Ian Lerche • Walter Glaesser

Environmental Risk Assessment

Quantitative Measures, Anthropogenic Influences, Human Impact

Ian Lerche
Walter Glaesser

Environmental Risk Assessment

Quantitative Measures, Anthropogenic Influences, Human Impact

With 125 Figures

Springer

AUTHORS:

PROF. IAN LERCHE
INSTITUT FÜR GEOPHYSIK UND GEOLOGIE
UNIVERSITÄT LEIPZIG
TALSTR. 35

04103 LEIPZIG
GERMANY

PROF. WALTER GLAESSER
INSTITUT FÜR GEOPHYSIK UND GEOLOGIE
UNIVERSITÄT LEIPZIG
TALSTR. 35

04103 LEIPZIG
GERMANY

E-mail: lercheian@yahoo.com
walter.glaesser@ufz.de

ISBN 978-3-642-06572-9 e-ISBN 978-3-540-29709-3

Springer is a part of Springer Science+Business Media
springeronline.com

Softcover reprint of the hardcover 1st edition 2006

Cover design: E. Kirchner, Heidelberg

Preface

The world is a dirty place and getting dirtier all the time. The reasons for this ever-increasing lack of cleanliness are not hard to find, being basically caused by the actions of the six billion people who inhabit the planet. The needs of the people for air, water, food, housing, clothing, heating, materials, oil, gas, minerals, metals, chemicals, and so forth have, over the centuries, given rise to a variety of environmental problems that have been exacerbated or been newly created by the industrialization of the world, the increase in population, and the increase in longevity of the population.

The costs of cleaning even fractions of the known environmental problems are truly enormous, as detailed in the volume *Environmental Risk Analysis* (I. Lerche and E. Paleologos, 2001, McGraw-Hill). The chances of causing new environmental problems, and their associated costs of clean up, are equally challenging in terms of anthropogenic influences and also of the natural environmental problems that can be triggered by humanity. This volume discusses many examples of environmental problems that have occurred and that are still ongoing. The volume also considers the effects in terms of sickness and death of fractions of the population of the planet caused by such environmental problems.

The purpose of the volume is to show that, even with the natural and anthropogenic environmental problems we already know about, the procedures for investigating the problems and then suggesting both remediation methods as well as preventative measures are not at all obvious. Many options are possible; many risks related to health concerns and to further pollution concerns have to be considered before one can decide on a particular remediation procedure.

The applications chosen to illustrate these points are taken from a variety of areas and with different causes. In this way one can see that the environmental problems are major, worldwide, and often have no social, economic, or politically acceptable solutions, even when scientific solutions are available.

Each chapter is self-contained and has a summary at its beginning so that one can obtain the purpose of the chapter before commencing to read the chapter. Readers who are not interested in particular subjects may then easily skip to remaining chapters. The disadvantage to this form of communication is that some figures have to be repeated in different chapters. However, such keeps the chapters self-contained and has the advantage that the reader does not need to keep jumping to other chapters to obtain the relevant figures.

Where possible, the case histories discussed are chosen from situations where we have performed the field work so that we are clear on what was done and the resolution with which data were obtained. In some situations such was not possible, and reliance on published information had to be accepted without the chance to check reliability. But that, too, is part of the uncertainty in environmental problems and their potential remediation.

The level of the book is set such that students actively involved in learning how to analyse environmental problems should have little difficulty in understanding the case histories. For professionals in the field, seriously involved in remediation efforts, this volume (plus the companion volume referred to above) provides a detailed set of procedures to analyse the scientific consequences of environmental problems. The volume should also be of use to decision-makers in both government and private industry who are actively involved in balancing the social, economic, political, scientific, and health issues for the best benefits of the population.

This work has been partially supported by the DAAD through their award of a Visiting Professorship to I.L. at the University of Leipzig. The University of Leipzig is also thanked for its contribution to this support and Professor Werner Ehrmann is particularly thanked for the courtesies and support he and his group at Leipzig have made available during the course of this work.

Walter Glaesser
Ian Lerche
Leipzig

Contents

Chapter 1

Natural and Anthropogenic Environmental Problems

Summary

Naturally occurring processes continually alter the Earth's atmosphere, topography, biomass and biofaunal loads, and their distributions around the world. When these processes adversely impact the environment relative to the perceived needs of humanity, they are considered environmental problems. Remediation can be performed but preventative measures seem difficult to provide. Anthropogenic processes are those produced by Man's activities in exploiting and modifying the environment. In general, there is usually a negative effect on the environment caused by such anthropogenic processes.

Remediation can be performed and, moreover, preventative measures can also be provided. This chapter presents an overview of the influence of both natural and anthropogenic processes on the environment. Remediation methods are considered as part of the entrepreneurial activities of mankind, while preventative measures are viewed from the self-preservation perspective of humanity. Finally, some basic rules of engagement are given when one is handling environmental problems, which enable a logical, ordered approach to be taken subject to the expediencies of cost/benefit/health and intrinsic need. Arguments can also be given to evaluate quantitatively the relevant conditions for prioritizing remediation and avoidance procedures.

1 Introduction

Humanity needs several factors to be present at the surficial domain of the Earth in order to survive. In order of precedence, determined as the absence of a factor leading to rapid death, these factors are: air, water, food, shelter, clothing, and energy. Regarded as the sine qua non for environmental conditions to maximize the well being of mankind, these factors are precisely those that humanity does the best job it can to <u>minimize</u> rather than maximize. We <u>are</u> our own environmental problem.

Within this broad classification of the needs of humanity, it is relevant to examine how natural and anthropogenic effects impact the survival and well-being of humanity, and how we can learn to control, modulate and/or remediate what we determine to be deleterious effects for humanity. One problem that is pervasive is who decides, and how one decides, precisely what constitutes a deleterious effect, and whether it persists on a short timescale or a long timescale as far as human involvement is concerned.

2 Natural Environmental Problems

As far as can be determined, evolution of the air, water, biomass, and landmasses of the Earth have been continuous throughout the total history of the planet. Until the advent of Man, the response of living creatures to such natural events was essentially passive: adaptation, mutation, and species extinction. The appearance and diaspora of Man across the globe, plus the ever ongoing demands of a place to live, places to grow food, and modification of the local environment by replacement of local plant material by cities, roads, houses, industry and farms, has led to more aggressive actions by the natural processes of nature on the desired habitat of mankind - a so-called "natural" environmental problem.

Basically, the natural processes of both rapid and slow geological evolution continue and mankind can perform only remediation after the fact but cannot pre-ordain or control to any significant extent the natural processes themselves. Perhaps the most important processes to list that influence the environment where mankind lives are: (i) climatic variations (e.g., El Nino); (ii) floods; (iii) droughts; (iv) earthquakes; (v) landslides and avalanches (snow, mud, and submarine turbidities); (vi) forest fires; (vii) volcanic explosions; (viii) food pests; (ix) meteoritic impact; (x) sea level fluctuations; (xi) hurricanes/typhoons/monsoons/tornadoes; (xii) ice floes and ice sheets.

Some of these processes influence mankind immediately (such as earthquakes, hurricanes and floods), whereas others have a longer-term impact over years or centuries (such as droughts ongoing over tens to hundreds of years), while a large meteoritic impact (as in the case of the Cretaceous-Tertiary boundary event and the associated extinction of the dinosaurs) would have an immediate, and likely non-recoverable, influence on the survivability of man kind as a species. So, natural environmental events range across the full spectrum of time and space scales, and from local, recoverable events, to irrecoverable long-term or short-term events.

While one can do little to control the natural processes, the problem humanity faces is to provide remedial action to those unfortunate enough to be impacted by any such event. Remedial action in such cases involves saving survivors, recouping the influenced area if possible, population redeployment if the impact is non-recoupable (such as in drought-stricken areas), and emergency medical, shelter, and food and water supplies. While such remediation actions help humanity, they do nothing to ameliorate the basic natural processes. And humanity has a long-term record of reoccupying those land areas that have been subjected to on-going or periodic natural environmental processes.

For instance, Kobi, Istanbul and San Francisco are all sited on extremely earthquake prone regions, as known from historical records of earthquake occurrence. Yet, after each such earthquake event, mankind rebuilds on precisely the same areas again and again. Presumably, the reasons for this apparent illogical activity are that there are more compelling reasons for reoccupation than for desertion (e.g., excellent port facilities; strategic control of a narrow strait; excellent farm land; no more land available to a constrained population).

There seems little that one can do to safeguard fractions of humanity from such natural processes; we are faced with the eventuality of having to learn to live with natural environmental processes and their impacts on mankind. Remediation action after each such event seems all that we are able to do. The situation is extremely different, however, for anthropogenic environmental events and these are considered next.

2.1 Anthropogenic Environmental Problems

As remarked in the introduction, there are several basic requirements mankind needs in order to survive: air, water, food, shelter, clothing, energy, and a place to call home. In addition, the collective activities of mankind result in industrial development, mining, modulation of land uses, and transportation requirements. Humankind is extremely efficient at taking the basic requirements for survival, together with its own collective activities, and converting them both to waste and to conditions untenable for human survival in the long-term.

We first provide some illustrations of these sorts of environmental problems, and then consider both preventative and remedial courses of action. We then look at the political and social implications for prevention controls and remediation of anthropogenic environmental problems.

2.1.1 *Examples of anthropogenic environmental problems*

Air

The fundamental requirement for existence is air to breathe. And yet mankind seems bent on polluting this absolutely fundamental existence requirement. Automobile exhaust pollution in major cities of the world (Los Angeles, Bangkok, Sao Paulo, Beijing) is rapidly making such locations less than viable sites where humanity can survive without major adverse impacts on human health and longevity. From an industrial perspective, mankind seems determined to spew ever-increasing amounts of SO_2 and CO_2 into the atmosphere, leading to acid rain (from sulfur dioxide converting to sulfuric acid) that kills forests and pollutes good farm land, while the industrially produced carbon dioxide enters the atmosphere and so provides one of the major components contributing to the greenhouse effect on a global scale. The release of massive amounts of fluorcarbons into the atmosphere has the major negative effect of destroying the ozone layer, and so of allowing more solar UV light to penetrate to the Earth's surface leading to major skin cancer problems, to say the least.

As a race, mankind would seem determined to minimize the fundamental factor it needs to survive- air to breathe.

Water

After air, water is the next fundamental factor needed for human survival, both for internal consumption for physiological maintenance and also for food supply control, be it for watering land-based plants and animals or for fresh water and salt water farming of fish, crustaceans, seaweeds, etc. And yet, here again, mankind would seem determined to pollute lakes, rivers and oceans of the world in a variety of manners, less than conducive to their use as sustainable prerequisites for human existence.

The main sorts of primary contaminants are chemical, biological, physical, and nuclear, which may be accidentally introduced into the fluid environment or introduced with clear determination and purposes. Primary chemical contamination of the fluid environment is as varied as the chemicals produced by the industrial activities of mankind, ranging from heavy metals through arsenic (from gold mining mainly) through oil (from pipeline breakage and tanker ship disasters), through paper pulp residue (as in the southern part of Lake Baikal, Russia), to horrendous mixes of residual chemicals used in the paint and lacquer industries. This list can be extended almost endlessly depending on mining activities and industrial activities around the world.

Primary biological contamination of the fluid environment can occur from a variety of sources: human and animal waste; hospital detritus; biological weapons residue (such as anthrax stored on the Aral Sea Island); decayed organic material from residual foods and unused animal parts, and so on. The impact of this primary biological contamination on pure water supplies can be enormous, both in terms of human toxic contamination and also in terms of secondary concentration of biological contamination through the food chain. Such an effect is, of course, also a major problem with chemical contaminants.

Physical contamination arises when, for example, used building materials are indiscriminately dumped in a water supply. The presence of asbestos, of lead piping, or of the fine detritus can pollute water to such an extent that fish die, oysters become toxic, and the water has so much physical detritus in suspension it becomes undrinkable. Other aspects of physical contamination can arise by performing different combinations of operations on water supplies: either dump different mixes and types of physical contaminants into a given water supply, of change the water flow conditions anthropogenically (dams, levees, canals, etc). Both of these impacts can produce deleterious results on water quality.

Two other major forms of water pollution are provided by mankind: nuclear and oil. In the case of nuclear pollution one has merely to look at the Kara Sea region, full of rusting hulks of Soviet nuclear submarines with fuel-laden reactors still onboard, and also nuclear cores from land-based power plants, in order to have a prime example (but, regrettably, not the only instance) of a major disaster waiting to happen.

Oil pollution of rivers, groundwater, and oceans seems to be an almost daily occurrence. The recent leakage (July 2000) by Petrobras of oil into the Iguanu river running south from Brazil to Argentina, the Exxon Valdez disaster, the Amoco Cadiz tanker shipwreck off the coast of France, the leakage from submarine oil transport pipelines, and from offshore exploration and production platforms, together with the massive sustained (up to around 25%) leakage into the tundra region and groundwater supply of the ancient and decaying Soviet oil pipeline system, all provide examples of the ability of mankind to foul the very waters needed to sustain plant, fish and animal growth, and also the waters directly needed by mankind in order to survive.

Humanity seems determined to increase the amounts of contaminants it unleashes into its global water supplies.

Foodstuffs

In order to survive humanity needs to supply itself with foodstuffs. Current estimates would indicate that the world as a whole has about 25-30 days reserve supply of food. Yield enhancements of crops have been ongoing for centuries, and the harvesting of fish and animals as long as civilization has been around.

But, as the world population has risen to its current level of around 6 billion persons, the pressure to increase food production has steadily risen. Because grains are subject to a wide variety of bacteriological, fungal and pestal infections, two major prongs of attack have been made to overcome these problems and so increase yields. First, one can spray growing grains with bactericides, fungicides and pesticides in order to limit infestations. The price that is paid here is that the various bactericides, fungicides and pesticides become directly absorbed by the grains, and also contaminate the earth where they are then absorbed by plants using root feeding. One of the major open concerns at the present day is whether direct human ingestion of such treated grains has a deleterious effect on human health and longevity. Second, one can genetically modify plant materials to make them intrinsically resistant to the various forms of infestation, thereby automatically increasing yield. What is under intense debate at the present time is the long-term effects on humanity of ingesting such genetically modified foods.

In respect of proteins supplied by animal flesh, perhaps the decades-long major problem of BSE in English cows (with the attendant transfer to humans in the form of the brain wasting Jakobs-Creutzfeldt disease) should serve as a very salutary warning that the diseases caused by food supplies to animals can be transmitted to humanity in excruciatingly nasty forms, leading to death and also to further vector transport of such diseases by infected humans.

In attempts to stimulate yields of crops and fish farms, fertilizers and pesticides are used in abundance throughout the world. Run-off of such residual contaminants by rain, or by crop irrigation waters, pollutes the waters flowing further downstream, including the effect on humanity, fish, plant, and wildlife that depend on such waters for existence. Perhaps the example of DDT, a long-lived chemical used as a pesticide, which causes thinning of eggshells in birds and so leads to a steady loss of bird embryos, is an example of the long-term serious ecological repercussions that can be caused by application of fertilizers and pesticides. One the one hand crop yields are indeed increased, and on the other hand major ecological damage is done. An uneasy and very uncomfortable situation exists of attempting to feed the world but at the expense of poisoning so me fraction of the bio-types inhabiting the world-including Man. Without some alternative in terms of adequate food production for the burgeoning world population, it would seem that humanity will continue with this exceedingly uncomfortable and dangerous situation.

As a further example of the ability of Man to contaminate his own food supply, consider the example of the release of radioactive strontium from the Windscale (now Sellafields) nuclear reactor in NW England. This radioactive strontium permeated the ground, was absorbed and concentrated by grasses, which were then eaten by cows, and so the radioactive strontium found its way, in ever more concentrated form, into produced milk. Because strontium is of the same chemical family as calcium, its properties and affinities are similar. Accordingly, the milk drunk by babies and small children, which the human body uses to produce bone tissue in the young, resulted in bone material in affected children with highly radioactive strontium as an integral component. Needless to say, the incidences of bone and tissue cancers in such children were out of all proportion relative to other children who did not receive contaminated milk.

Humanity seems to be extremely efficient at attempting to deprive itself of the very foodstuffs it needs in order to survive.

Land Use

The ever-increasing world population needs more space for habitation and for agricultural usage. In counterpoint, mankind seems determined to make less land accessible and useable for species survival. For example, the Chernobyl nuclear power plant disaster and subsequent long-term radioactive pollution of the land in both the near and distant vicinities of the massive radioactive release, should caution humanity about the need to decommission quickly such unsafe reactors to avoid the risk again of having such a problem. And yet, in countries without the wealth to install safer replacement reactors, continued use of Chernobyl style reactors persists, with the attendant high risk to man, land, air and foodstuffs being exacerbated as such reactors age.

Again, one needs only to look at the example of the denudation of the Amazon basin rain forest to recognize a serious threat to many factors on which humanity relies. Because trees absorb carbon dioxide and produce oxygen, the destruction of the Amazon rain forest will eliminate one of the largest single regions of oxygen production. In addition, about 95-98% of the nutrients are in the trees, so that once the trees are gone, there is nothing to hold the top soil in place, leading to massive soil erosion and transport down the Amazon river to the delta mouth. Thus, while the Amazon burn-off will produce soil with rich nutrients for a few years, over a decade or less the soil ceases to be productive and a wasteland is created. Further problems are the loss of potential new drugs and chemicals from plants in the forest, the total erasure from the face of the earth of animals that have specialized ecological niches adapted to a rain forest environment, and the continued denudation by water cannons and open pit mining as procedures for extracting gold and other minerals in a cost effective but environmentally irresponsible manner. The loss of societal human cultures that also call the Amazon forest home, is also a major factor to be reckoned with, and their loss diminishes the diversity and richness of mankind's disparate evolution as a species.

Severe threats to the land on which mankind lives are provided by the acts of waste production from human and industrial activities, something that mankind is amazingly adept at accomplishing in a remarkably short time. On the industrial side there are the problems of where to deposit chemical. Biological, nuclear, and general waste products from human activity. Classic examples of such indiscriminate depositories are the Love Canal, USA, chemical depository on which, twenty years later, houses were constructed with the concomitant massive increase in all sorts of human illnesses to the point that the Love Canal housing estate had to be abandoned and massive compensation paid to the survivors of the affected families.

On the nuclear waste depository side is the USA national low-level waste site at Barnwell, South Carolina, where the influence of subsurface fracturing and faulting, plus the potential for destabilizing earthquakes on the integrity of the waste depository is raising major concerns. The problem of the safe disposal of high-level radioactive waste is an area of pressing concern around the world, both in terms of the very definition of "safe," the length of time such long-lived highly toxic radioactive waste must be kept under control in any depository, and even the question of long-term (1000 year) safe record keeping for future generations. The billions of dollars the US government has already invested in exploring the feasibility of the Yucca Mountain site as one such possible safe places, is a measure of the urgent need to resolve such problems safely and quickly-the Kara Sea dumping by the Soviets has al ready been mentioned as another lethal problem-an accident waiting to happen.

General-purpose landfill sites and oil pollution of the subsurface, together with the disposal of animal and human sewage waste are becoming even more critical problems as the world population doubles every 20-30 years. There is a limit to the natural recycling rate of such organic wastes, and many have argued that the limit has been reached if, indeed, it has not already been surpassed.

The problem of disposal of biological waste (be it from hospitals or as the result of bacteriological warfare products and by-products) is, undoubtedly, one of the major problems facing the 21st Century as more and more biologically toxic agents are generated, stored, or released (on purpose or by accident). There seems to be no end in sight to such pollution of the ground on which we rely for our living accommodation and major food supplies worldwide. Humanity is, without a doubt, creating and causing its own environmental disaster in this respect as well.

Mineral and thermal pollution

The ever-increasing search for minerals and metals of economic worth has always had a major role to play in changing the environment. The marble quarries of the island of Thassos, and the gold silver and copper industries of the past on the island, have all led to changes of forested areas to ugly eyesores of waste rock deposits next to the mines and quarries. This scenario is repeated around the world time, and again as mankind has quarried and tunnelled for rock, metals, coal, uranium etc. Two or three major environmental pollution problems stand out on this regard, with a host of ancillary problems as corollaries.

Perhaps most exemplifying the cumulative problem of tunnel mining over the centuries is the general region around Bochum in the Ruhr Valley of Germany, where co al and metal mining has been carried out for centuries. Many mines were not recorded, or the records were lost, many were illegal. At the present day, the underground is a maze of known and unknown tunnels, interconnected and cross-connected. As a result of long-term groundwater infiltration, and so causing a general lack of cohesiveness of mechanical support in the residual rock strata, major sinkholes appeared in 1998-1999 in a heavily populated area. Immediate resolution of the problem called for cement to in-fill the huge sinkholes. But the general problem is not so simply solved. There remain miles of such tunnels in one of the most heavily populated and industrial sectors of Germany. There is absolutely no idea available of the likely occurrence of another such event, when it may occur, or where. The potential environmental impact is truly enormous.

In addition to physical collapse damage, both open pit (quarry) mining and tunnel mining lead to exposure of waste rock to the elements. The problem of leachate production of acid mine drainage waters is particularly severe in many areas of the world where coal has been the main economic component being sought. The production of heavy metal residues in river, lake and ocean waters as a consequence of mine waste piles being indurated is not a problem to be dismissed lightly either.

There are, also, soil erosion problems that arise both from mining activities and also from irresponsible forest clearing-cut and slash-and-burn techniques. As remarked already, the Amazon basin rain forest provides a prime example of such uncontrolled exploitation.

As an added element to the environmental problems created by mining activities, there are the major contamination problems caused by oil spillage already referred to. Also, the seemingly inexhaustible appetite of mankind for more and more power (usually in the form of electricity generation) leads to thermal pollution problems of rivers, lakes and oceans from the cooling systems, as well as to generalized thermal pollution from energy-hungry devices. The current estimate of 5% of the USA electricity energy output for computers, with a rise to around 12% predicted over the next decade, serves to indicate the compelling need for energy frugality if we are not to thermally pollute water reserves to the point where nothing can survive in such high temperature waters. How this environmental problem will develop, as the world demands ever more power is not at all clear.

Population as an environmental problem

Mankind is its own worst enemy in respect of environmental pollution and contamination. As already said, there are natural events that contaminate the environment humanity deems desirable to continued species survival, but the cumulative effect of such events is small compared to the scale of anthropogenically generated environmental problems.

The biggest problem is caused by the sheer number of human beings, currently around 6 billion with an increase estimated to double that number in a decade or two. This large number of people requires air, water, food, shelter, clothing, and land. In return, humankind produces human waste, fouling of the air, water and land it uses, and also, because of the close proximity of the mass of humanity (almost 30% live in major cities worldwide), contagious diseases. Such diseases can be spread by rodents (plague), by mosquitoes (yellow fever, Nile fever, malaria, etc.), by human-to-human contact or nearness (Ebola, typhus), by human sexual activity (HIV, AIDS, syphilis), and by other transport vectors. Thus, in addition to the anthropogenically produced environmental problems, humanity is quite capable, indeed remarkably so, at spreading debilitating and life threatening diseases. And yet humanity seems bent on increasing its population past the point of sustainability that the Earth can maintain.

We are indeed our own worst enemy in the nominal pursuit of high environmental quality; there are more of us every day to pollute and contaminate the very environment we need to survive; and we are excellent at being polluters on the individual and industrial levels.

War

Mankind has always warred with itself. Ever since recorded history began there are records of war. The reasons for wars are as many and as varied as mankind has been able to invent, ranging from the need for more slaves, more minerals, more oil, more land, through the desire of one religious, political, business or national group to foster its own brand of system upon another people or, indeed, just the massive aggrandizement of raw naked power. It would seem that man kind is determined, and always has been, to perform genocide and racial murder if the need can be fabricated to convince, by might or right, a particular group to participate.

As a consequence of this inherent historical conflict, mankind has become ever more ingenious over the centuries at generating weapons of greater and greater killing power. Indeed, a goodly fraction of humanity's endeavors have always been in the area of military weapons development.

The end result is that, as a race, today we possess weapons that can trivially annihilate and obliterate humankind from the face of the planet; a remarkable achievement indeed! Many of the weapons, such as hydrogen bombs, are so dangerous that their very testing in the atmosphere is internationally banned because of the global radioactive fallout problems-a major environmental problem. Many others, such as enhanced biological anthrax, cannot be tested at all outside the confines of a secure laboratory because of their uncontrollable nature. And yet, mankind continues to lay waste people and land with variations of these and other weapons.

Perhaps one of the best recorded events from ancient times is the sack of Carthage by the Roman Empire, where not only was the city put to the torch, but the agricultural fields around the town were sown with salt, making them incapable of producing enough crop yields to support the Carthaginians, leading to a diaspora of the survivors of the sack who were not enslaved. Indeed, this event is a clear example of chemical warfare.

In the modern era, after the use of mustard gas in the First World War, such chemical weapons were internationally banned. But the use of Zyklon-B in the concentration camps of Germany during the Second World War, the more recent gassing of the Kurdish population in Iraq by its own government, the use of chemical defoliant Agent Orange in Vietnam by the USA, all show that international treaties on chemical weapons last only as long as a nation considers them expedient.

One further consequence of war is the legacy left afterwards of landmines strewn everywhere. Because land mines are cheap to produce, and contain minimal metal content, they are used as a deterrent difficult to detect by an opposing force. But after a war, there are often no records available of where land mines were deployed, their deployment patterns, or the number deployed. Finding such fields of sudden death is often by accident-and normally by human beings being maimed or killed. The land is then not useable until all such land mines are cleared. And in poor countries this clearing is often long delayed or never undertaken before yet another war breaks out with a repeated sewing of death and destruction.

It is a sad mark of the intrinsic nature of Man that mankind uses its wealth much more to further death of its own species than to improve the environment for the benefit of the species.

2.2 Resources and environmental issues

Natural resources. depletion and waste

Resources that humanity brings to its own domestic and industrial uses are of two basic types: intrinsically renewable on a time frame much shorter than human existence (such as fishing, forestry, agriculture, animal husbandry); intrinsically not renewable on such a timescale (such as oil, diamonds, metals, minerals, marble). But the fact that some resources can be intrinsically renewed does not mean that they will be. For instance, human greed has led to massive over-fishing offshore Newfoundland and in the North Sea, to name but two instances. Thus, a potentially renewable resource can be so over-utilized by humanity as to destroy the resource!

For intrinsically non-renewable resources, such as oil for example, there are two major problems to consider: Is the currently available supply large enough to meet the demand? Are the known reserves and the rate of reserve replenishment by exploration large enough to keep the current mode of civilization in operation and for how long?

In the cases of both renewable and non-renewable resources one is faced with the problems of responsible management of the resource. The problems of depletion by over-utilization or by limitations on the amount of the resource need to be incorporated in the management, so that one is well prepared, well ahead of time, for the eventual loss of the resource. Exacerbation of the resource supply and demand occurs when one or more nations control a supply that other nations need-as in the case of OPEC and the demand of the western world for ever more oil. Indeed, wars have been fought to keep the supply intact and to keep the economy of the western world in operation. Might does make right in such situations. Can one really imagine that massive intervention of the western military powers would have taken place in the Middle East during the so-called Gulf War of 1991 if there had been no oil present to protect? Or the invasion of Egypt a half century ago to keep the Suez Canal open to oil tanker traffic when Nasser attempted to stop such a vital flow of oil? One would, at the very least, be naive to think that such interventions were for purely altruistic reasons.

At the same time, resources are often wasted. For example, the main Russian oil pipelines from the West Siberian Fields (and earlier from the Rumanian Ploetsi fields and the Azerbaijan oil fields) often had sustained leakage rates of more than 10-20%-- a serious waste of a potentially valuable commodity.

The quest for more resources, both renewable and non-renewable, in order to maintain and improve the quality of life for mankind has not had a history of being in the best general interest of mankind. Instead, the resources have often been controlled by a few individuals of a few countries, with a more rapacious desire to become rich and/or powerful at the expense of others. Humanity would still seem not to have learnt much, if anything, from the evergreen lessons of history.

Human resources and waste

Throughout the world there are basically two different levels of human existence, although one can split the groupings into ever-finer fractions. At one level, a goodly fraction of humanity spends all of its energy just in order to survive on a daily basis, and still goes to bed hungry at night. At another level, the remaining fraction of the world population satisfies its daily survival requirements in a small fraction of a day, and so has time for other pursuits, such as technical innovation, information developments, research and applications, art, culture etc. In short, one has a small percentage of the world inventing and designing methods and procedures for improving their own lot and leaving less developed nations even further behind.

The inventiveness of mankind in all fields of endeavor is astonishing; from satellites to the planets, man travelling to the Moon, Shakespearean plays, computer technology, biological and medical advances, and on and on. But the callousness of mankind to the less fortunate fraction of humanity is equally astounding-and always has been. Not only will the poor and unfortunate be always with us but also they will become poorer and even more unfortunate relative to those who sit in the domains of the highest quality of life.

Such is both a very poor use of human resources (How many potential geniuses sit in such impoverishment with never the hope of realizing their potential in full measure? And so humanity is deprived by its own actions, or lack thereof, from benefits that could otherwise accrue to mankind), and also a waste of human resources in that, once the impoverished fraction could spare some time from the unrelenting toil of providing daily sustenance, the released human potential would be truly astounding.

The tapping of the total of human potential resource capability, together with the minimization or elimination of blunted ambition and enthusiasm, could herald a blossoming of human endeavor not seen since the Golden Age of Greece. But the probability that such a scenario will prevail seems even further in the future than the Golden Age is in the past.

Entrepreneurial developments

The broadest classifications of collective human endeavor are four in number: business, political, religious, and national. Within these broad groupings environmental problems and potential solutions have to be considered. Not all groupings make for ease in containing or eliminating anthropogenically produced environmental problems; indeed the groupings can themselves be the main source of such problems, and often have no intrinsic desire to provide solutions. Consider some illustrative points.

2.2.1 *Business*

The fundamental purposes of business are to make as much profit as possible at whatever is undertaken, and also to grow. Within these two basic paradigms, one must also remember that each major business endeavor produces waste-be it in the form of carbon dioxide emissions, cyanide from gold mining, or whatever, and the waste must be disposed of. One must also keep in mind that businesses have accidents (Bhopal India toxic gas release from a Union Carbide plant; oil refinery explosions on a recurring basis; etc.).

From the point of view of a business without any external controls on its operations, it is most cost effective just to dump the waste material. But with the modern realization that we all live on a very small planet indeed, there are increasingly stringent controls being enacted by governments on local, regional, national, and supranational levels to force businesses to police their waste and to clean up previous waste production where possible. Businesses, in turn, end up passing these additional costs directly onto product prices for consumers where possible, and taking a tax write off of the costs also.

The ultimate motive is still the cost/profit drive. And the ultimate environmental problem still remains: Where is the waste to be disposed of? Environmental external control does not solve the basic problem.

2.2.2 *Political*

Methods of political governance of populations have been as many and as varied as humanity has been able to devise; and corruption of, and by, political officials has also been as equally varied over the course of history. The activities of governments, some secret and some open, have also led to major environmental problems, which often have come to light only years later, if at all. For instance, the Superfund Clean Up in the USA has, over the years, been used for partial remediation of military bases with chemical spills into the ground, for remediation of nuclear plants which were often not reporting correctly their accidents, for biological weapons testing facilities remediation, etc. When environmental problems of such magnitude are caused by a government, then there is little chance in a variety of less than open national governments of even finding out what the problems are, never mind attempting remediation. All too often, as in the case of the USA, such matters are buried under the aegis of "National Security", or are performed at national defence sites, off limits to public oversight control and even to many working at such sites. Qui custodiat custodiensis? remains still an unanswered, and likely unanswerable, question when it comes to environmental problems created by or on behalf of government bodies.

2.2.3 *Religious*

As remarked already, mankind has always had a fatalistic attraction to waging war, leaving ecological devastation in the wake. One of the major causes of war over the centuries has been, arguably, the nominal claim of promoting one religion over another. Examples that readily spring to mind are the Crusades of the Middle Ages, the Protestant-Catholic conflict in Cromwellian times in England and Ireland (and still ongoing today in Northern Ireland), the Arab-Israeli wars of the last half century, the Serbian-Croatian conflict of the last decade, to name but a very few.

With the advent of modern weapons and bombs of significant destructive power, plus the increase in terrorist and guerrilla warfare of the last few decades, the environmental aftermaths of war in the name of religion are become even more diffusely widespread throughout the world than they were in the past.

It would seem that those who practice religion should do so with their own zeal but should not be dogmatic about ramming their particular brand of religion down the throats of others, often at the point of a gun, who worship a deity in different ways. Regrettably, however, the damage to the various peoples of the world, their livelihoods, buildings, land, countries, etc., in the name of religion seems unlikely to be controlled in the near future. Presumably, such religious wars will continue to be fought, disaster strewn behind as the consequence, to the detriment of mankind and the environment, as cultures are annihilated and land laid waste.

2.2.4 *National*

The common needs of a nation of people and their mutual agreement on societal forms of controls and rights are, perhaps, the quintessential backbone underpinning how the nation evolves. The interests of the nation as a whole become the paramount concern, with the needs of other nations, subscribing to alternative common good agreements and to alternative forms of societal controls and rights, being of secondary consideration. The conflicts of such different perceptions of worth of humanity have led to major wars over the centuries, to the detriment of the different perceptions, and producing ravages to humanity, countries and cultures as the bitter fruit.

As with any war, national effort goes into diverting natural resource raw materials to the instruments of war, so that not only is there a loss to society of the benefits such resources could bring, but the national effort to improve the environment for the benefit of the nation is thwarted as all goes up in smoke and ashes. And the depletion of such non-renewable resources, together with their waste on the pursuit of killing one' s neighbor, is hardly a responsible measure of the best that mankind has to offer in managing such resources.

But the interests of security of an individual nation, and of the apparent need to convert other nations to a similar societal grouping (by force if necessary), would surely seem to be the Achilles heel in any composition of a nation. Certainly, the continuing warring conflicts of mankind provide a strong historical database indicating that such is the case. Apparently, this depletion and waste of natural resources will always occur. Apparently, the aftermaths of war in the form of the Four Horsemen of the Apocalypse will continue to ride mankind. Apparently, we learn nothing from historical sanguinary lessons. Where are you now Genghis Khan?

2.2.5 *Capital*

Recognition that an environmental problem exists is the first stage in attempting to address the issue, irrespective of who instigated the problem. But dealing with the problem in a remediation sense requires money. This money can come from two major sources: private industry or government.

In the case of private industry, money to remediate self-made environmental problems or to minimize production of such problems will generally be spent under one of four broad conditions: (i) when the government penalties for not so doing are more burdensome a charge to the company than are the remediation costs; (ii) when a fine can be levied for not taking sufficient action (Petrobras was fined $10 million for the recent Iguanu river oil contamination plus the clean-up costs); (iii) when the public perception of an environmentally responsible company helps boost sales to such an extent that they exceed in profit the environmental costs; (iv) when the company needs to recycle waste material to recover components of the waste for re-use because the cost of so doing is cheaper than buying fresh components. In that case recycling is in the best interests of a corporation. Always the cost/profit motive is dominant.

In the case of governments, several factors come to the fore in addressing remediation of environmental problems. First, the only form of money a government has is through the taxes it collects from corporations and individuals. Hence, any money a government commits (other than from penalties of fines imposed) to remediate environmental problems must come either from the general revenue surplus fund, from current operational budget, of from a surcharge to taxpayers. Second, identifying the culprit in a particular environmental problem may not be easy: a corporation may long since have become defunct but its environmental pollution legacy survives; the government may (and often is) the main pollution culprit over the years; the environmental standards imposed and met in an earlier era may no longer be relevant when viewed in the light of later scientific and technological information. Third, the government rarely performs remedial action itself Instead, it often prefers to let a contract to private industry to perform remediation, occasionally to the same corporation that produced the problem in the first place! Then some form of independent monitoring of the terms of the contract is required to ensure proper remediation is carried out.

Whichever way the problem is addressed, the point remains that capital is required. Too little capital invested will not solve an environmental problem but will often exacerbate one; too much capital invested leads to opportunities for graft and corruption. An effective capital balance would seem difficult to achieve given the venial pecuniarity of mankind.

3 Remediation of Environmental Problems

In this overview of natural and anthropogenic environmental problems, it is not appropriate to consider specific technical methods for remediation of every specific type of environmental concern. Such an undertaking is best left to those who are specialists within given areas. But several factors stand out as sharp focuses to consider in general; these factors are listed here.

3.1 Population and behavior

Because humanity as a whole is remarkably gifted at producing environmental problems that diminish the quality (and quantity) of the very environment mankind requires in order to survive, it would appear that three dominant characteristics need to be addressed by mankind.

First, and foremost, mankind must be both educated and encouraged (with a reward/penalty system most likely imposed) to minimize the anthropogenic environmental problems. Failure to do so means that we will increasingly pollute the planet in all aspects, a less than healthy long-term species survival trait for mankind.

Second, changes to the infrastructures of societies and to the living standards of mankind as a whole should be encouraged at a global level, so that some of the vast untapped human potential can be channelled to further environmental improvements. Failure to do so, without altruistic aid from the better-developed nations, will increase the gap and just lead to more environmental problems.

Third, the major future problem is surely the explosive growth of human population, which creates even more strain on the world natural resources, and which seriously erodes the global quality of air, water, food, shelter, clothing, and energy that mankind needs to survive. If we do not learn to control our own human population, then at least one or more of the Four Horsemen of the Apocalypse will eventually do so for us, in less than a humane manner.

3.2 Particular environmental remediation concerns

The major environmental problems split into two broad categories: avoidance, control and/or remediation of those problems that have a deleterious impact on our current and near future use of the environment (such as oil spills, heavy metal water pollution, carbon dioxide in the atmosphere, etc.); treatment and disposal of environmentally damaging materials that have already been created and which would have profound long-term consequences for the global environment if allowed loose (such as nuclear waste, biological organisms, long-lived chemical toxic pollutants, etc.).

Each requires an intermix of national and supra-national government protocols, accords and laws for enforcement; each requires responsible action on the part of individuals, businesses and governments; and each requires coordinated remediation on a global scale involving all nations. One cannot, for instance, declare a river pollution problem "solved" in a given nation just because the water born pollution is transported to a neighboring nation by the river. Such chicanery may serve well as political rhetoric but it does nothing at all to solve global environmental problems.

Perhaps we are making a start on the awareness sensitivity level of particular issues (such as the greenhouse problem, or the nuclear waste and nuclear accident problems), but we have a long way to go to remediate all such major environmental problems to the benefit of mankind.

3.3 Prevention/Remediation

One of the classic English sayings is "An ounce of prevention is worth a pound of cure". This saying also pertains to the environment. It is better not to create an environmental problem than it is to have to remediate one. But until we clean up the environmental mess we have already made, we will have to spend enormous amounts of money to remediate. At the same time, prevention of future potential environmental problems by global agreement, national laws, independent agencies, corporate and individual responsibility must surely become the sine qua non if mankind is to flourish in a healthy environment and not wither and die in self-made environmental pollution.

4. Rules for addressing environmental problems

In the ancient Greek days, advice given to doctors was "First and foremost, do no harm". That adage can equally well be applied to remediation of environmental problems today.

1. First, and foremost, study the particular problem thoroughly before proposing a course of remedial action.

2. Be ready with different potential courses of action to be used in parallel or sequentially as the need is seen.

3. Before commencing a major remediation effort, do a pilot study to determine that there are no untoward nasty environmental surprises caused by the suggested remediation procedure itself.

4. Do not do an incomplete or poor remediation treatment. It will just have to be done all over again properly at even more expense.

5. Humankind as a whole must be trained to stop or minimize its anthropogenically generated environmental problems. DO NO HARM!!

Chapter 2

Restoration of lignite mining sites in the former GDR: lessons to be learnt from Zwenkau

Summary

The interactions have been investigated between the near-surface sediments of the sediment dumping at the Zwenkau open pit (used for extraction of lignite) and the produced mining water. These environmental impacts are the legacy of the energy policy of the former GDR. The pyrite oxidation that takes place in the overburden sediments causes the formation of an oxidation front and the pH-value of sediments falls to about 2 to 3 very soon after deposition. The primary mineral contents are destroyed and a considerable number of elements are mobilized that remain locally in the sediment pores. Because of the very weak seepage, rainfall remains in the alluvial areas as dump lakes. The dump lakes are very acidic because the rainwater transports the mobilized elements from the surface. Autochthonous and allochthonous formations of secondary minerals have been observed and analysed. A continuously ongoing interaction of exogenous and endogenous effects hinders the hydrological and geochemical equilibrium stabilisation. What to do with such a "moonscape" is considered in the conclusions to the chapter.

1 Introduction

In consequence of the energy policy in the former GDR, extensive exploitation of the Central Germany lignite-mining district was undertaken with the result that a number of mining landscapes remain, which are typically marked by mining lakes, dumps and slagheaps. The energy policy in GDR times was one of avowed self-sufficiency in heating homes and providing energy for industrial plants. Given the negligible oil and gas resources of the GDR, the only viable alternative was lignite mining for such uses, and this resource was intensively mined.

The point here is that energy policies, such as that in the GDR to produce lignite for a variety of purposes, can have long-term effects on the environment, mostly deleterious. This chapter looks at the legacy of this interaction of energy policy and environment in terms of the results it has produced in the Zwenkau region.

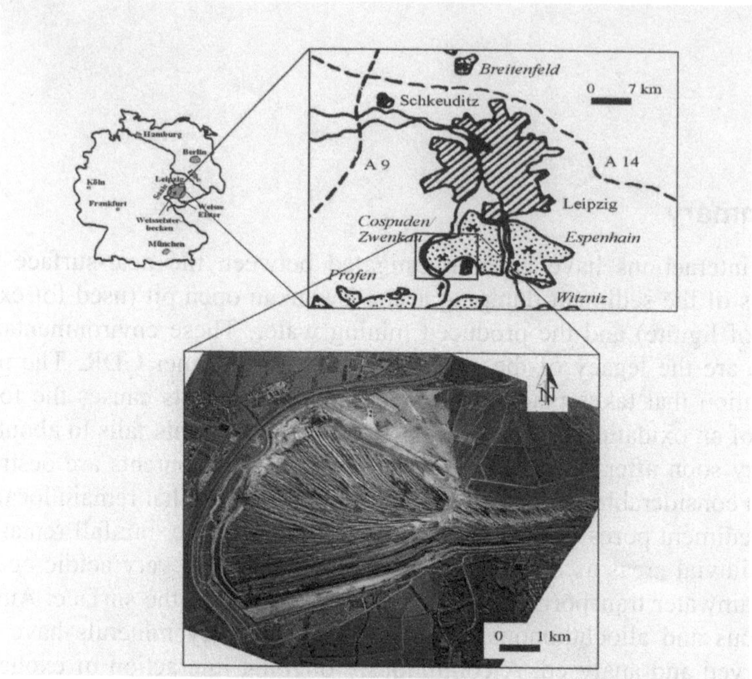

Fig.2.1. Location of the open mining pit Zwenkau (air survey from German Aerospace Center [DLR] – infrared-image, flight height 3,600 m, Kodak Aerochrom-Infrared 2443)

The overburden dumps of the open lignite pit Zwenkau, located south of Leipzig, and forming part of the Northern Weisselster-Becken (figure 2.1), were investigated. The advantage of the Zwenkau pit is the presence of freshly deposited sediment ribs, together with up to 17-year-old ribs in the overburden dumps, so that one has a history of the environmental effects recorded in the dumped materials. Unfortunately, the mining-induced alterations of the topographical, hydrological, and geological situations of these areas are mostly irreversible (Caruccio et al. 1988; Glaesser 1995 a, b,1997; Koelling 1990; Johnson and Thornton 1987; Mills 1985; Pflug 1998; Prein1993). Especially in the Central Germany lignite mining district, one of the major problems for the recultivation of the mining landscapes is acid mine drainage, i.e. the oxidation of the pyrite and markasite, which begins very soon after depositing sediments in the overburden dumps (conveyor bridge dumps). Because this oxidation is the beginning of different subsequent processes, the hydrological and geochemical equilibria are highly affected. Sulphate, iron, and hydronium ions are produced so that the pH-values of the sediments are drastically lowered to strongly acidic conditions (Ahonen and Tuovinen 1989; Alpers and Blowes 1994; Evangelou 1998; Moses et al. 1987; Nordstrom 1982; Wisotzky 1994). This pH decrease initiates the destruction of the primary minerals, an intensive mobilization of elements and, ultimately, the precipitation of secondary minerals (Wiegand et al. 2000).

The overburden dumps consist of a mixture of Tertiary sediments (marine fine sand, silt, clay) and Quaternary sediments (gravel, sand, and clay) as in the unmined land (Bellmann and Starke 1990). Rainfall precipitating onto the overburden dump flows along the rib slopes into the alluvial area (between pairs of ribs), often accompanied by the generation of erosional channels. Due to the binding character of the sediments, the water will remain without any significant infiltration, forming small dump lakes. This type of water is also called dump water. It is of interest to know how the characteristic properties of the sediments, such as structure, homogeneity, and mineral abundance, change over the period of storage after spilling. What kind of qualitative and quantitative interactions between the sediments and the dump water take place? How fast are the processes of mineral destruction and precipitation in terms of time and pH-values? And what does one do with such "moonscapes" once one has sorted out what is currently going on?

2 Methods

Ninety-six sediment samples were acquired from the top of the ribs and, depending on the location, further samples were also obtained from depths of up to 1 m (in some cases up to 2 m). Following the mining reports of the Mitteldeutsche Braunkohlengesellschaft mbH (MIBRAG), it was possible to assign ages to the sediments (starting from their deposition in the conveyor bridge dumps) as shown schematically in figure 2.2a. Each point was localized by the Global Positioning System (GPS) with an uncertainty of about 5 m.

At the UFZ and the University of Leipzig standard sedimentological investigations (e.g. pH-values, grain size, water absorbing capacity, elution) were undertaken. In addition, optical microscopic observations and chemical analyses (x-ray fluorescence [XRF], x-ray diffraction

[XRD]) of selected sediments were carried out. In addition to the sediments, various samples of water were acquired.

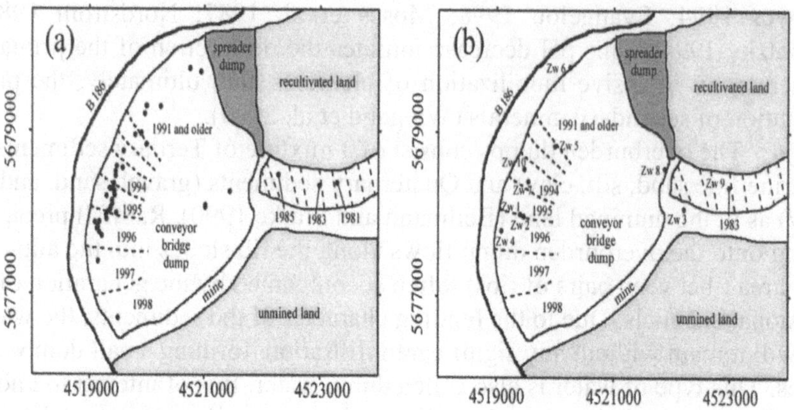

Fig. 2.2. Scheme of the lignite open pit Zwenkau including the estimated ages of the conveyor bridge dump and the locations where (a) sediments and (b) water samples were acquired

Figure 2.2b shows the locations of the dump lakes from which water was obtained at periodic intervals (about 3 months). Numbers 3 and 6 are groundwater drainage basins. The water samples were analysed immediately after obtaining the samples, while the chemical investigations (inductive coupled plasma mass spectroscopy [ICP-MS], ionic chromatography [IC]) followed later in the laboratory. These standard tests were undertaken so that one could identify optically, chemically, and with isotopic methods the dominant contaminant products in both the sediments and the connate waters. The major components obtained are discussed below.

3 Results and Discussion

Depending on the dumping technology, different mineral mixtures of the sediments from the unmined land were deposited in an arbitrary distribution so that, after spilling into the conveyor bridge dump, a strong primary heterogeneity was present. The surface is quickly fractionated after deposition because of the prevailing winds and rainfall – the fine silt and clay being transported into the alluvial area and coarse-grained gravels remaining on the tops or slopes of the ribs. This fractionation can be labelled secondary heterogeneity, and was observed in even the oldest ribs in Zwenkau. The conveyor bridge dump is characterized by a loose, unsorted structure with a high pore volume, a broad grain size distribution, and a small water absorbing capacity. Figure 2.3 shows the typical grain size distribution curves of the sediment samples analysed from the surface and to a depth of 1 m. There is no significant difference with depth or age.

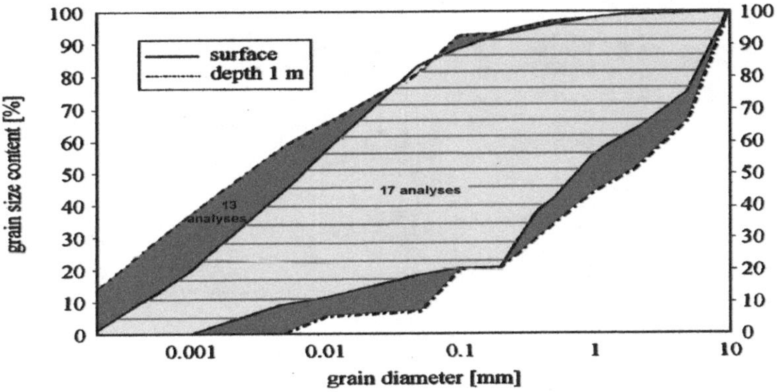

Fig.2.3. Typical grain size distribution curves of the overburden sediments obtained from the surface and to a depth of 1 m

The surface of the conveyor bridge dump remains dry even after strong rainfall. Only a small amount of water seeps into the sediments and only up to a penetration depth of about 1 cm. An investigation of the water absorbing capacity revealed that the mean water absorbing capacity of the sediments is about 30 %, which is extremely low. In addition, the filling of the pores by secondary minerals, to be discussed later, makes the sediments bind very tightly so that water cannot easily penetrate the surface.

Just a few days after deposition of the sediments an oxidation front is established within the near-surface region of the dumps, due to the oxidation of pyrites, which is found in the Tertiary sediments. A simplified equation for this chemical reaction is given by Wisotzky (1996):

$$4FeS_2 + 15O_2 + 14H_2O \Leftrightarrow 4Fe(OH)_{3\,(am)} + 8H_2SO_4 \qquad (2.1)$$

As a reaction product of pyrite oxidation, considerable iron ions and sulphuric acid are released. Because of the aggressive character of sulphuric acid a number of further reactions occur (figure 2.4).

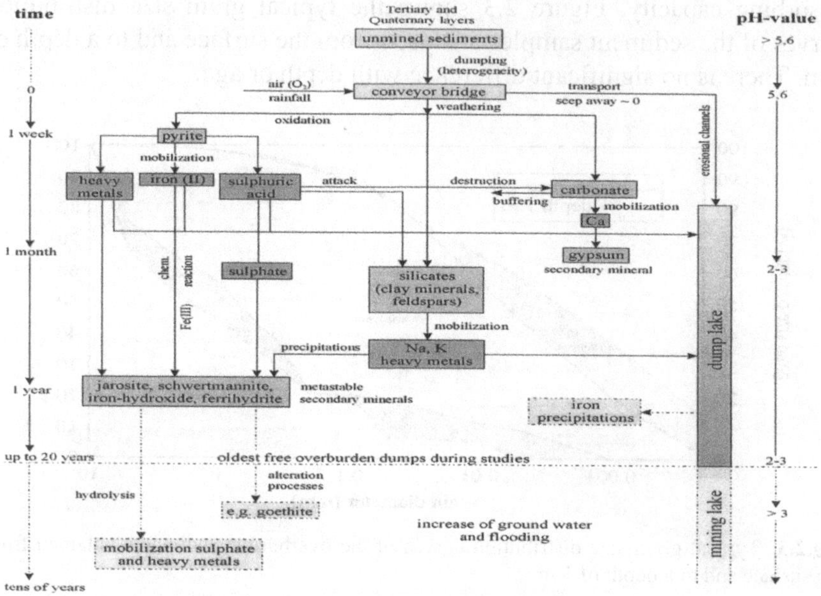

Fig. 2.4. Chronological scheme of alteration processes in the conveyor bridge dump of the open lignite pit Zwenkau

The whole process causes a drastic change of the geochemical situation. The pH-value of the sediments decreases to a level of about 2 to 3 within the first 20 weeks after deposition. At this point the iron buffer reaction (Prenzel 1985) stops the acidification:

$$Fe(OH)_3 + 3H^+ \Leftrightarrow Fe^{3+} + 3H_2O \qquad (2.2)$$

The pH-value then stabilizes, as indicated by an almost constant value, which can be found as well in the oldest ribs in Zwenkau. Figure 2.5 shows the alteration of pH with sediment age.

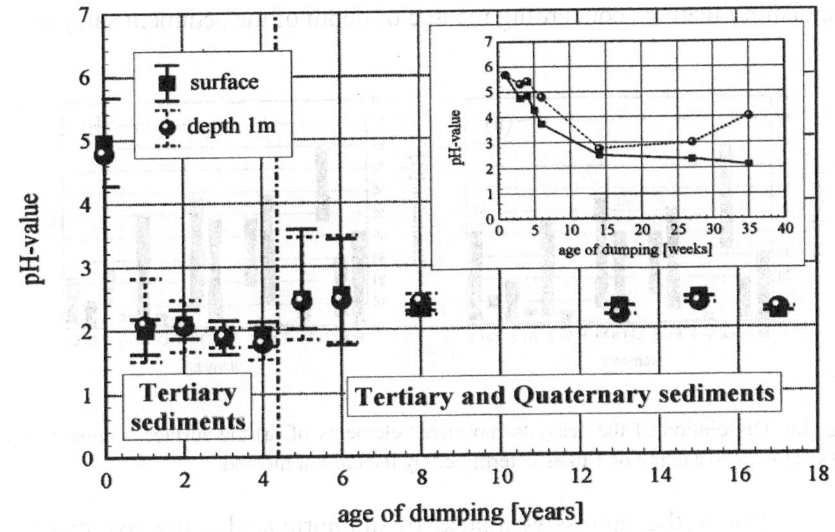

Fig. 2.5. pH-values of the dumped sediments versus sediment age

Figure 2.5 also shows that there exists a shift in the pH-value of about 0.5, consistent with the transition from the Tertiary and Quaternary mixed sediments to the pure Tertiary sediments. This step can be explained by the occurrence of pyrite that is typical for Tertiary sediments. In ribs consisting of both types of sediments the amount of pyrite is less than in other ribs, resulting also in a less marked decrease of the pH-value. In addition, the strong heterogeneity of the sediments causes local differences of the pyrite concentration so that the pH-value also shows strong local variations.

Depending on the acidification of the sediments, the primary minerals will also change (figure 2.4). Carbonates, if they existed in the unmined land, are completely destroyed by buffer reactions, except in the oldest ribs of the mining pit where carbonate concretions can be found sporadically. Other minerals, such as glauconite, feldspars, kaolinite, and muscovite / illite can partially withstand the virulent attack of sulphuric acid, but are slowly destroyed over the years. Only the major mineral quartz remains mostly unaltered.

The element signatures of the sediments, obtained by chemical analyses, show very strong variations. The conveyor bridge dump is of such a strong heterogeneity that it is impossible to find any significant distinguishing features concerning the age or depth of the sediment samples.

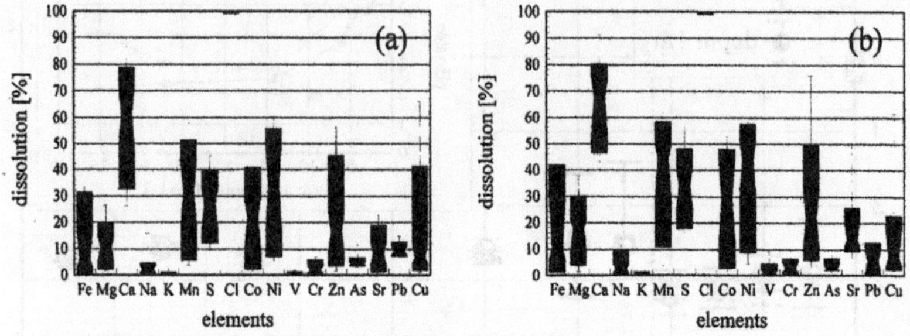

Fig. 2.6. Dissolution of the "easy-to-mobilize" elements of (a) the surface sediments and (b) sediments to a depth of 1 m as determined by the elution method

Due to the aggressive attacks of sulphuric acid, a massive dissolution (sometimes complete as in the case of Cl) of the "easy-to-mobilize" elements starts within the first few days after deposition of the sediments, as can be seen in figure 2.6 (see also figure 2.4). While rainfall transports most of the mobilized ions from the surface to the alluvial area (because of the extremely weak seepage) they remain at a depth of 1 m in the sediment pores and can be found there even in the oldest ribs after storage of 17 years. With a mean dissolution of about 20%, the "easy" mobilisation is very high.

The intensive input of ions into the alluvial areas is responsible for their very acidic character, so that even young dump lakes (about 1 month after deposition of the sediments) show pH-values in the range of 2 to 3 (figure 2.4). Because the mean amount of rainfall is less than the average evaporation in the Zwenkau region, the element concentrations in the dump lakes are even more increased. A comparison of the hydrochemical analyses of the rainwater and the dump water has also demonstrated the massive material input from the conveyor bridge dump. Due to these permanent alterations (dependent on the local concentration of pyrite and therefore also sulphuric acid, rainfall, evaporation, and precipitation) it is impossible to reach thermodynamic equilibrium in the dump lakes.

The dominating anion is sulphate as can be seen by the almost complete filled right half of each pie. In the case of cations (left half of the pies) different major concentrations can be observed according to the type of sediments present in the neighbouring ribs. Aluminium and iron dominate in the dump lakes surrounded by pure Tertiary sediments, while magnesium and calcium dominate in the case of dump lakes in the alluvial areas close to Tertiary and Quaternary sediments, which are present in ribs deposited before 1994. This behaviour is basically caused by the amount of pyrite (limited to the Tertiary sediments).

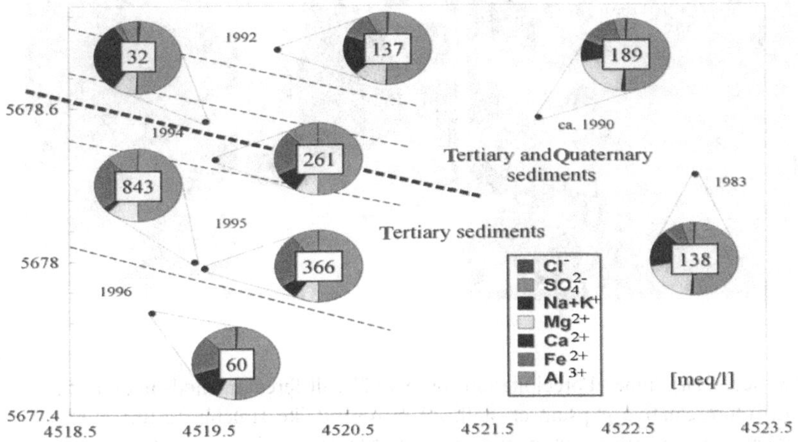

Fig. 2.7. Comparison of the element concentrations [meq/l] in the dump lakes of different ages – drawn in pie-chart fashion. The total concentrations of the dump lakes are given as [meq/l] in the middle of each circle (axes GAUSS-KRUEGER-Coordinates)

Away from locations in the conveyor bridge, it is possible to assign to the dump lakes the age of the surrounding sediments. Figure 2.7 shows a scheme in typical pie-chart fashion of the open pit has been noted already in the pH-value of the sediments (see figure 2.5).

Depending on the level of acidification, the age of the sediments and the geo- and hydrochemical situation, metastable secondary mineral associations were locally precipitated (see figure 2.4). Therefore, one has to differentiate between allochthonous and autochthonous formations.

At the surface of the conveyor bridge dump, as well as in the sediment pores, gypsum was observed about 1 month after sediment deposition. About half a year later iron hydroxide (and jarosite after 1 year) were present. Gypsum prefers to crystallize in rosette-like structures (figure 2.8a), while iron hydroxides form only amorphous mineral phases that appear as a cover surrounding the remaining primary mineral grains, as shown in figure 2.8b in the case of a pyrite grain. This cover retards the further weathering of the minerals, which also explains why pyrite minerals could be found in very old ribs. Generally, the sediment pores are filled by the formation of secondary minerals, a phenomenon that increases with age of the sediments.

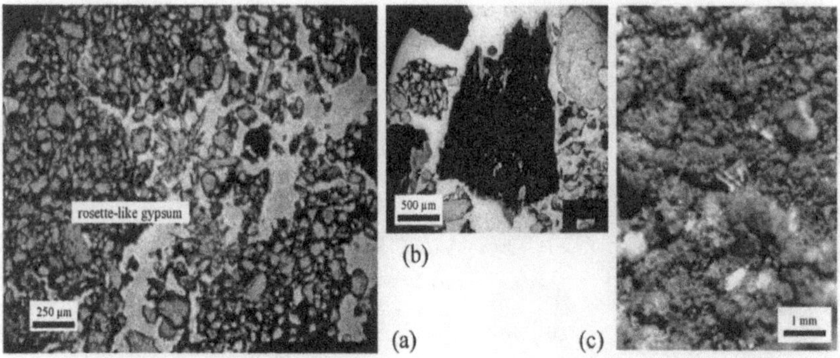

Fig. 2.8. Secondary mineral precipitations observed by different methods in optical microscopy: (a) representative gypsum crystallized in rosette-like structure in the pores of the overburden sediments (thin section microscopy); (b) pyrite grain covered by iron hydroxide (thin section microscopy); and (c) gypsum and iron hydroxide crystallization overlying the sediments in the wet areas of the conveyor bridge dump (direct light microscopy)

Apart from this autochthonous mineral precipitation, allochthonous secondary minerals can also be observed and analysed, such as gypsum, jarosite, alunite, melanterite / rozenite, iron hydroxide, ferrihydrite, and schwertmannite (figure 2.8c). These minerals usually occur in the wet areas of the conveyor bridge dump. The secondary mineral phases can also buffer parts of the mobilized elements by incorporation.

Considering the conveyor bridge in terms of exogenous processes (dumping, wind, rainfall) and endogenous processes (destruction and secondary formation of minerals), an interaction can be recognized (figure 2.4). The exogenous dumping is an initiating effect for a number of endogenous consequences: oxidation of the pyrite causes the release of sulphuric acid; mineral destruction is initiated by sulphuric acid, which then causes a massive mobilization of elements; and, finally, the formation of secondary mineral phases. Because of further occurring exogenous events over time, such as wind and rainfall, surface fractionation takes place, mobilized elements are transported to the dump lakes, with erosional channels being created. A considerable amount of new sediment reaches the surface so that the ongoing endogenous processes continue to be supported and new processes initiated. This interaction can be observed even in the oldest ribs in the Zwenkau conveyor bridge dump.

The major problem for the recultivation of the open pit Zwenkau is the strong potential of acidification contained in the conveyor bridge dump, estimated at about 30 kg H_2SO_4/m^3 (Cesnovar and Pentinghaus 2000). From investigations in other open pits in the Central German lignite mining district, which have been flooded already, it is known that further interactions (hydrolysis, mineral alteration; figure 2.4) will happen over the years (Geller et al. 1998). An important fact is that the owner of the lignite pit Zwenkau does not intend to cover the conveyor bridge by a spreader dump as is usually done. Rather it is planned to flood the whole area. In the case of natural flooding (natural increase of groundwater), contamination of the groundwater with iron, sulphate, aluminium, and heavy metals is inevitable. An arranged flooding is to be preferred because of its accelerated flooding behaviour and the additional amount of non-acidic water. Similar to the dump lakes, the large content of clay in the conveyor bridge dump, and the pore filling effect by secondary minerals, will both cause a temporary sealing of the dump surface, which hinders the acidic potential. Furthermore, the possibility of sediment sliding and slumping is present because of the mechanical instabilities of the conveyor bridge dump (Carstensen and Pohl 2000). Flattening the sediment slopes and planting reduce this effect (Kirmer and Mahn 1996).

4 Technical Conclusion

In this study the consequences of the energy policy of the former GDR in terms of lignite mining, and the resulting damage to the environment, have been viewed from the long term interactions between the sediments of the conveyor bridge dump of the lignite open mining pit Zwenkau and the produced mining water. Due to dumping of the unmined sediments, an extremely heterogeneous sediment mixture is created. Very soon after deposition, oxidation of the pyrite minerals begins, an oxidation front is formed, and the pH-value decreases to about 2 to 3 within the first 20 weeks. Sulphuric acid, as a product of pyrite oxidation, destroys the primary minerals. Large amounts of elements are mobilized and primarily remain in the pores of the overburden sediments. Precipitation takes place of autochthonous secondary minerals, such as gypsum, jarosite, alunite, melanterite / rozenite, and iron hydroxides. In this way the pore volume of the sediments is filled, which causes a reduction of the water absorbing capacity of the sediments. Rainfall cannot infiltrate the sediments and will shed on the slopes of the ribs, creating dump lakes in the alluvial areas. On its way to the alluvial troughs the rainwater dissolves mobile elements and transports them to the dump lakes, also drastically and rapidly decreasing the pH-values to about 2 to 3. Allochthonous mineral formation was also analysed. Optical microscopic methods showed that secondary iron hydroxide in the sediments covers the remaining primary minerals, and so retards further weathering. A continuing interaction is seen of the exogenous and endogenous effects, even for the oldest ribs in the conveyor bridge dump, which is why hydrological and geochemical equilibria cannot be established.

5 A Legacy for the Future?

In the case of Zwenkau , it would seem that the pit is to be artificially flooded to provide a lake, perhaps suitable for future recreational purposes. However, there is no clear understanding of the long-term repercussions of this action. Perhaps initially the flooding will act to dilute the acidic conditions, but what will happen in the long-term as leaching continues (and presumably then causing oxygen-starved bottom waters) is a complete unknown, and one that needs careful monitoring.

What one does with such open mining pits in general is the subject of major debate. Complete sediment infilling is one proposal for then one can have a sport ground, or a meadow. But what effect such infilling will have on subsurface acidic groundwater, or on people playing on such areas, or on future subsidence, or on crop growing (if such is undertaken) or ingestion by food chain animals or humans, seem not to be at all clear for their effects. Presumably some of this uncertainty is the cause for the major debate concerning what to do. We remain ignorant of our fate in this respect at least- and that is perhaps the strongest environmental lesson to be learnt from the energy policy of the GDR and the inherited legacy we face today.

Chapter 3

Carbon Dioxide Development and the Influence of Rising Groundwater in the Cospuden/Zwenkau Dump: Observations and Inferences

Summary

Investigations at the Cospuden/Zwenkau dump show that as sources for CO_2 in the lignite open pit mining dumps, both lignite particles as well as carbonate dissolution come into question. The lower dump material was laid down as the waste material after lignite mining, and was set down in the form of ribs parallel to the exhumation direction; it makes up about 2/3 of the dump mass. The surface of this first component was let sit fallow for many years and so had an intensive air admittance before the overlying cover material, made up of local materials, was superposed. The residual large quarries, due to coal excavation, were flooded and converted to lakes. The change in the lake levels with time, and the horizontal influx of lake waters into the dump, means that carbon dioxide is altered from free phase gas to gas in water solution with concomitant changes in dissolution and out-gassing capabilities.

The mining spoils that were redumped had the consequence that the original formations were destroyed, as were the various water transport pathways through the geologically deposited rocks, and the redeposition was a mix of unconsolidated spoil material. Because of aeration (ground water withdrawal), naturally occurring reducing anaerobic regions were changed to aerobic oxidizing conditions. Near the formations impeding gas exchange at the dump surface a considerable amount of carbon dioxide would seem to be dissolved in water. Measurements of the carbon dioxide flow at the dump surface show no increase in out-gassing rate relative to a control area outside the dump region.

Isotopic composition of the carbon dioxide from many probes through the dump speaks out against the origin from various sources. The negative $\delta^{13}C$ values (between –24.5 to –26.5 ‰) imply a dominant organic origin rather than one from carbonate dissolution.

Out-gassing measurements show that, despite the higher carbon dioxide content of the lignite open pit dump in comparison to ordinary earth, no enhanced emission rate is seen. For the organic and carbonate contents in the dump materials, one can calculate the maximum possible carbon dioxide development. From one tonne of lignite (water content 52%, coal material content 48%) there results about 582 m^3 CO_2 from burning (corresponding to about 1144 kg). Paying attention to the fact that the dump is composed of about 2/3 underlying dump materials and about 1/3 of overlying dump material, there is about 29 kg of coal material per tonne of mixed dump material. From this figure one can obtain about 106 kg (54 m^3) of carbon dioxide. Because the extraction of each tonne of lignite required three tonnes of mining "spoils", the carbon dioxide amount that could be produced by direct burning of lignite in the atmosphere can be further increased by about 30% as a consequence of oxidation and carbonate dissolution processes in the dump itself.

1 Introduction

In respect of the genesis and distribution of CO_2 in aerated soils, both naturally occurring and anthropogenically influenced, there have been many investigations and publications. The general result of such investigations is that the main source for CO_2 is the respiration of organic plant materials in their root regions. In order that these microbiologically driven processes provide CO_2 contributions in the surficial regions and also the atmosphere, there has to be a significant influence of temperature.

The CO_2 generated in the soils is mainly transported to the atmosphere by diffusion-type processes, but convection, air pressure variations, climatic, and hydraulic variations must also be taken into account (Matthess 1990; Merkel et al.1992). Transport in the subsurface by CO_2 solubility variations in undersaturated water is also a partial contributor. As a consequence, the CO_2 content in the aerated soils is strongly dependent on the local conditions, but generally lies under 1%. In soils that are used for agricultural and forestry businesses the CO_2 content can rise as high as 3%. A further increase in the CO_2 content, at the expense of oxygen content, can be due to the loss of plant material (Langhoff 1971; Aslanboga et al. 1979). Investigations of the composition of trapped gases in the mining dump of the Rhine lignite area have yielded massive increases in the carbon dioxide content (up to 20%) in association with very low oxygen content. Carbonate in the dumped material is viewed as the main source for the high CO_2 content (Wisotzky 1994). Because of high sulphuric acid development, which follows from oxidation processes caused by pyrite connections to ground water and sulfide reactions (Bierns de Haan 1991), there is then a CO_2 release from the carbonate material (Singer and Stumm 1970; Stumm and Morgan 1981; Taylor et al. 1984).

Tertiary sediments in central Germany, which in the main consist of deposition during the Rupelian and which provide the sealing layers and also the dump material in the central German lignite region, are considerably higher in pyrite and carbonate content (marine carbonates and fresh water carbonates) than those in the Rhine lignite area by a factor of 100 or more (Schreck et al. 1998; Dohrmann 2000; Wiegand 2002). Accordingly, these sediments provide the basis for the development of a much greater, and thereby environmentally relevant, carbon dioxide content.

This geological state then leads one to ask the fundamental question of whether the developed carbon dioxide is, in the main, transported to the atmosphere or under what sort of conditions would the carbon dioxide so generated remain in the dump? In this chapter we consider this basic question.

2 Methods and Measurements

The first measurements of the gas composition in the Cospuden dump, taken at a depth of 10 m, yielded the expected high concentration of carbon dioxide. Isotopic analyses of carbon, which originally were thought to distinguish the various carbonate types as contributors to the CO_2, yielded the result that the CO_2 in the investigated region was not dominantly caused by carbonate leaching, but rather was brought about by oxidation of organic material. Because of the lack of vegetation, only finely disseminated, unoxidized, lignite particles, available to microbiological activity, can be the source of the carbon dioxide (Hiller et al. 1988).

In order to make a first estimate of the growth potential of the dump materials for carbon dioxide production, a drilling core was investigated for its content of organic (C_{org}) and carbonatic (C_{carb}) carbon. These two classes of carbon were considered as the main potential sources for the carbon dioxide. The drilling was done from the present-day surface of the dump through to the original base of the open-pit mine. For technological reasons the dump is composed from two structural components: an underlying bridging support dump, and an overlying covering deposited dump that parallels the land relief.

The first component was immediately laid down as the waste material after lignite mining, and was set down in the form of ribs parallel to the exhumation direction; it makes up about 2/3 of the dump mass. The surface of this first component was let sit fallow for many years and so had an intensive air admittance before the overlying cover material was superposed, made up of local materials. Because various sorts of geological material were used for this deposited part of the dump, the result is that there are different geochemical compositions. Thus the underlying bridging support dump material is made up of an extremely homogeneous mixture of carbonate, phosphorite and pyrite rich Tertiary marine materials (sand and silt), while in contrast the covering dump material is composed mainly of lamella-shaped deposition of Quaternary blanketing materials (till, melt water sands, river terrace gravels), with a Tertiary section built up between the two components composed of river sands, silts and, occasionally, a considerable fraction of lignite (Bellmann et al. 1977). The core profile showed a high C_{org} in the region of the covering dump and, with one exception, no C_{carb} content, while in the underlying bridge dump material low C_{org} but considerable C_{carb} contents were the rule. On this basis results from more probes for both the dump materials and also the subsurface air could be broken out into these two types of dump components. In total 40 drillings for the underlying bridge dump and 32 for the covering dump were undertaken. In order to obtain such a large number of quantitative drilling results, both drill cores and pressure driven cores were obtained. The drill probes provide a complete coverage of the dump profile, while the pressure driven cores were obtained for each type of dump material, respectively.

For the investigation of the gas composition, closable gas sondes were installed at various depths in each type of dump material. Following the massive flooding of 2002, which came extremely close with non-connate waters to the residual Cospuden "hole", the pressure water table in the dump rose considerably and much more quickly than had been anticipated. This rise occurred only in the communicative water pathways in the dump, but did not oversaturate the total dump. As a consequence, more sondes had to be installed in the area of investigation and new sondes installed at shallower depths. Stainless steel was used for the stationary gas sondes finish, in order to prohibit any changes in gas composition as a consequence of corrosion effects. Because of the change in the capillary tube size (an internal diameter of 2.16mm), the flushed volume obtained around the probe area is then minimized in terms of the danger of contamination from atmospheric air.

For the measurement of the out-gassing rate, right-angled frames with U-shaped profiles (made from stainless steel) were placed on the surface of the dump. The measurement locations were free of vegetation and were approximately $0.25m^2$ in area. To obtain measurements of the out-gassing rates, measurement boxes (so-called "Lemberger Boxes" (State Bureau for Environment of Baden-Wuerttemburg 1992)) were placed in the water filled areas. From the permanent thoroughly mixed central region of the box, repeated probes were taken and the CO_2 content verified. From the increase in concentration during the measurement time it was then possible to compute the out-gassing rate. Away from the measurement areas of the various dump types, two comparison measurement areas were chosen. At the lysometer control station Brandeis (near Leipzig), the State Environmental Company, under the business direction control of the Sachsen Ministry for Environment and Agriculture, placed 1995 more lysometers that were filled with material from the Espenhain dump. That monolithic excavated block has a surface area of about 1m^2 and is about 3 m long. In terms of material composition and isotope composition, the material is extremely similar to dump materials at Cospuden/Zwenkau. Continuous investigations could be made of the composition of the gas phase at several of the lysometers. Extraction of gas at three different depths was possible. In addition, seeping water amounts and their pH values could be obtained. In order to hold the pressure values in the lysometers at constant values during the period of the gas collection, a special probe technique was available. The extraction of gas was accomplished by diffusion in helium charged vessels. In this way it was possible to establish a weekly probe cycle, which helped lead to a much better understanding of the measurement results.

2.1 Results

2.1.1 Investigations of the Dump Materials

The typical range of the C_{org} and C_{carb} contents through the profiles that spanned the total dumpsite are shown in Figure 3.1a. Organic carbon material was found in all investigated probes, with the higher concentration in the overlying dump material.

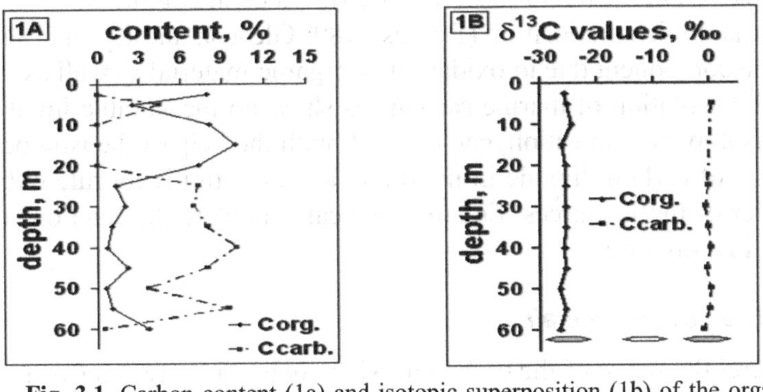

Fig. 3.1. Carbon content (1a) and isotopic superposition (1b) of the organic and carbonatic materials for a 60 m long profile through the dump

Carbonatic carbon material is to be found only in the material from the underlying dump, with one exception. This distribution pattern was also found in the other investigated probes and pressure driven cores, and has its origin in the different geological components of the dumped substrate material.

The changeover between the two types of dump materials is clear to see over a mean thickness of about 20 m (a variation of between 9 m to 29 m dependent on the rib positions). Cores from the surface of the underlying dump material demonstrate that the weathering effect in most cases led to no carbonate content. The dissolution with sulfuric acid, arising from the rapid pyrite decomposition by weathering, is apparently complete in this region. From the complete data set one can calculate the following values. The overlying dump material contains, on average, 6.2% organic carbon (a range of 0.7% to 14.7%) and only exceptionally is carbonatic carbon material (the residual from the glacial till) found. In the underlying dump material, however, there is a mean organic carbon worth of 1% (a range of 0.4% to 3.7%) and a carbonatic carbon value of 1.4% (range of 0% to 10.1%).

Figure 3.1b shows the results of the isotopic investigations. The $\delta^{13}C$ values of the carbonate lie between +0.5 and –1.5 ‰, and so inform on their marine origin. The $\delta^{13}C$ values of the organic material vary between –24.5 and –26.5 ‰, which are typical values for humic material (plants, turf, coal) (Hiller et al. 1988). The bars in the lower area of the picture represent the variation spread of all investigated probes. Isotopic signatures that could have indicated fresh water carbonates were not found. The $\delta^{13}C$ values should typically lie between the two basic material groups in the unfilled regions. The results show that for the development of carbon dioxide two different reservoirs exist, which are significantly different in their isotopic composition (Deines 1980; Gleason and Kyser1984). That both the component due to oxidation of organic material as well as that due to acid dissolution of marine carbonates show no measurable involvement in the isotopic composition, can be used, with the help of the isotope investigations of carbon dioxide in the dump air, to definitively rule out one or the other of these sources. Equally, one can calculate the fractions of each source in a mixture.

2.1.2 The subsurface air

In general the range of the carbon dioxide content is between 6% and 20%, but in some cases can be above this range (figure 3.2). The open symbols represent gas from the overlying dump, while the closed symbols, crosses and stars represent gases from various regions of the underlying dump material. The carbon dioxide content in the subsurface air is not obviously dependent on the state of the dump material. Nevertheless, it seems that temperature influences the oxidation process. The open circles that are connected with a broken line represent gases from 1m deep (the overlying dump). By 5 m deep (stars with a continuous line, and in the overlying dump material) it is no longer so obvious to see the effect. Gases from 10 m deep are marked by crosses and open rectangles.

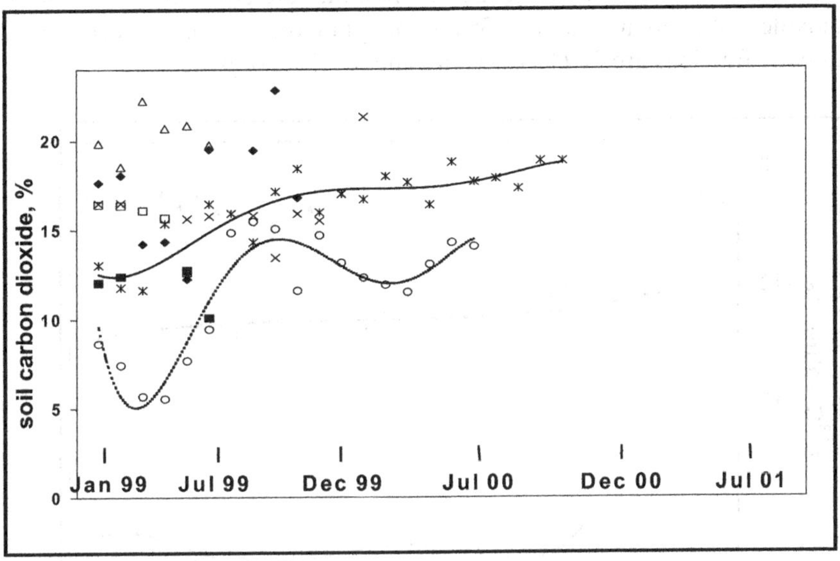

Fig. 3.2. Carbon dioxide content for the subsurface air

In total, during the time of the investigation, the carbon dioxide content shows an increasing tendency. One possible cause is the steadily increasing pressure surface of the waters in the dump; on the other hand, consolidation of the dump materials during the drilling and core recovery could also lead to such results.

If one considers the isotopic compositions of the gases, they speak out against the origin from various sources (figure 3.3). In the lower part of the diagrams one finds gas from the overlying dump (open symbols). The negative $\delta^{13}C$ values (between –24.5 to –26.5 ‰) imply a dominant organic origin (see also Cerling et al. 1991; Haas et al. 1983; Thorstenson et al. 1983). The smaller negative $\delta^{13}C$ values in the upper region of the diagrams (filled symbols) were measured on gases from the underlying dump material. Because fresh water carbonate can be excluded, these values must be interpreted as representing a mixture with a strong component of carbon dioxide originating from marine carbonates, and a smaller fraction that originated from oxidation of organic carbon materials. In contrast to the overlying dump material, both source components are present in the underlying dump material. The $\delta^{13}C$ values that lie in between (crosses and stars) represent gases taken from other underlying dump materials. In this case the isotopic composition indicates a greater fraction of an organic source. With the help of an isotopic balance equation one can work out the individual fractional components. The first mixture contains 33.5% carbon dioxide that is organic, and 66.5% of inorganic origin. For the second mixture the fractions are 50.6% (organic) and 49.4% inorganic.

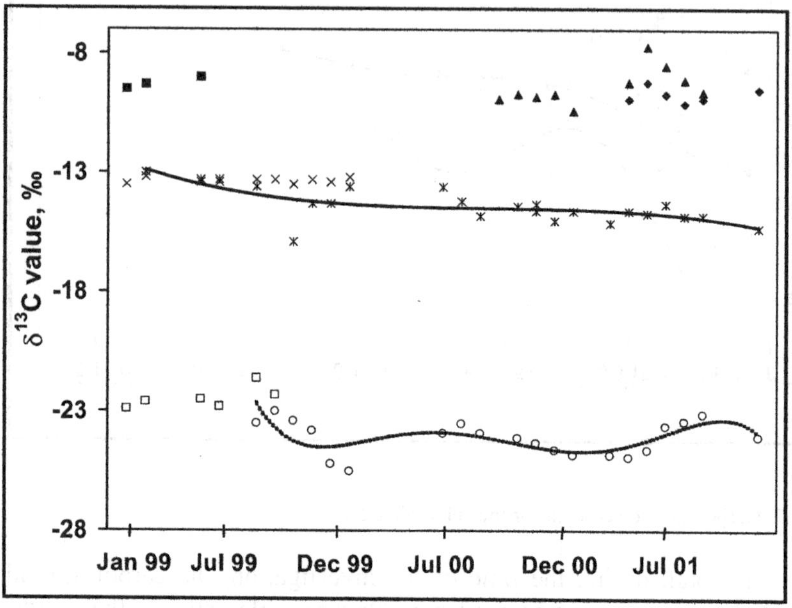

Fig. 3.3. Carbon dioxide content for the subsurface air

In contrast to the carbon dioxide content, the $\delta^{13}C$ values vary surprisingly little for an individual dump material, implying that the various carbon dioxide contents do not result as a consequence of variations in the fractions contributed by each source but rather arise from different source strengths or from dispersal effects.

A smaller decline in the $\delta^{13}C$ values during the time of the investigation is, with extremely high probability, due to the decline in the carbonate replacement; the cause can be exhaustion of the carbonates but may also be due to pyrite exhaustion.

2.1.3 Measurements of the Out -gassing Rate

Because of the high content of carbon dioxide and the precursor lignite open pit mining dump, the question of the development of carbon dioxide after the end of mining was of particular interest. Measurements of out-gassing rates should indicate if these dumps emit more than the average amount of carbon dioxide into the atmosphere. The results of the measurements are shown in figure 3.4.

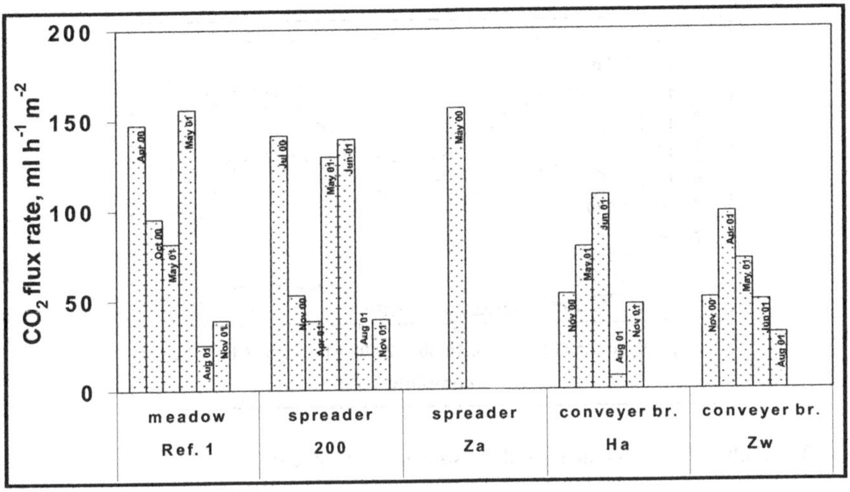

Fig. 3.4. Carbon dioxide out-gassing rates for various measurement levels

All measurement areas clearly show a yearly time-dependence. In addition, from the surface of the underlying dump there is a smaller flow rate than from the overlying dump material. As the most important operating factor, the humidity is predominant. During the time the dump material lay on the ground it built a surficial region of some thickness composed of fine-grained sands and shale minerals. This process was increased by water (Wiegand 2001). Before the measurement campaign in August 2001 there was a long rainy period, which indicated extremely small flow rates of gas. A surprising result is the fact that, despite the higher carbon dioxide of the dump, practically the flow rate of the similar measurement horizons was not exceeded. Data from literature for forest regions lie between 80 to 800 mlh^{-1}m^{-2} (Normann et al. 1997). Near the formations impeding gas exchange at the dump surface a considerable amount of carbon dioxide would seem to be dissolved in water. While lake waters entering the dump contain about 100 mg/l of HCO_3^-, the waters in the dump contain 900 mg/l and even more of HCO_3^-.

2.1.4 Lysometer Investigations

Figure 3.5 shows the carbon dioxide content in three different depths of the monolithic block (undisturbed dump material taken from the quarry Espenhain) taken at the lysometer station Brandeis in Sachsen. Here, too, the contents of CO_2 increase with increasing depth, as in the dumps.

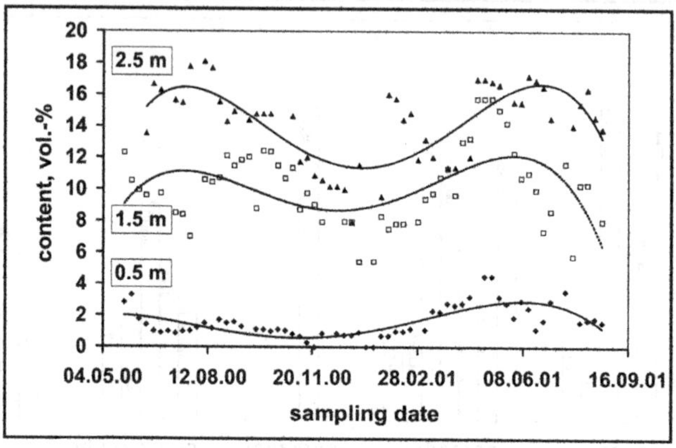

Fig. 3.5. Carbon dioxide content for the lysometer atmosphere

In contrast the oxygen content decreases. Because the dimensions of the lysometers are not very big, the effects of temperature and carbon dioxide growth are extremely noticeable. Owing to the gas exchange processes, variations in the neighborhood of the surface are lessened. Air pressure, rainfall, and seepage water amounts were also measured, but without any significant effects being at all noticeable. At the beginning of the measurements, depth dependent differences in the isotopic composition of carbon dioxide were obtained (figure 3.6). The lysometer materials had a composition between 0.79 % and 9.08 % C_{org} (with a mean value 3.13 % for five measurements, and $\delta^{13}C$ = -25.5±0.3 ‰) and only small amounts of carbonate content, so the isotopic composition favors strongly the organic source. Small contributions from the carbonate source push the measurement direction of $\delta^{13}C$ towards the positive direction. Low pH values in seepage waters from this region during this time period make carbonate replacement possible. During the period of the probe measurements the $\delta^{13}C$ were more and more negative, which would seem to imply an end to the carbonate replacement possibility. Higher pH values in seepage waters in the lysometers confirm this suggestion.

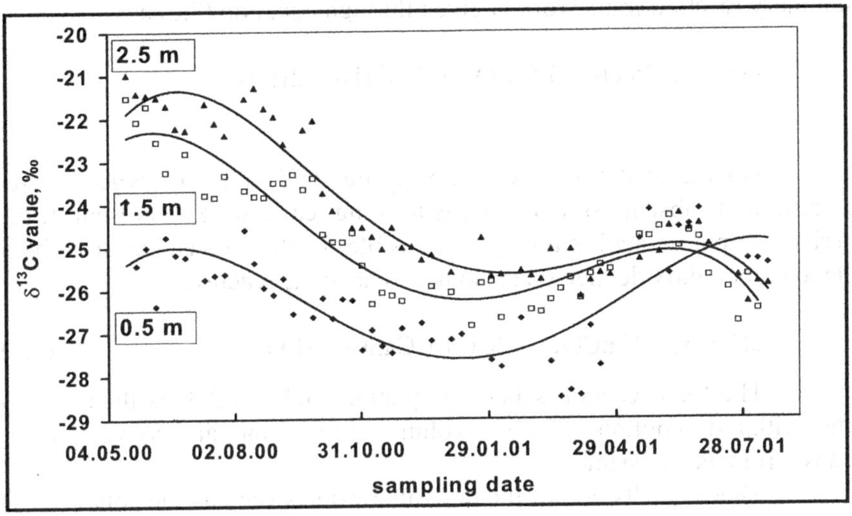

Fig. 3.6. Superposition of the various carbon dioxide isotope measurements

2.2 Discussion

The investigations carried out show that as sources for CO_2 in the lignite open pit mining dumps, both lignite particles as well as carbonate dissolution come into question. In particular, the overlying dump contains a higher content of coal-like materials. In the oxygen bearing regions oxidation processes start, eventually leading to carbon dioxide and water through the equation:

$$C_nH_{2n}O_n + nO_2 \rightarrow nCO_2 + nH_2O \tag{3.1}$$

A change in the gas volume is not associated with this reaction; gas exchange follows mainly through diffusion processes. Oxidation reactions are exothermic; a vertical temperature distribution curve from the investigation region shows a temperature increase of about 4°C between 5 m and 11 m depth (Schreck et al. 1998). This temperature maximum coincides with the high C_{org} content of the drilling profile of figure 3.1a.

In order to change carbonate in the underlying dump to carbon dioxide there has to have been a previous reaction that changes pyrite (containing sulphatic sulphur) to sulfate oxide and so to sulphuric acid. A summary reaction framework is given through equation (3.2) as:

$$FeS_2 + 3.75\ O_2 + 3.5H_2O \rightarrow Fe(OH)_3 + 2H_2SO_4 \qquad (3.2)$$

For the oxidation of one mol pyrite, 3.75 mols of oxygen are required, in the absence of another gas forming reaction being present. If the derived sulphuric acid is totally responsible for the conversion of carbonate to carbon dioxide then one also has to have the reaction:

$$2H_2SO_4 + 2CaCO_3 \rightarrow 2CO_2 + CaSO_4 + H_2O \qquad (3.3)$$

The total reaction scheme (equations 3.2 and 3.3) is then associated with a diminution of the gas volume in the dump, and convective gas transport is then possible.

These results do not include other effects such as the solubility of carbon dioxide in water, or the use of sulphuric acid for other reactions. Such would, in turn, lead to a further reduction of the gas volume. The development of gas overpressure as a consequence of the above reactions is excluded in unsaturated regions. Because of the dump structure there is an emergency flow path for water out of the neighboring residual lake in the dump. In this way there is the possibility that particular areas within the dump have water outflow but gas trapping. An increase in the hydrostatic pressure leads to compression of such gas bubbles. However, with increasing pressure there will also be a greater amount of carbon dioxide dissolved in water so that any overpressure can be partially or completely compensated.

For the organic and carbonate contents in the dump materials, one can calculate the maximum possible carbon dioxide development. From one tonne of lignite (water content 52%, coal material content 48%) there results about 582 m^3 CO_2 from burning (corresponding to about 1144 kg). Paying attention to the fact that the dump is composed of about 2/3 underlying dump materials and about 1/3 of overlying dump material, there is about 29 kg of coal material per tonne of mixed dump material. From this figure one can obtain about 106 kg (54 m^3) of carbon dioxide. Because the extraction of each tonne of lignite required three tonnes of mining "spoils", the carbon dioxide amount that could be produced by direct burning of lignite in the atmosphere can be further increased by about 30% as a consequence of oxidation and carbonate dissolution processes in the dump itself.

The out-gassing measurements show that, despite the higher carbon dioxide content of the lignite open pit dump in comparison to the ordinary earth, no enhanced emission rate is seen. Near the sealing layers at the surface, solution effects can also play a role. What cannot be predicted is the carbon dioxide gas-combining rate in secondary minerals that can occur due to later changes in the CO_2 saturation capability of the dump.

From the above reactions one can also estimate the relative reaction speeds. In contrast to the pyrite oxidation, which occurs in a relatively short time, and the consequent carbonate dissolution that is also rapid, oxidation is a much slower process, even when catalysed by microbial activity. It follows, therefore, that the dump is not a particularly intensive generator of carbon dioxide but, rather, that over a long period of time stored carbon dioxide is emitted.

That the measurements of the gas phase allow an investigation over a long time period means that it is worthwhile to undertake further research with lysometers. Because the previous investigations have shown that it is possible to estimate the residual carbonate dissolution using carbon isotope measurements, simulation of the effects of water level increase using gas and isotope analyses should be undertaken. Of particular interest would also be to investigate when (and if) a change in the reduction conditions could occur due to methane generation.

The work undertaken and the results obtained have provided the basic character of the carbon dioxide generation potential and its escape in open pit mining dumps. Thus one can surmise that there is an additional component to CO_2 production as a greenhouse gas apart from that arising from the burning of fossil fuels (in this case lignite) to provide energy. Within the framework of this project, no qualified conclusions have been presented.

3 Discussion and Conclusions

Lignite mining in central Germany was (and in a much smaller manner, even today, is) associated with major changes in the landscape. In order to deliver energy from the excavated lignite, first the lignite must be separated from the residual "spoils" material. Above and in between the lignite seams and benches there were mainly sandy, shaley or gravel beds that were, therefore, excavated with the large mining equipment used, and then later redumped in the open pit region where the lignite had been removed. After variable but long time periods ranging to tens of years, these dump landscapes were then covered with an overlying dump material in order that one could later use the areas for commercial land or forestry purposes.

The residual large quarries, due to coal excavation, were flooded and converted to lakes.

The mining spoils that were redumped had the consequence that the original formations were destroyed, as were the various water transport pathways through the geologically deposited rocks, and the redeposition was a mix of unconsolidated spoil material. Because of aeration (ground water withdrawal), naturally occurring reducing anaerobic regions were changed to aerobic oxidizing conditions. The total hydro-geochemical conditions were changed to aerobic oxidizing conditions. The oxidation effects were not confined to the surface and near-surface regions. Because of the lack (after redeposition) of the original transport controlling formation structures, the oxidation penetrated deep into the dump. The oxidation processes led to a change to sulphuric acid of pyrite, which was deposited in Tertiary marine sediments. A known consequence is the acidification of water around and in the dump. If such occurs in waters neighboring the dump, such as lakes formed from the open quarries, then there is a lowering of the pH.

A further result is the increase in carbon dioxide content in the unsaturated regions of the body of the dump. Procedures such as carbon content and carbon isotopic investigations can be used to show that there are two sources. One source is from marine carbonates that are to be found in marine sediments of the lower Oligocene (Rupelian) and are dissolved by the development of sulphuric acid. A second source is organic material in the form of coal-like particles that begin to change slowly to carbon dioxide at reaction centers.

In order to estimate the carbon dioxide potential of a dump area, many drill profiles and piston cores (over 70) were used. A tonne of spoils material contains approximately 29 kg of coal-like material, from which 54 m^3 carbon dioxide can be released. In order to extract a tonne of lignite (that was burnt to convert about 50% of its substance to carbon dioxide that was then released to the atmosphere) three to four tonnes in total of spoils material had to be excavated as well. As a consequence, the redeposited spoils material generates carbon dioxide that can produce up to about a 30% increase in the total carbon dioxide budget release to the atmosphere.

Measurements of the carbon dioxide flow at the dump surface show no increase in out-gassing rate relative to a control area outside the dump region. Therefore, there are many possible different causes for this lack of an effect: partial sealing of the dump surface by shale material; solution of carbon dioxide in neutral rising dump waters, or limiting speed of the oxidation reactions. These effects mean that the dump is not an intensive, spontaneous carbon material source, but rather is slow and diffuse in character. A long-term effect cannot be ruled out because of the high carbon dioxide content of the dump.

Chapter 4

Carbon Dioxide Development and the Influence of Rising Groundwater in the Cospuden/Zwenkau Dump: Quantitative Models

Summary

The variation of carbon dioxide in the Cospuden/Zwenkau dump is due to a variety of reasons, some having to do with the different types of residual mining spoils used as underlying and overlying dump materials, and some having to do with the dynamical evolution of conditions both within the dump and externally, as time progresses.

The purpose of this chapter is to show how one can quantitatively handle such variations of external and internal conditions, even when one does not know with precision whether, for instance, a leak exists and the fraction of gas it can leak. Nevertheless one can construct model behaviors that allow one to explore the probable consequences of such likely leaks. In addition, the variation of lake level pressure, of the seasonal swing of temperature, and the variation in the availability of the supply of carbon dioxide, are all factors that one can investigate to determine their influence on the saturation of carbon dioxide in the dump waters, the amount of free phase gas one anticipates may be in the dump, and the potential loss of carbon dioxide from the dump to the atmosphere. Results of such investigations are presented here so that one can determine the influence of each factor in the potential release of carbon dioxide and also the residual free phase gas still trapped in the dump as well as the amount still in solution with water. The general procedures are applicable to any dump involved in carbon dioxide generation and /or release.

1 Introduction

As noted in Chapter 3, the rise of carbon dioxide content in residual mining dumps of the former DDR has consequences for land usage, for possible release of acid mine waters, as well as for release of carbon dioxide to the atmosphere, thereby increasing the greenhouse gas emission rate.

Chapter 3 explored the experimental results obtained from probes of the dump at Cospuden/Zwenkau, to the south west of Leipzig in Saxony, Germany. This chapter is concerned with providing quantitative models to allow a sharper insight into the manner in which carbon dioxide can be distributed within a dump, either as free phase gas or in water solution, and to examine the effects of temperature variations as well as rising water pressure due to flood effects on the carbon dioxide content. In addition, the effects of probable leaks in the dump covering material in terms of the influence on carbon dioxide release to the atmosphere are considered.

In order to deliver energy from the excavated lignite, first the lignite must be separated from the residual "spoils" material. Above and in between the lignite seams and benches there were mainly sandy, shaley or gravel beds that were, therefore, excavated with the large mining equipment used, and then later redumped in the open pit region where the lignite had been removed. After variable but long time periods, these dump landscapes were then covered with an overlying dump material in order that one could later use the areas for commercial land or forestry purposes. The residual large quarries, due to coal excavation, were flooded and converted to lakes.

The mining spoils that were redumped had the consequence that the original formations were destroyed, as were the various water transport pathways through the geologically deposited rocks, and the redeposition was a mix of unconsolidated spoil material. Because of aeration (ground water withdrawal), naturally occurring reducing anaerobic regions were changed to aerobic oxidizing conditions. The total hydro-geochemical conditions were changed to aerobic oxidizing conditions. The oxidation effects were not confined to the surface and near-surface regions. Because of the lack (after redeposition) of the original transport controlling formation structures, the oxidation penetrated deep into the dump. The oxidation processes led to a change to sulphuric acid of pyrite, which was deposited in Tertiary marine sediments. A known consequence is the acidification of water around and in the dump. If such occurs in waters neighboring the dump, such as lakes formed from the open quarries, then there is a lowering of the pH.

A further result is the increase in carbon dioxide content in the unsaturated regions of the body of the dump. Procedures such as carbon content and carbon isotopic investigations can be used to show that there are two sources: One source is from marine carbonates that are to be found in marine sediments of the lower Oligocene (Rupelian) and are dissolved by the development of sulphuric acid. A second source is organic material in the form of coal-like particles that begin to change slowly to carbon dioxide at reaction centers. In order to estimate the carbon dioxide potential of a dump area, many drill profiles and piston cores (over 70) were used. A tonne of spoils material contains approximately 29 kg of coal-like material, from which 54 m^3 carbon dioxide can be released. In order to extract a tonne of lignite (that was burnt to convert about 50% of its substance to carbon dioxide that was then released to the atmosphere) three to four tonnes in total of spoils material had to be excavated as well. As a consequence, the redeposited spoils material generates carbon dioxide that can produce up to about a 30% increase in the total carbon dioxide budget release to the atmosphere.

Measurements of the carbon dioxide flow at the dump surface show no increase in out-gassing rate relative to a control area outside the dump region. Therefore, there are many possible different causes for this lack of an effect: partial sealing of the dump surface by shale material; solution of carbon dioxide in neutral rising dump waters, or limiting speed of the oxidation reactions. These effects mean that the dump is not an intensive, spontaneous carbon dioxide material source, but rather is slow and diffuse in character. A long-term effect cannot be ruled out because of the high carbon dioxide content of the dump.

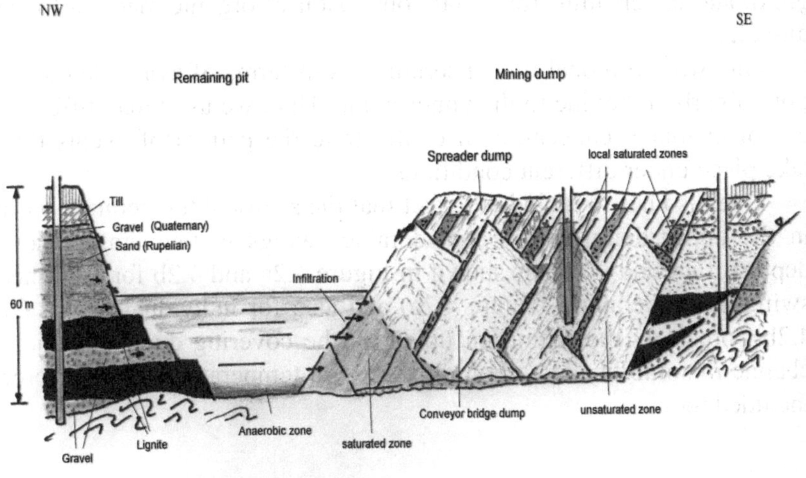

Fig. 4.1. Cross-section of the dump at Cospuden/Zwenkau

Figure 4.1, showing a cross-section through the dump at Cospuden/Zwenkau. The triangular regions show the lower dump material (Tertiary) that was allowed to lie fallow and exposed for several tens of years, while the shaded upper regions show the low permeability covering dump material (Quaternary). To the left of the dump is the residual lake produced by anthropogenic flooding of the residual quarry. The rise in the lake level due to natural flooding by about 7 m in 1996 is also shown. Further flooding events due to major rainfall are not shown, but even as late as 2002 there was a rise in the lake level to almost the dump surface due to the torrential rains that were the norm for that year.

The corresponding lake level rise was some 25 m. Note that the upper dump material is approximately 15 m thick but varies from around 5 m thick at the peaks of the lower dump structures to around 25 m thick at the valleys between the peaks.

Interest in this chapter centers on the evolution possibilities for carbon dioxide in the upper dump. As shown in Chapter 3, the supply of carbon dioxide to the upper dump from below, and the generation of carbon dioxide in the upper dump, are sufficiently high that one can maintain a steady-state balance of supply at all times. Precisely how large a supply is available is not known except, as shown above, one is able to provide an approximate upper limit from total conversion of organic material to carbon dioxide.

One way to model this uncertainty is in terms of available concentration of carbon dioxide to the upper dump. Here we use three different values of available concentration to illustrate the pattern of events that can take place under different conditions.

One also has to include the fact that the seasonal temperature variation in the upper dump shows considerable swings in temperature down to depths of around 20 m, as shown in Figure 4.2a and 4.2b for both the total swing in temperature (figure 4.2a) and also for individual depths (figure 4.2b), precisely the region occupied by the covering dump material. The change in solubility of carbon dioxide with temperature must therefore be included too.

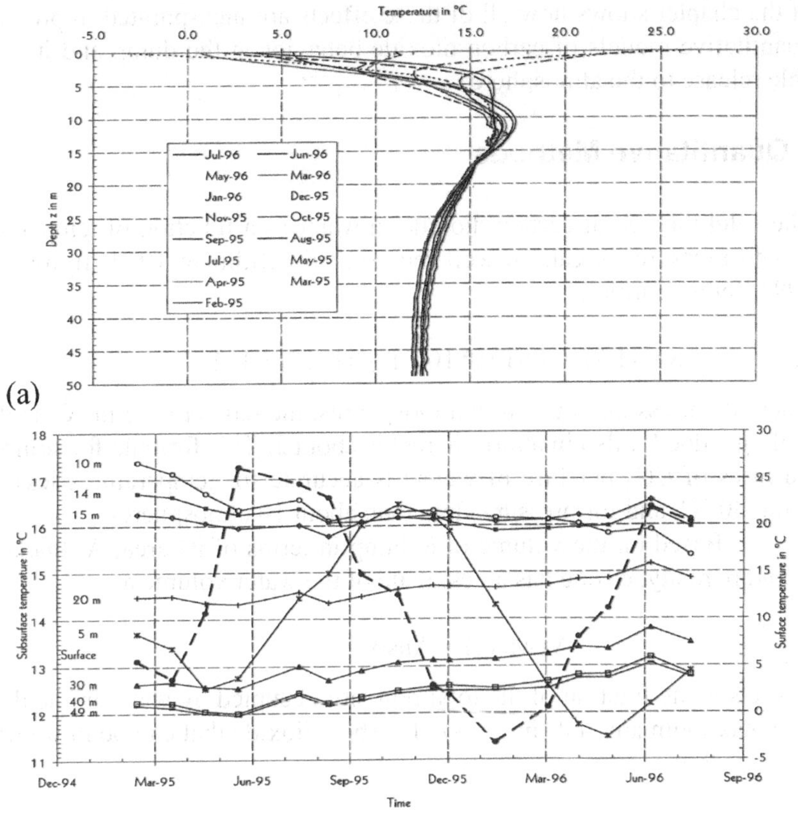

(b)

Fig. 4.2. a. Variation of temperature with depth and season of the year through the Cospuden/Zwenkau dump; b. Variation of temperature at individual levels through the dump

Of concern is the probability of a leak in the overlying dump material with a release of some, or all, of the carbon dioxide that is either in the form of free phase gas in the dump or in water solution. In addition, if there is a leak then the formation pressure in the overlying dump will be reduced so that the solubility of carbon dioxide will be lowered at a given temperature, thereby releasing gas that, in turn, can escape through the leak. This factor must also be included.

Finally, because the lake water level changes with time, and because there are connective hydraulic pathways into the dump, then the formation pressure will also change with time. In turn, this change implies that the carbon dioxide solubility will vary from this effect alone. Such a change in the system must also be included in quantitative models. The next section of the chapter shows how all of these effects are incorporated in providing quantitative models of carbon dioxide behavior in the dump and its probable release to the atmosphere.

2 Quantitative Methods

The solubility, S, of carbon dioxide in water as a function of temperature, T, and pressure, p, can be well represented (Milliman 1974; Iijima et al. 1993) by the formula

$$S = (1.72\text{-}0.06T+9*10^{-4}\,T^2)*(1+2*(p\text{-}1)/P) \tag{4.1}$$

where the pressure is given in atmospheres, the temperature in °C, and the scaling value P (also in atmospheres) is about 2. This formula for solubility (in litres of CO_2 per litre of water) is accurate for temperatures less than about 30 °C, and for pressures less than about 10 atmospheres.

Based on the volume of a dump in terms of its area, A, thickness, h, and porosity, φ, one has an estimate of the water volume as

$$V(\text{water}) = Ahs\varphi \tag{4.2}$$

where s is the fraction of the total porosity occupied by water in the dump. The maximum amount (in litres) of carbon dioxide that can be in water solution is then

$$V_{max} = V(\text{water})S \tag{4.3}$$

Any carbon dioxide above this limit must then be in the form of free phase gas within the dump. As the temperature and pressure in the dump change with time or space, so, too, does the solubility so that carbon dioxide can either go into solution with the water or must be out-gassed from the water to be in the free phase gas budget component.

Because the overlying component of the dump is not fully saturated there is freely available space for the gaseous component of the carbon dioxide to be emplaced.

The available concentration, C, of carbon dioxide is the proxy used to describe how much carbon dioxide is allowed to be in the system. If the available concentration is less than the solubility then all the available CO_2 goes into water solution; if C>S then only the fraction S goes into water solution with C-S being free phase gas.

If there is a probability, p, of a leak in the overlying dump material then the fraction, f, of free phase gas that can be leaked is allowed to be any value up to the maximum available free phase gas total. In addition, once one has a probability of a leak then, presumably, the pressure in the fraction of the dump containing carbon dioxide can also change to a lower value, which then allows carbon dioxide to out-gas because the solubility is decreased. So a component of this "new" free-phase gas is then also subject to the leakage conditions of the original free phase gas and can escape the system.

As the water level in the neighboring dump lake rises during flood conditions, then the formation pressure in the dump also rises in domains connected hydraulically to the lake conditions. As a consequence, more carbon dioxide from the free phase gas component can go into water solution, to be released later when either the lake water level drops or when the temperature in the dump rises, thereby lowering the solubility limit.

A simple Excel program has been constructed to include all these behaviors. A representative sample version of the program is given in Appendix 1 to this chapter.

Based on the numerical procedure there are a variety of interesting examples to explore. These are sequentially investigated below in two groups as follows. Group A effects in the absence of a potential leak: (i) variable available carbon dioxide; (ii) seasonal temperature effects; (iii) lake level rise. Group B effects in the presence of a potential leak: (i) leak efficiency; (ii) lake level variation effects; (iii) available carbon dioxide; (iv) seasonal temperature effects. We consider each in turn.

3 Model Behaviors

3.1 Group A Models

3.1.1. Variable Available Carbon Dioxide

As a first simple illustration consider the situation where the temperature in the dump is at the yearly average of about 12 °C. The pressure in the dump, if sealed, is around 2 atmospheres (corresponding to an average material density of about 2 g/cm^3 at an average depth of 15 m) while the area of the dump is about 100 m (width) x 1000 m (breadth) =10^5 m^2, with the average depth to the overlying dump material base at around 15 m. The saturation limit for carbon dioxide in water under these conditions is some 2.23 l/l. For the three situations of available carbon dioxide concentration of 1 l/l, 3 l/l, and 10 l/l one can then compute the carbon dioxide in water solution and the amount in free phase gas. Note that the maximum possible available carbon dioxide concentration in the Cospuden/Zwenkau dump is around 27 l/l if all the residual lignite has been converted to carbon dioxide (Glaesser et al. 2004). Note also from Glaesser et al. (2004) that the observations show carbon dioxide content (by volume) ranging from a low of a few percent near the surface to around 18% at depth, so that the three values of 1, 3 and 10 l/l adequately cover this range. The three illustrations therefore take the position of less than total maximum carbon dioxide production in the dump because of the rise in carbon dioxide measurements with time during the course of the investigation over a few years.

For the three cases there are several obvious consequences. First, for the situation of an available carbon dioxide concentration of 1 l/l, all the carbon dioxide stays in water solution because the solubility is greater than the supply. Thus a water-dissolved amount of 10^5 liters of carbon dioxide is present in the overlying dump. Second, in the situation where one raises the available carbon dioxide concentration supply to 3 l/l, then a small fraction of the carbon dioxide is in the form of free phase, while, third, for the situation of 10% carbon dioxide most is in the free phase gas form, as shown in figure 4.3.

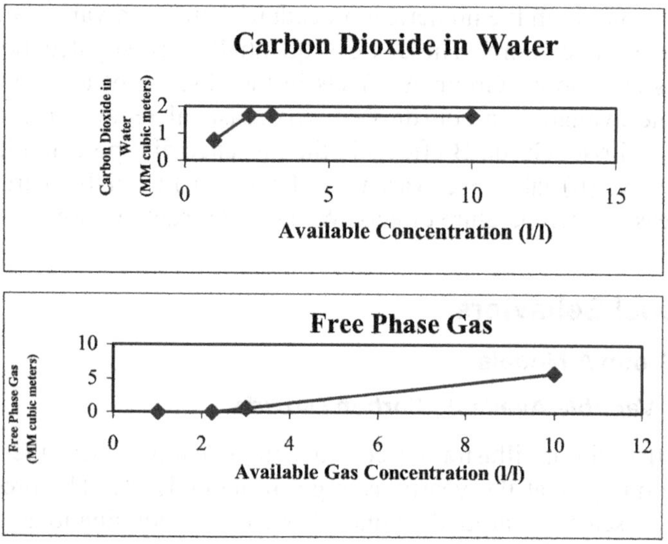

Fig. 4.3. Variation of the free phase gas component and the carbon dioxide in water solution as functions of the available carbon dioxide concentration when the dump is sealed

3.1.2. Seasonal Temperature Effects

The temperature in the shallow part of the dump, less than about 20 m from the surface, shows considerable swings with the time of year, ranging at the surface from around 0 ° C in winter through to around 25 °C at high summer. In order to more than cover the total swing of temperature variations at depths to around 15 m, here we allow the surface range of temperature to prevail. In that way we exaggerate the effects of temperature variability in terms of carbon dioxide in water solution or free gas phase as the year cycles. We also deal with just the situation of a carbon dioxide content of 3 l/l because it is the case most sensitive to changes in the solubility, which lies around the 2.2 l/l mark for an average temperature of 12 ° C. Shown in figure 4.4 are the variations of carbon dioxide in water solution and as free phase gas as the temperature varies. As expected, the amount of carbon dioxide in water solution drops significantly (by a factor of almost 2) as the temperature rises from the mid-winter value of around 0 ° C through to the summer high value of around 25 ° C. Correspondingly, the amount of free phase carbon dioxide gas climbs from a value of zero in mid-winter to just over 1 MMm^3 (MMm^3 = million cubic meters) in high summer.

3.1.3 Variable Lake Level Effects

As shown in figure 4.1, the variation in the lake water level, is about 10 m or more in 1995 and, in the summer of 2002 when there was massive flooding in Eastern Germany, was almost 35 m, reaching

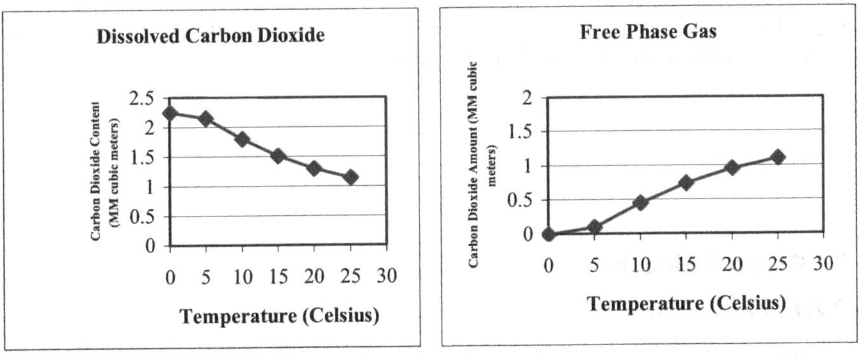

Fig. 4.4. Variation of the free phase gas component and the carbon dioxide in water solution as functions of temperature when the dump is sealed

nearly to overflow stage at the dump surface. The infiltration of this lake water horizontally into the dump meant a considerable increase in the pressure, almost 1 atmosphere increase in 1995 and about 3 atmospheres increase in 2002 over the "nominal" value of about 2 atmospheres. The influence of such a pressure change on the phase in which the carbon dioxide can exist is now examined.

As a canonical illustration we again choose the available carbon dioxide concentration to be at the value 3 l/l and deal with a fixed temperature of around 12 ° C because the influence of the entering cold lake water will keep the seasonal temperature changes to a lower swing magnitude. Figure 4.5 shows the variation of free phase gas and gas in water solution as the pressure in the dump rises due to lake level increases. For values of pressure less than about 3 atmospheres, the carbon dioxide in water solution steadily rises until, at 3 atmospheres and above, the water is completely saturated at the value 2.25 l/l. Correspondingly, the free phase gas amount drops steadily as the pressure increases until, at the 3 atmosphere mark, all the free phase gas has gone into solution with the water.

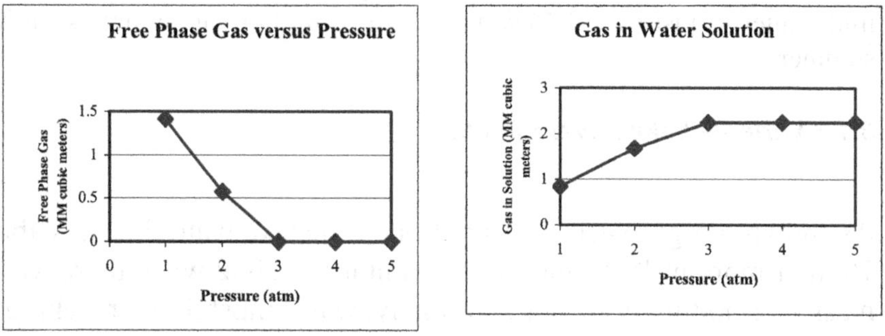

Fig. 4.5 Variation of the free phase gas component and the carbon dioxide in water solution as functions of pressure when the dump is sealed

3.2 Group B Models

3.2.1 Leak Efficiency

In order to illustrate the effects of a potential leak for carbon dioxide from the dump, in this first set of illustrations we take the original formation pressure in the dump to be at 3 atmospheres, representing a compromise between lake level rise and overburden load; we also take the dump to be at a fixed temperature of 12 ° C, and the final pressure, if there is a leak, is set at 1 atmosphere, representing direct contact to the atmosphere. Because there is no guarantee that there is a leak, we allow the probability of a leak to vary, but the maximum fraction of free phase gas that could escape through the leak is fixed at 50%. For instance, at a leak probability of 0.1, the probable fraction of gas leaked is then 0.5*0.1, which is just 5%, while

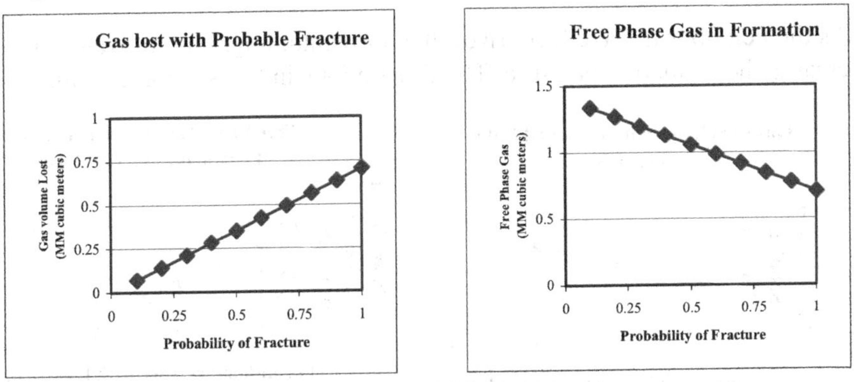

Fig. 4.6 Variation of the probable free phase gas and probable gas loss as functions of the leak probability for parameters given in text (Group B, example a)

at a 90% chance of a leak the fraction of free phase gas that could be leaked is 0.5*0.9, a total of 45%.In this situation we deal with the probable amounts of gas leaked, the probable amounts of free phase gas remaining in the formation, and the probable amounts of gas still retained in water solution. We also set the available carbon dioxide concentration at 3l/l. Two contributions to the gas leakage are clear. First, there is leakage of the original free phase gas from the formation; second, because the formation pressure is lowered to 1 atmosphere from the original 3 atmospheres, then the solubility of carbon dioxide is lowered and so gas is exsolved from the water. This extra free phase gas is then subject to the same leakage conditions as the original free phase gas. Figure 4.6 shows the probable amounts of free phase gas in the formation as well as the probable amounts of carbon dioxide leaked as functions of the leak probability. Note from figure 4.6 that the probable free phase gas in the formation drops from around 1.3 MMm^3 to just under 0.75 MMm^3 as the probability of a leak steadily increases, while the probable amount of carbon dioxide that could be leaked to the atmosphere steadily increases (almost linearly) to a maximum of around 0.75 MMm^3 as the leak probability increases, representing the 50% limit for a 100% chance of a leak. Thus the efficiency of the leak and the probability of a leak both influence the amount of probable carbon dioxide that could escape.

3.2.2 Lake Level Rise

As the neighboring lake level rises, the horizontal migration of water in creases the formation pressure. The illustrations in this section consider

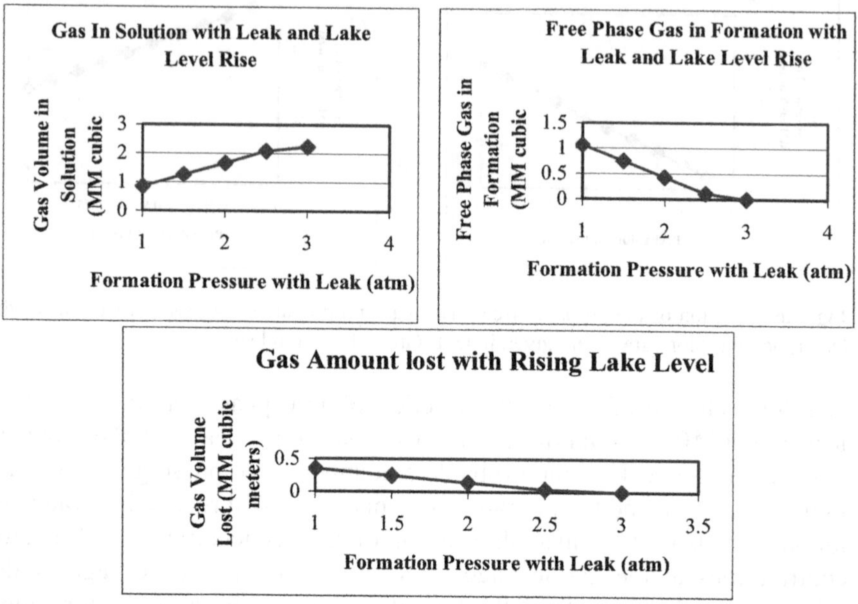

Fig.4.7. Variation of the probable free phase gas, the probable gas in water solution and probable gas loss as functions of the lake level pressure for parameters given in text (Group B, example b).

the original formation pressure set at 3 atmospheres, the dump temperature set at 12 ° C, the available carbon dioxide concentration set at 3 l/l, the maximum free gas fraction leakage at 50%, and the probability of a leak is set at 50% as well. The final formation pressure, if there is a leak, is now dependent on the lake level. As the lake rises so, too, must the external pressure from the original final pressure of 1 atmosphere, which is the value if the leak is open to the atmosphere. Accordingly, the solubility of carbon dioxide in water will be higher in the higher lake level situations, and so more carbon dioxide will remain in solution, with less available for the free phase gas component, which is the portion subject to possible escape to the atmosphere or to the lake waters.

Figure 4.7 shows this sort of situation for the probable residual gas in water solution, the probable free phase gas remaining in the formation, and the probable gas loss from the dump. Note that once the final pressure on the dump equals the original 3 atmospheres there is no gas loss because the conditions have not then changed. At lower values of lake level, as recorded by the pressure conditions, there is probable gas loss of around 0.35 MMm^3 at a final pressure of 1 atmosphere, which drops to zero as the final pressure reaches 3 atmospheres. For the probable free phase gas remaining in the dump, figure 4.7 shows that it is a touch above 1 MMm^3 at a final pressure of 1 atmosphere, gradually decreasing to zero as the final pressure rises above the 3 atmosphere mark. Correspondingly, the carbon dioxide in water solution rises steadily as the water pressure rises, representing the increase in solubility with increasing pressure, until saturation is reached at 2.225 l/l at the 3 atmosphere mark.

Thus a rising lake level inhibits (by a significant factor) carbon dioxide release even in the presence of a probable leak. In turn, by considering the three parts of figure 4.7 to represent a lowering lake level, and so a lowered pressure, it follows that a falling lake level promotes carbon dioxide release by precisely the same factor.

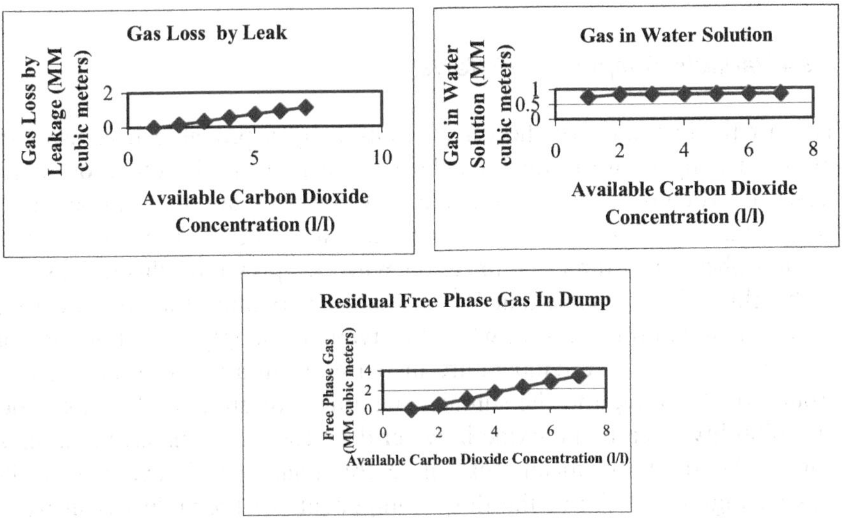

Fig. 4.8. Variation of the probable free phase gas, the probable gas in water solution and probable gas loss as functions of the available carbon dioxide concentration for parameters given in text (Group B, example c).

3.2.3 Available Carbon Dioxide

This illustration shows the effects of varying the amount of available carbon dioxide in the presence of a probable leak.

The parameters fixed are the original pressure at 3 atmospheres, the dump temperature at 12 ° C, the final pressure, if there is a leak, at 1 atmosphere, the probability of a leak is again held at 50% and the maximum free gas that could be leaked is held at the fractional value of 50% as well. The variable in this illustration is the supply of available carbon dioxide concentration. Shown in figure 4.8 are the probable gas in water solution, the probable gas in the formation, and the probable gas lost to the system as the available supply of carbon dioxide is varied. Note that the water quickly saturates with carbon dioxide so that, once the available carbon dioxide concentration crosses around 2 l/l, thereafter the dump waters are saturated at all higher values. Correspondingly, the free phase carbon dioxide gas in the dump rises and so too does the amount of carbon dioxide that can be released to the atmosphere. Both rise linearly with increasing carbon dioxide availability, with the probable loss to the atmosphere being about 0.25 of the free phase carbon dioxide in the dump because the leak probability and the maximum fraction that could be lost through the leak are both set at 0.5, so their product is precisely the 0.25 recorded in figure 4.8.

3.2.4 Variable Temperature Effects

This final illustration of the effects caused by a probable leak uses the swing in temperature throughout the year to indicate the retention or release of carbon dioxide from the dump. Again we retain the original dump pressure at 3 atmospheres, and the final pressure, if there is a leak, is set at 1 atmosphere. For direct comparison with the previous illustrations, the probability of a leak is held at 50%, as is the maximum fraction of carbon dioxide that could be leaked, while the available supply of carbon dioxide is held at 3 l/l. As the temperature increases from a winter time value of around 0 ° C through to the summer maximum of around 30 ° C, so, too, the solubility of carbon dioxide in water decreases. In turn, such a diminution in the solubility means that more free-phase gas is available to be leaked. Figure 4.9 shows the three components of probable gas in water solution, probable free phase gas in the dump, and probable gas loss as functions of temperature. Note from figure 4.9 that the probable gas in water solution decreases from winter to summer from about 1.3 MMm^3 to 0.5 MMm^3. For the free phase gas in the dump and the gas loss from the dump, one sees from figure 4.9 that there is an increase in probable gas loss from around 0.25 MMm^3 at winter temperatures to around 0.42 MMm^3 in summer, with also a shift in the free gas component in the dump from around 0.75 Mmm^3 to around 1.25 Mmm^3, representing the exsolution of gas from water due to temperature increase plus the probable loss through the leak. The point of this example is to show that temperature effects can also play a significant role in altering the gas loss for other conditions being held fixed.

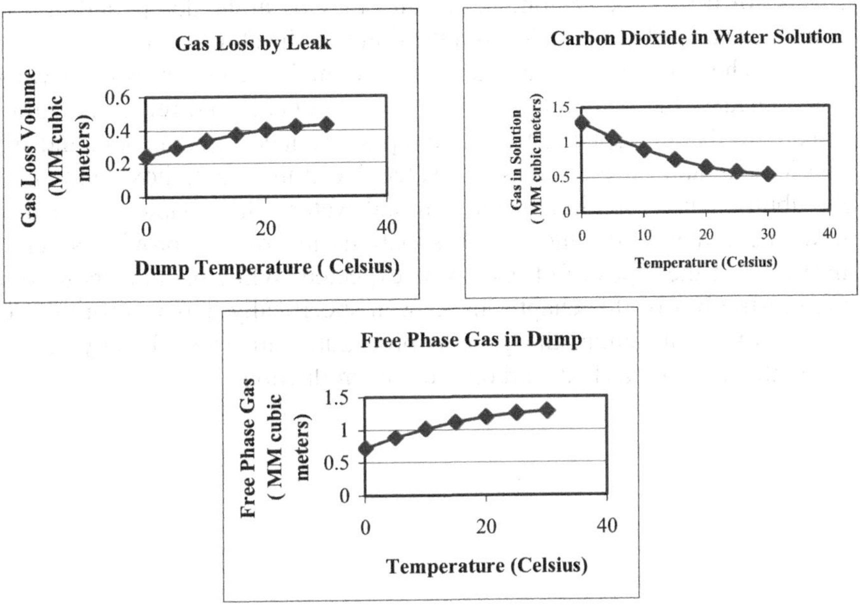

Fig. 4.9. Variation of the probable free phase gas, the probable gas in water solution and probable gas loss as functions of the dump temperature variations with season for parameters given in text (Group B, example d)

4 Discussion and Conclusions

The variation of carbon dioxide in the Cospuden/Zwenkau dump is due to a variety of reasons, some having to do with the different types of residual mining spoils used as underlying and overlying dump materials (Glaesser et al. 2004), and some having to do with the dynamical evolution of conditions both within the dump and externally, as time progresses.

The purpose of this chapter has been to show how one can quantitatively handle such variations of external and internal conditions, even when one does not know with precision whether, for instance, a leak exists and the fraction of gas it can leak. Nevertheless one can construct model behaviors that allow one to explore the probable consequences of such likely leaks. In addition, the variation of lake level pressure, of the seasonal swing of temperature, and the variation in the availability of the supply of carbon dioxide, are all factors that one can investigate to determine their influence on the saturation of carbon dioxide in the dump waters, the amount of free phase gas one anticipates may be in the dump, and the potential loss of carbon dioxide from the dump to the atmosphere.

The necessity for such quantitative model behaviors is clear when one considers the potential of a dump for producing and releasing carbon dioxide to the atmosphere, for the dump to produce carbonic acid and sulphuric acid, and for the release of waters laced in carbon dioxide to either the abutting lake or directly into ground water. This chapter has shown how one can indeed handle such situations in order to provide a set of measures of the types of effects to be expected. While parameters used in the illustrations of this chapter have been specifically drawn from the Cospuden/Zwenkau dump, the general procedures are valid for any sort of dump that has, or has had, carbon dioxide production..

Appendix 1.Carbon Dioxide Program

Area being considered=	20000	sq.m
Thickness being considered=	10	m
Porosity of formation=	0.5	fraction less than unity
Temperature of formation=	15	Celsius — Keep less than 33C!!!!
Formation pressure =	3	atmospheres
Solubility of CO_2 in water =	3.0255	cubic m/cubic m — Water Volume= 100000 cubic meters
Available concentration of CO_2 =	1.08	cubic m/cubic m of water
CO_2 volume in formation water=	108000	cubic meters
Free phase CO_2 in formation =	0	cubic meters

FREE PHASE GAS LEAK

Probability of a leak=	0.3	fraction less than unity
If leak then estimate of fraction of free phase gas leaked=	0.2	fraction less than unity
Probable free gas leak fraction=	0.06	
Probable Residual free gas in formation=	0	cubic meters
Probable free phase gas leak amount =	0	cubic meters

SATURATED WATER GAS LEAK

If leak, then reduction of pressure to value=	1	atmos
Solubility if leak=	1.0085	litres/litre of water
If leak new CO_2 volume in water=	100850	cubic meters
If leak CO_2 volume released=	7150	cubic meters
Probable gas leak fraction =	0.06	
Probable residual gas in water=	100850	cubic meters
Probable water gas lost =	429	cubic meters
Probable free gas in formation=	6721	cubic meters

TOTALS IF THERE IS A LEAK

Probable gas in water solution	1E+05	cubic meters
Probable free gas in formation	6721	cubic meters
Probable total gas loss	429	cubic meters

Chapter 5

Environmental and Economic Risks from Sinkholes in West-Central Florida

Summary

Data from the last twenty years for sinkhole occurrences in West Central Florida are used in conjunction with population and housing data, including house prices, to assess the risk of a house being swallowed by a sinkhole, together with the likely economic loss. The top five relocation cities in each of Hillsborough, Pasco, and Pinellas counties are investigated in detail to determine relative risk as well as absolute risk. Because of the massive urbanization taking place in the area, the sinkhole risk is increasing due to water related problems as well as because of the underlying karst topography. Pinellas County is better placed in this regard than Hillsborough or Pasco Counties because of its lower karst component. The city with the highest likelihood of risk is Tampa, probably due to its massive urbanization over the last twenty years. Coastal cities most at risk from sinkhole effects on housing are New Port Richey and Hudson, but they are a very distant second and third compared to Tampa. Economic damage risk estimates to housing are around $5MM per year for Tampa and are sure to increase in the near future.

1 Introduction

The Tampa Bay area in West-Central Florida is one of the largest relocation destinations in the United States. Resources of the region include the world-famous beaches, balmy year-round temperatures, and theme parks. However, this region is also prone to numerous natural environmental risks, including hurricanes, tornadoes, and sinkholes. The whole region is underlain by carbonates, and the presence of sinkholes in carbonate karst domains is well known.

The purpose of this chapter is to examine the three-county region of Tampa Bay, namely Hillsborough, Pinellas and Pasco counties, and create an environmental risk analysis for sinkhole occurrence based on the last 20-year interval of sinkhole data in the region, subdivided by risk per county and per city, and with 15 cities chosen for the study (top 5 relocation cities from each of the three counties). An economic risk analysis considers median housing value of each city, used as a weighing factor to determine monetary environmental risk of each location. The first part of this study examines sinkhole formation in Florida; the second part determines risk analysis based on data for the region.

2 Sinkhole Formation and Occurrence

2.1 General Remarks

Sinkholes are closed depressions in the land surface formed by the dissolution of near-surface rocks, or by the collapse of the roofs of underground channels and rivers. Sinkholes are very common in Florida due to the landscape being mainly a karst terrain, which is one defined by water dissolving the overlying limestone bedrock, and characterized by sinkholes, caverns and springs. The main cause is dissolution of underground limestone by slightly acidic water. Broadly speaking, the main sequence of events leading to a sinkhole is:

1.Rain falls through the atmosphere, absorbs carbon dioxide, and forms a weak carbonic acid;

2.This acidified water moves through the vadose zone and reacts with organic matter, becoming more acidic;

3.This acidically enhanced water slowly dissolves the limestone, causing voids or cavities into which overlying sediments can collapse;

4.The end result is a sinkhole.

West-Central Florida has two distinct aquifer systems:

1.Floridan Aquifer System: a large confined aquifer system extending over most of Florida and the southeastern U.S;

2.Surficial Aquifer System: a discontinuous unconfined system that represents the local water table in the study region. This second aquifer system is more influenced by changes in rainfall, run-off, and land use than the Floridan, and therefore is the main catalyst for sinkhole development in the study area.

Florida has three main types of sinkholes:

1. Limestone Solution Sinkholes: Limestone, if exposed at the surface, is more vulnerable to physical and chemical breakdown. Solution

erosion causes a gradual depression, and potentially (and intermittently) lakes;

2. Cover-Subsidence Sinkholes: These occur where limestone is covered by 50-100 feet of permeable sand. Dissolving limestone is replaced by sand that slowly creeps down into the void;

3. Cover-Collapse Sinkholes: These occur when a cavity in the limestone grows so large that the weight of the overlying roof of sand and clay can no longer be supported by the residual (if any) limestone. This type of sinkhole accounts for over 95% of Florida sinkholes, and is, by far, the most catastrophic in terms of the sinkhole sizes and depths produced. Typical sinkhole depths are around 10-50 feet, while equivalent radii of the sinkholes range from 10's to 100's of feet, with a rough median value at around 30 feet. Figure 5.1 shows a representative example of such a sinkhole.

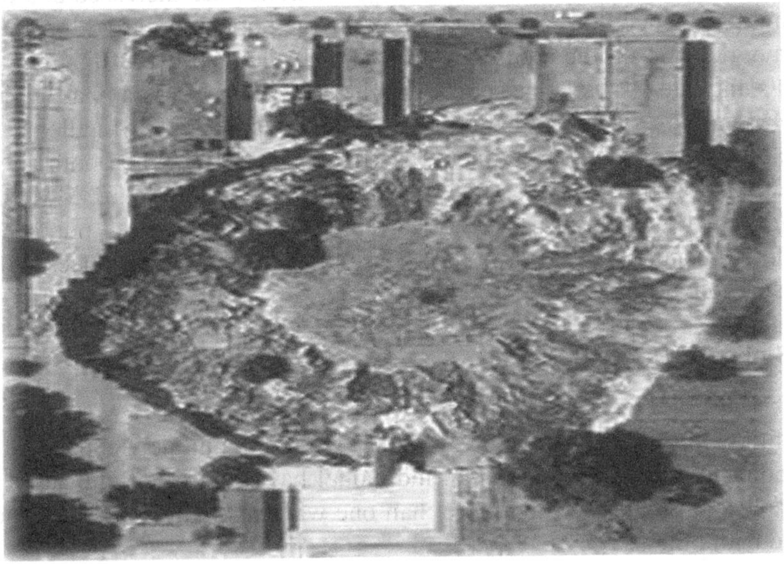

Fig. 5.1 View of a sinkhole swallowing a house in Central Florida

2.2 Activities Accelerating Sinkhole Expansion

2.2.1 Natural Causes

Triggering by heavy rains or flooding, which made the soil "roof" over a limestone cavity very heavy, to the point that it eventually collapses, is the most common natural cause. Late spring is the time of the occurrence of most sinkholes due to this flooding. Droughts can also lower the ground water levels, reducing the buoyant support of a cavity roof and promoting collapse.

2.2.2 Anthropogenic Causes

Large-scale urban development is responsible for increased sinkhole activity. Factors include: changing or loading the surface with retention ponds; office buildings and homes; changes in drainage patterns; drilling vibrations; and heavy traffic.

The number and size of Florida's sinkholes has increased at a disturbing pace since the 1970's, due to explosive population growth placing a heavy demand on the aquifer system, and accelerated by droughts in the 1970's and late 1990's. One source of relief just opened on Tampa Bay is a 25 MM gallons /day desalination plant, with a second 25 MM gallons/day operation slated to begin producing freshwater in 2007. When both are in operation, together they should offset almost half of the 110 MM gallons/day of water that Tampa Bay currently uses. The avowed purpose is to allow the aquifer to refill, in turn possibly decreasing the chances of future sinkhole occurrence.

2.3 Sinkhole Warning Signs

The following warning signs of sinkholes, while not uniquely connected to sinkholes as the only cause on an individual basis, are often indicators of sinkholes, particularly when more than one warning sign occurs in coincidence with others.

1. Fresh exposures on fence posts or other structures;
2. Slumping, sagging trees, or other objects; doors and windows that fail to close properly;
3. Ponding where rainfall has not collected before;
4. Wilted vegetation, caused as moisture normally supporting vegetation in the area is, instead, drained into a sinkhole developing below the surface;
5. Structural failure: cracks in walls, floors, and pavement.

2.4 Sinkhole Data and Inferences

The three county area under consideration has different areas and sinkhole data for each county (Figure 5.2). The demographic data for each of the cities in the three county area, including the housing units, land area, and density of population are shown in the figures spread throughout this chapter. The definition of "city" is the legislative district so that, for instance, all the individual parts that constitute Zephyr Hills are one city.

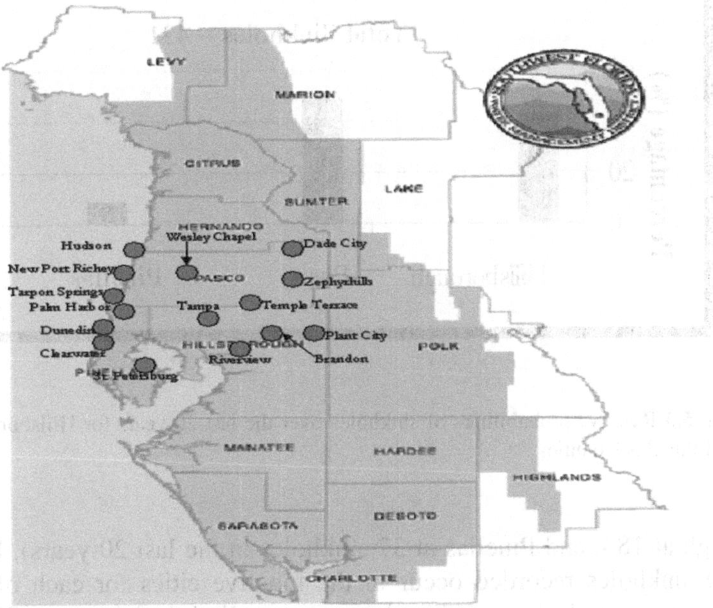

Fig. 5.2 Map of the region being investigated together with the locations of the 15 cities (Base map courtesy of the Southwest Water Management District, Florida, with cities overlay added).

. These data will be used later in the economic risking section of the chapter. Detailed sinkhole data were obtained from the Southwest Florida Water Management District, which oversees all aspects of water use and regulation in the Tampa Bay region (www.swfwmd.state.fl.us).

Median housing statistics were obtained from the U.S. Census (2000), specifically its online search engine American FactFinder (http://factfinder.census.gov).

The sinkhole count for each of the counties is presented in Figure 5.3. As shown on Figure 5.4 a, b, and c, the relative percentages of sinkholes per county indicate that the dubious winners are Pasco and Hillsborough counties in terms of total sinkholes (Pasco County at 210, Hillsbor

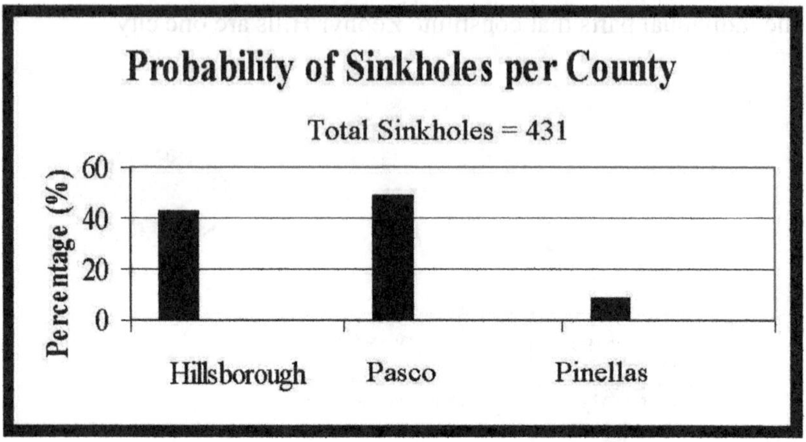

Fig. 5.3 Relative probabilities of sinkholes over the last 20 years for Hillsborough, Pasco, and Pinellas Counties

ough at 184, and Pinellas at 37 sinkholes in the last 20 years). But not all the sinkholes recorded occur in the top five cities for each of the three county areas. In fact, within the corporate limits of the total fifteen cities there are only 29 sinkholes recorded for Pinellas County, 190 for Pasco County, and 137 for Hillsborough County.

The remaining sinkholes recorded (8 for Pinellas County, 20 for Pasco County, and 50 for Hillsborough County) presumably occur outside the corporate limits of the cities and so are less damaging in terms of property in the rural areas compared to the highly developed urban city areas.

For each individual county one can plot the relative probability of sinkhole occurrence for each of the five cities by taking the sinkhole count in each city and dividing it by the total for all five cities within that county. This sort of plot is presented in figure 5.4 for the three counties, showing that cities in Hillsborough and Pasco Counties are, in general, at much higher relative risk than those in Pinellas County. Thus, if one had made the decision to relocate to a city in one of the three counties, then one does not have to contend with the sinkholes in the other two counties, just in the county of relocation.

On the other hand, the <u>absolute</u> risk depends on the total number of sinkholes in each of the counties within the corporate city limits. Accordingly, in figure 5.5 we show the absolute probability of a sinkhole occurrence for all 15 cities. Note that Tampa (in Hillsborough County), and Hudson and New Port Richey (in Pasco County) have much higher sinkhole risks than any of the other 12 cities, being at 28% risk, 37% risk and 42% risk, respectively. Thus if one were contemplating relocation, then avoidance of the three cities in relation to the remainder is one way of increasing the chances that one's house will not be swallowed by a sinkhole. Presumably house insurance against sinkhole loss is higher in these three cities than in the remaining twelve.

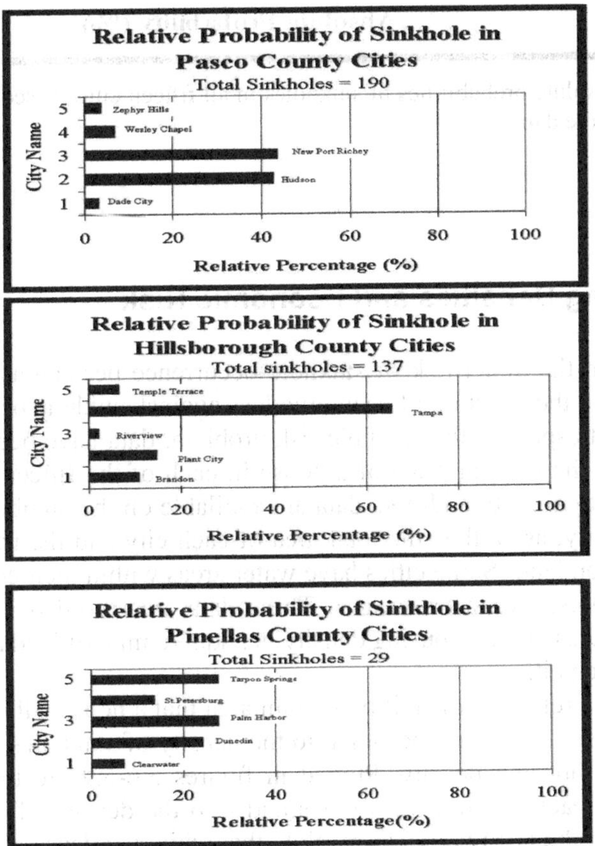

Fig. 5.4. Relative probability of a sinkhole in cities in (a) Pasco County; (b) Hillsborough County; (c) Pinellas County

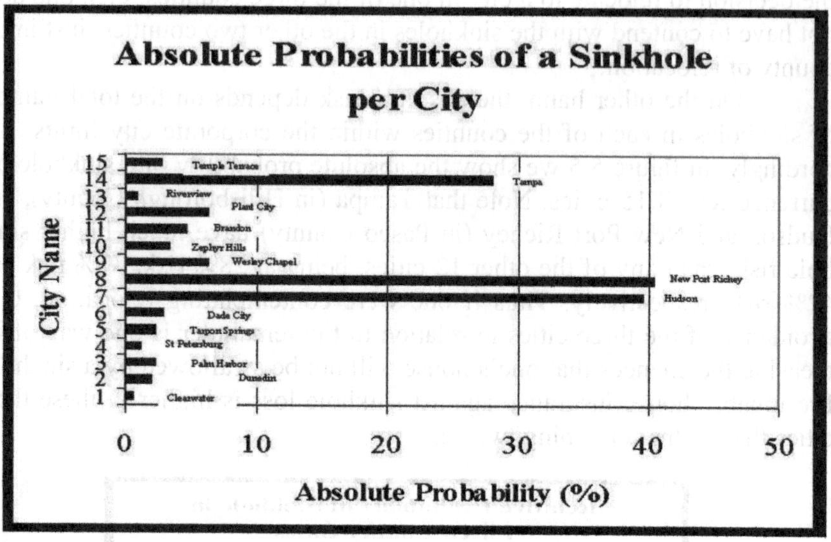

Fig. 5.5. Absolute probabilities of sinkholes in all fifteen cities based on the last 20 years sinkhole data

3 Housing Densities and Economic Risk

Apart from the direct risk of sinkhole occurrence per county or per city, there is also the question of economic loss against sinkhole occurrence. To handle that aspect of the sinkhole risk problem, data have been assembled relating to the median price of a house in each of the fifteen cities in the three county area. In addition, data are available on the number of housing units per city, as well as the total area of each city and the fraction of the city lying on land (Some cities have water areas within their corporate limits as harbors, rivers, or ocean.). These data are recorded in figures 5.6 through 5.9 as are the housing density per square mile of land area for each of the fifteen cities.

The relative probability per unit area that a house will be subject to sinkhole risk must be proportional to the density of sinkholes per unit area and to the housing density. Plotted in figures 5.6-5.9 are the density of housing in each of the fifteen cities and also the density of sinkholes reported over the last twenty years. Note that while the density of housing is variable by around a factor of four or so, the density of sinkholes has a much greater variability. Plotted on figure 5.10 is the product of sinkhole

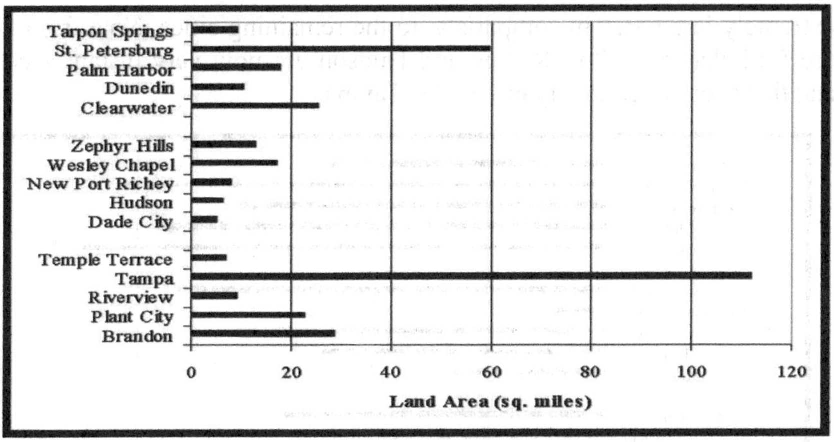

Fig. 5.6. Areas (in sq. miles) of the individual cities in the investigation

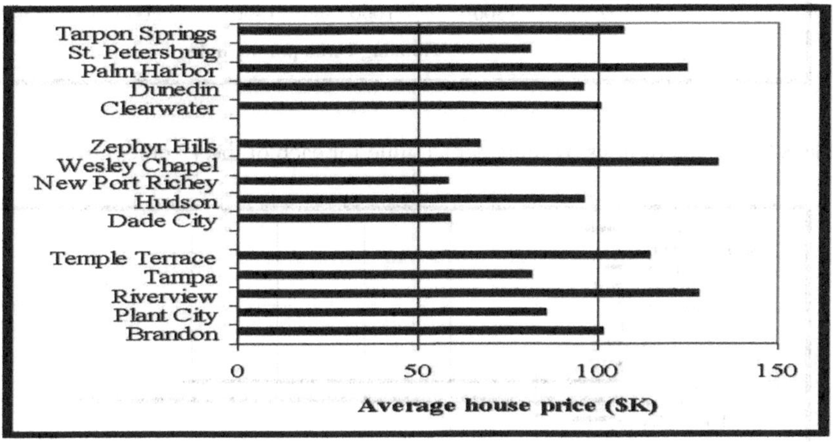

Fig. 5.7. Average price (in $K) for housing units in each of the 15 cities based on year 2000 information

density and housing density, showing that, per unit square mile, the highest probability density of a house being influenced by a sinkhole is for Hudson and New Port Richey. On the other hand, for each city as a whole it is not just the density of likely occurrences that is important, but also the absolute number of likely house disasters. This variable is proportional to the size of the city, so that even a high-density effect can be lowered in terms of city influence for small area cities. As shown in figure 5.11, when the city area is allowed for, then Tampa is the city with the highest likelihood of sinkholes damaging houses over the twenty-year period because of its extremely large area in comparison to the remaining cities. Note from figure 5.11 that New Port Richey and Hudson are now very distant second and third compared to city effects for Tampa.

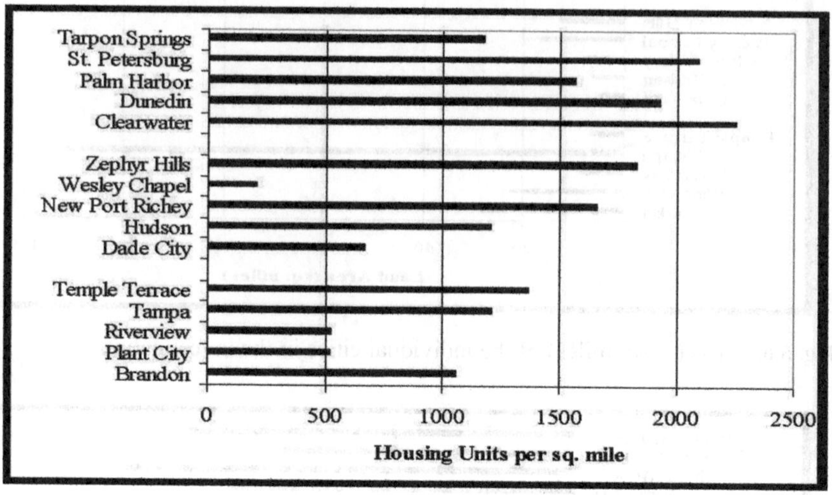

Fig. 5.8. Density of housing units per sq. mile for each of the 15 cities

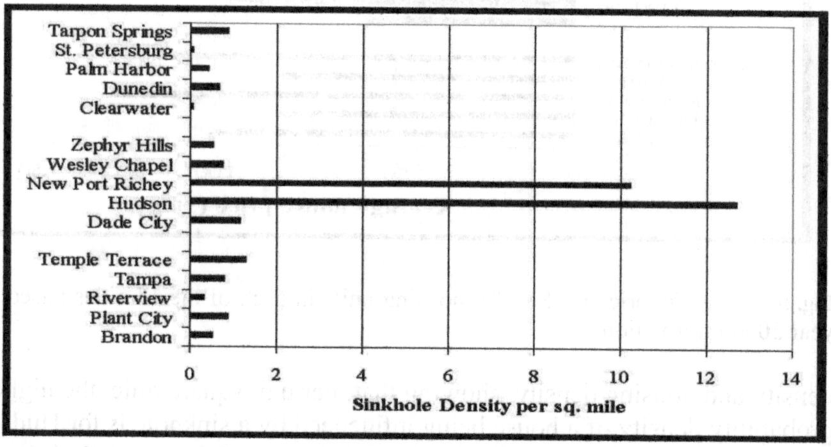

Fig. 5.9. Sinkhole density per sq. mile for each of the 15 cities based on the 20 year information

Economically, the cost of each house involved must also be included in the economic risk analysis. Using the median price of housing per city, as shown in figure 7, one can then compute the likely economic costs per city per year from sinkhole damage. This sort of information is presented in figure 5.12, showing that Tampa faces the biggest bill on average per year (of around $5MM), again with New Port Richey and Hudson coming in very distant second and third.

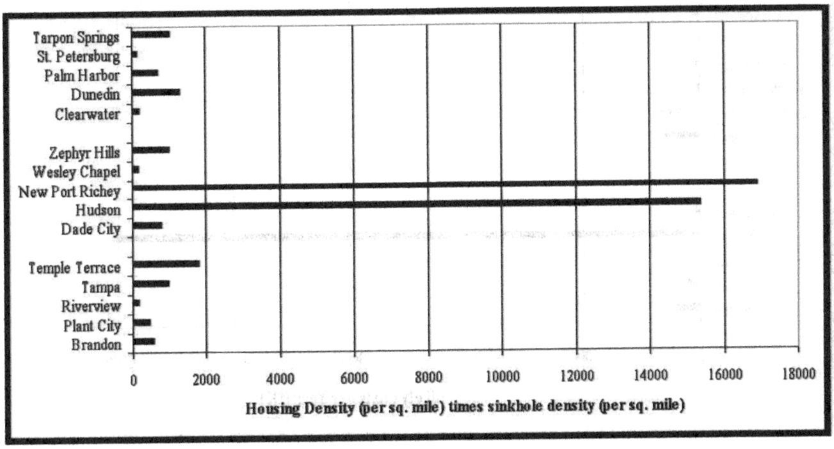

Fig. 5.10. Product of housing density and sinkhole density showing the relative chances of a house being impacted by a sinkhole for each of the 15 cities

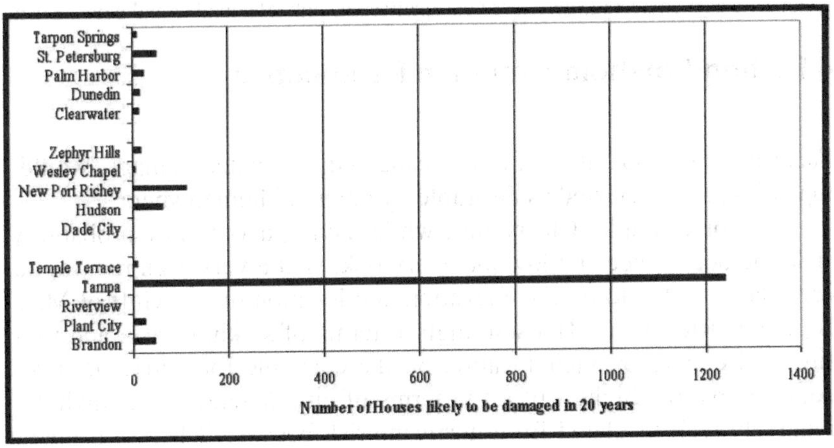

Fig. 5.11. Number of houses likely to be impacted by a sinkhole over a twenty year period for each of the 15 cities using the city area as well as the housing density

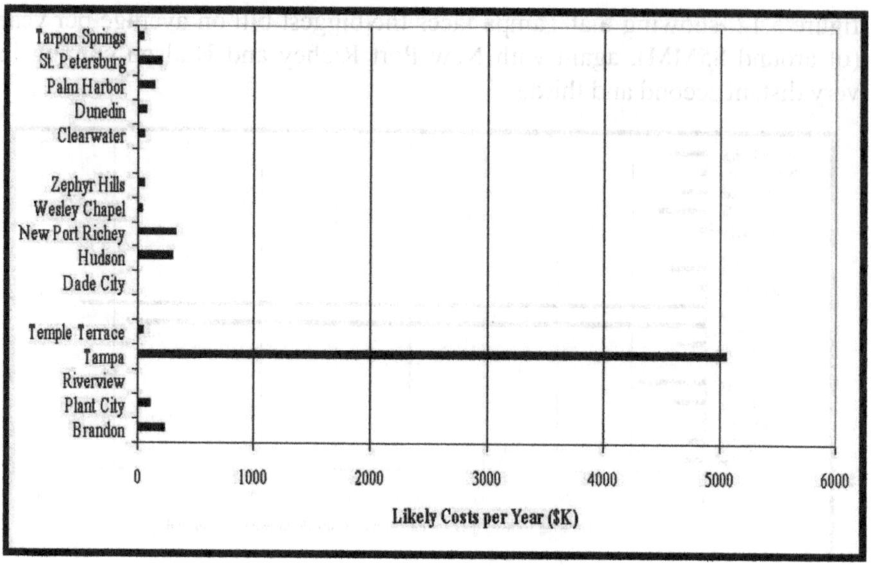

Fig. 5.12. Likely costs per year for each of the fifteen cities as a result of sinkhole damage to houses based on average costs per house, city area, sinkhole number and housing density

4 Putting Sinkhole Events in Perspective

Factoring the probability of a hurricane and associated damage, would the top cities still be viewed as desirable by potential homeowners?

For example, Clearwater, while having a very low probability of sinkhole occurrence and low monetary risk, is at a very high risk for hurricane damage due to its low elevation and location on the Gulf of Mexico. New Port Richey and Hudson, high in terms of sinkhole risk, add to their financial danger by their location on the coastline for hurricane damage. And Tampa, the highest of all in terms of city damage from sinkholes, is relatively well insulated from hurricane risk because it lies well inland in comparison to the coastal high-risk cities.

The point to be made is that one cannot look only at sinkhole risk when one is contemplating relocation. One must take into account all factors of a given region before making an informed decision. On the other hand the overall danger from sinkholes in Pinellas County is much lower than in Pasco or Hillsborough Counties. A possible reason is that Pinellas County is comprised of coastal sands and smaller amounts of karst, versus central and western Pasco County and central Hillsborough County, which are both large karst platforms and account for numerous sinkholes. Pinellas

County also is an isthmus, underlain by saltwater. Therefore, aquifer fluctuations (both Surficial and Floridan) may have a lesser influence on sinkhole formation because saltwater can intrude into the fissures left in the limestone by lowered groundwater levels.

Sinkhole distribution was skewed in Hillsborough and Pasco Counties due to the previously mentioned karst platform that extends roughly down the center of these counties. The urbanization of Tampa, in particular, may have significantly increased the probability of sinkhole occurrence and collateral damage to houses in that city.

Chapter 6

Risks of Damage from Flooding Rivers: Correlation of Weakened Dyke Structures?

Summary

This chapter shows how to incorporate uncertainty on whether two dykes, considered as likely to collapse during high water in a river after weeks of torrential rain, are correlated or independent. Prior to actual collapse of at least one dyke there is no way to be sure of the state of the two dykes in relation to each other. We show that this uncertainty automatically adds an uncertainty to the mean assessed cost estimate should there be a collapse, and that this uncertainty can dominate over uncertainties brought about by imperfect knowledge of the failure probabilities and potential costs of each of the individual dykes, as well as the distributions chosen to represent such uncertainties, and the nominal values of the parameters and their skewness of dynamic range around the nominal values. The main aim is to show which factors dominate in assessments of uncertainty of potential costs should there be a dyke collapse. In this way one can focus on narrowing the relevant parameter uncertainties causing the largest contributions without the need to spend time and effort in addressing other parameter uncertainties that would help but little, if at all, without the main focus on the important contributors to uncertainty of the mean cost assessment.

1 Introduction

In the summer of 2002, there was torrential rain for many weeks in Poland and Germany. As a consequence, the Elbe River and many of its tributaries flooded beyond what had been considered the ultimate high water stage. Despite sandbagging the banks of the rivers there was massive flooding, due mainly to the low and flat nature of the landscape in the eastern part of Germany and the southern part of Poland. The presence of many dykes prevented some flooding but by no means all. The estimate of total damage to houses, industry, businesses and crops is in the several billions of dollars range. Indeed, many of the insurance companies that had to bear a considerable brunt of the damage costs have raised their general insurance rates to householders, often by more than 25% in one year, to slowly recover their insurance outlay. In the aftermath there has been interest in determining if individual dyke breaks were correlated or independent, with the hope that if such could be determined then one might be able to correctly plan for the future in terms of dyke support, rebuilding, or secondary dyke construction.

The purpose of this chapter is to show the differences that can ensue when one considers independence or correlated behavior of dykes, so that some idea can be obtained of precisely what it is that one should be aware of. The damage costs in either event are considerable, as are the dyke rebuilding or shoring up costs. Any advance on methods and techniques designed to address portions of the problems encountered have the potential to save lives and livelihoods in the future, as well as lower the costs. It is with this background picture in mind that the methodology in the next sections has been developed.

2 Damage and Probability

Perhaps the breaching of dykes by the flooding is best captured in figures 6.1 and 6.2 showing portions of the Mulde River (a tributary of the Elbe River) in Saxony after it had breached dykes that had protected old open cast pits, which had produced lignite coal during the DDR period. The filling of the pits and the concomitant overflow produced major damage with long-lasting implications. First was the actual flooding itself that brought fresh water into contact with the residual open pit tailings including high sulfur-bearing lignite, leading to the classical mine waste acid production. Second was the destabilization of the pit walls and so water-driven erosion that led not only to surrounding land surface crumbling but also led to ground water leaching through the opening of new high permeability pathways. The ground water in the Saxony region of the Mulde River is itself highly contaminated from being used in DDR times as toxic dump locations. Thus one opened up the possibility of release of previously "sealed" contaminants to anthropogenic toxicity. Third was the dynamical impact of the waters on further water-sodden dykes, constructed as secondary support systems, which then failed under the hydrodynamic pressure, causing considerable flooding (see figures 6.1 and 6.2) with damage to agricultural land and also farms, villages, roads, towns, bridges, and infrastructure, such as sewage systems, electrical cables, etc.

In many cases where buildings were heated with oil, damage to the oil containers resulted in secondary pollution being created by oil release. In short the damage produced hit not only the financial sector but also the use of land and the toxic hazard to humans and animals.

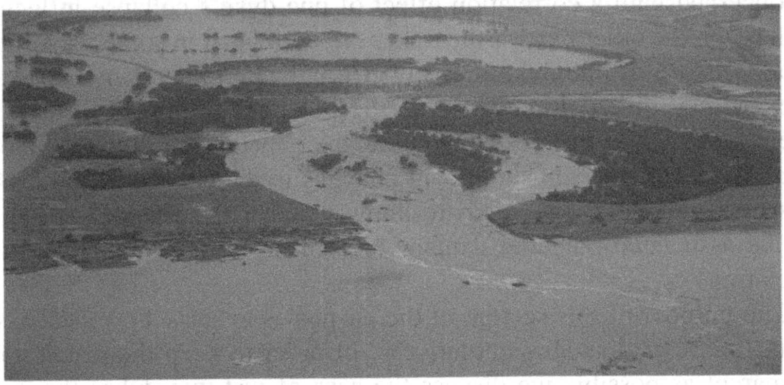

Fig. 6.1. Aerial picture of the breakthrough of the Mulde River at high water stand into a former open pit mine area at Goitsche on 15 August 2002

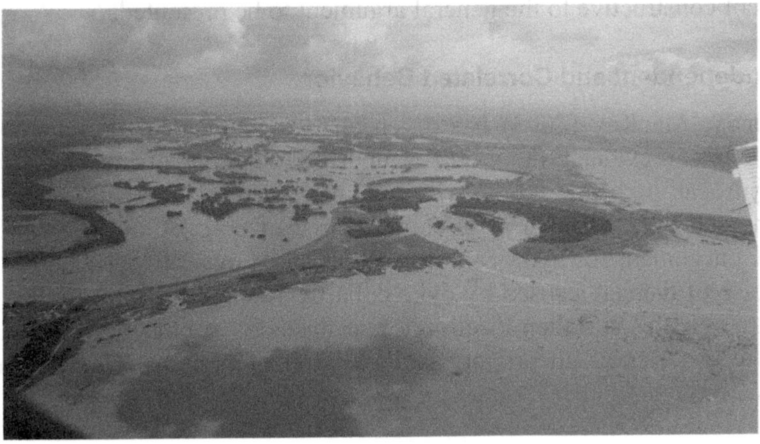

Fig. 6.2. Similar to Figure 6.1 but taken from a different aerial perspective to show the larger region covered by the floodwater breakthrough

One aspect of this general problem is the question of whether there is independence or correlation of dyke collapses. The reason that this problem is important to answer is that if the dykes are independent then the potential collapse of one dyke does not influence the potential collapse of a neighboring dyke. Both potential collapses are independently assessed and so remedial action to strengthen the dykes can proceed without having to be concerned with a correlation effect of one dyke's collapse influencing another. On the other hand if the dykes are correlated in some way, the potential collapse of one dyke may either accelerate the collapse of another dyke, or may divert floodwaters so the potential collapse of the second dyke is lessened. In this correlated case, the remediation and strengthening of one dyke can provide significant changes in the chances of collapse of the second dyke-and one needs to know whether the chances of collapse and the associated potential damage are enhanced or lessened by such a correlation.

The purpose of this section of the chapter is to show how such assessments can be evaluated quantitatively. In order to keep the discussion as transparent as possible we discuss the case of just two dykes throughout this chapter, although extension to as many dykes as needed can be done (which is technically involved but relatively simple to perform) at the expense of clarity of presentation. Such technical complexity is, in our opinion, not constructive to the general argument to be presented.

2.1 Independent and Correlated Behavior

Recently, MacKay (2003) has given an extremely lucid presentation of how one can evaluate the worth of undrilled horizons one suspects could be hydrocarbon bearing, both when one has no correlation between the horizons and also when there is some degree of correlation. The MacKay (2003) argument for hydrocarbons can be taken over almost wholesale to the case of damage caused by dyke collapse. We show here how this procedure operates and then discuss how it needs to be embellished to allow for the vagaries of uncertainty in both parameters and intrinsic assumptions.

2.1.1 MacKay's Argument applied to independent dyke collapse.

Considers two dykes, A and B, and assign them the following parameters: chances of collapse (p(A)=30%;p(B) =20%); mean potential cost if there is a collapse (M(A) = \$60 MM;M(B) = \$80 MM), from which one can calculate the chance-weighted cost for each dyke as CWC(A) = p(A)xM(A)=\$18 MM;CWC(B) = p(B)xM(B) = \$16MM. Because of the assumption of no dependence between the dykes, the chance of at least one of the dykes undergoing collapse is

$$P(\text{at least one}) = 1-(1-p(A))(1-p(B))= 0.44 \qquad (6.1)$$

An estimate of the mean costs (IM) expected if one considers both dykes with their independence of parameters one from the other is then simply given by

$$IM = (CWR(A) + CWR(B))/P(\text{at least one}) = \$77.3\ MM \tag{6.2}$$

2.1.2 MacKay's Argument applied to Fully Correlated Dykes

In this situation, MacKay (2003) makes a clear distinction between reciprocal dependence and conditional dependence. Full reciprocal dependence requires dyke B to collapse or stand (stand or collapse) according as dyke A collapses (stands)-assuming dyke A collapses first. There can then be no freedom of choice for the probability of dyke B collapsing; it must be identical to dyke A, i.e. p(B) = p(A). Thus if one has assessed a different probability for dyke B to collapse than for dyke A, prior to any collapse, then reciprocal dependence cannot be in play. Furthermore, prior to collapse of either or both dykes, there is absolutely no way one can know, as opposed to surmise, that the two dykes are indeed reciprocally correlated. The use of such a reciprocal correlation in standing dykes is then itself suspect. In such a situation, the two dykes can be only conditionally dependent. Assuming dyke A is put under floodwater pressure ahead of dyke B, and assuming further that p(A)>p(B), then the overall chance of at least one dyke collapsing is just p(A). Alternatively, if p(A)<p(B), then the overall chance of at least one dyke collapsing is just p(B).

An estimate of the correlated mean costs (CM) to be expected for both dykes is then

$$CM = (CWC(A)+CWC(B))/\max(p(A),p(B)) \tag{6.3}$$

For the numerical values used by MacKay (2003) p(A) = 0.3 while p(B) = 0.2 so equation (6.3) yields CM = $113.3 MM, which represents a significant increase in the estimate of $77.3 MM for the independent dyke collapse case but with lower probability of overall collapse(30% versus 44%). MacKay also points out that one can use "Monte Carlo sampling techniques to selectively add success-case outcomes from each zone" and that one can estimate the P(10) and P(90) range of costs using appropriate statistical tools.

3 Modifications to the Independent versus Correlated Situations.

There are several modifications that are appropriate to make to the basic arguments advanced by MacKay (2003). First, note that the basic collapse probabilities for each dyke are taken to be precisely known, whereas in fact it is most often the case that there is a range of values possible depending on how the assessment was made. Such uncertainty should surely be allowed for in providing estimates of likely costs. Second, the estimated values of mean costs for each dyke are also uncertain before a collapse occurs- and often even after a collapse! The uncertainty of such values should surely be included in the assessments.

Third, when one allows for these two groups of uncertainties then one traditionally performs some form of Monte Carlo calculations allowing each uncertain parameter to be chosen from an underlying distribution. One needs to determine to what extent the choice of distribution made influences the estimates of anticipated costs, and whether the relative uncertainty brought about by such choices is large or small in comparison to the uncertainty brought about by the dynamic range of each parameter.

Clearly, if the choice of distribution is significant in this regard then one has to have some way of better controlling that uncertainty before one bothers to try to control the uncertainty brought about by the dynamic ranges of the uncertain parameters. On the other hand, if it turns out that the choices of distribution have but little impact on the cost uncertainty in comparison to the uncertainty brought about by the parameter ranges, then one needs to concentrate on narrowing the dynamic ranges first. And in this regard one also needs to know which parameters are causing the greatest contribution to the uncertainty so one can concentrate on them first without spending inordinate time and money on improving other parameter ranges that do but little to improve the overall uncertainty in potential cost estimates.

Fourth, note that in application of MacKay's procedure it is assumed that it is known, ahead of one or both dykes collapsing, that the dykes are either uncorrelated or are fully conditionally correlated. In fact that information is usually not available, except by argument from analogy using prior historical precedent from earlier dykes in the same floodplain domain and with the assumption that the precedent applies to the two standing dykes. A more appropriate strategy is to take it that there is a probability, p_c of conditional correlation, and so a probability $1-p_c$ of independence of the dykes, when one can write the assessment of mean costs, R, assuming at least one dyke will collapse, as

$$\begin{aligned} R &= p_c(CWR(A)+CWR(B))/\max(p(A),p(B))+(1-p_c)\,(CWR(A) + \\ &\quad CWR(B))/[1-(1-p(A))(1-p(B))] \\ &= p_c CM + (1-p_c)\,IM \end{aligned} \tag{6.4}$$

There is an automatic uncertainty on this mean assessment because of the probability p_c. The variance on R is given through

$$\sigma(R)^2 = p_c(1-p_c)(CM-IM)^2 \tag{6.5}$$

Unless CM = IM, then the mean assessment carries a positive variance with it because of the uncertainty in whether uncorrelated (independent) or fully correlated conditions between dykes A and B is appropriate. And this variance exists even for statistically sharp choices of the parameters and so must be allowed for when comparing the variances brought about by parameter uncertainties and also uncertainties due to different distribution choices for all parameters including p_c itself.

The way one treats the system is then a variant of the Monte Carlo problem again. One allows for an uncertainty in the probability p_c that the dykes are correlated and so explores to what extent the behavior of the cost estimate is dominated by the choice or dynamic range of p_c or the choices for its distribution in comparison to the variance due to the fact the correlation probability is neither zero nor unity (see Lerche and MacKay (1999) for details of the method used).

Note also that a measure of the usefulness of the mean cost estimate R can be provided through the volatility, v, given by (Warren 1978)

$$v = \sigma/R \tag{6.6}$$

A value of $v \ll 1$ indicates that, for specific choices of the other parameters entering the estimate of potential cost, there is but little uncertainty on R, which can therefore be regarded as a trustworthy measure of the expected costs; a value of $v \gg 1$, however, implies considerable uncertainty on the estimate R, so that it does not provide a trustworthy measure of the expected costs. Alternatively, in terms of an equivalent gaussian cumulative probability, P(h), that one will have costs in excess of hR, where h is arbitrary but positive, one can write

$$P(h) = I(h)/I(0) \tag{6.7a}$$

where

$$I(h) = \int_g^\infty \exp(-y^2)dy \tag{6.7b}$$

with $g = -(1-h)/(2^{1/2}\ v)$, which illustrates the way the volatility controls the probability of assessing likely costs with an uncertainty on the mean value. A fairly good analytical approximation to equation (6.7a) is

$$P(h) = \{1\text{-}\tanh(h/(2^{1/2}\, I(0)v)/[2\tanh(1/(2^{1/2}\, I(0)v)]\} \quad (6.7c)$$

In short, there are a significant number of addenda one needs to make to the basic MacKay argument in order to determine the best sort of information concerning cost estimates to extract from the possibility that the dykes are either independent or fully conditionally correlated. These addenda can be conveniently split into four groups for evaluation: (i) Contributions to the uncertainty brought about by the probability the dykes are correlated when all other parameters are determined precisely; (ii) Contributions to the cost uncertainty when all parameters (including the correlation probability) have dynamic ranges of uncertainties but have fixed distributions from which a Monte Carlo scheme chooses individual values; (iii) Contributions to the cost uncertainty when the parameters can have different distributions; (iv) Contributions due to changes in the dynamic ranges of each parameter and of the central values chosen for each parameter around which the distribution choices and dynamic ranges are made. These problems are considered next.

4 Uncertainties and Correlations

In this section we systematically go through the four problems listed above and illustrate how they influence the evaluation of estimated costs.

4.1 Uncertainty in the Probability of Correlated Behavior

Start with the easiest of the four problems: variation in only the correlation probability, p_c, with all other parameters fixed at the nominal numerical values given by MacKay (2003) and reported here in Section 2. Then

$$R = 77.3 + 36p_c \quad (6.8a)$$

and

$$\sigma(R)^2 = p_c(1\text{-}p_c)(36)^2 \quad (6.8b)$$

yielding a volatility v given by

$$v = 36[p_c(1\text{-}p_c)]^{1/2}/(77.3 + 36p_c) \quad (6.8c)$$

The volatility takes on its largest value when p_c =0.405, and then v(max) = 0.19. The uncertainty on the mean estimates resources is then ±19% with R = \$91.88 MM thereby providing a range of uncertainty at one standard error ranging from \$74.45 MM to \$109.33 MM.

The basic outcome is then simple to state: unless one can, ahead of a collapse, somehow guarantee that there will be a larger, or smaller, value than about 40% for the probability of correlated behavior, then one automatically has at least a 19% standard error on the mean estimated costs for the two dykes if at least one of them were to collapse. The highest standard error is at a value of 50% for the correlation probability, and yields an uncertainty of ± \$18 MM on the mean value of \$95.3 MM, and a corresponding volatility of nearly 18.9%.

In addition to this basic problem there are also more uncertainties brought about by the uncertainty in the parameters used to assess the potential costs for the two dykes. These uncertainties not only bring about an uncertainty to the estimated costs but also change the mean costs to be expected, and they also bring even more uncertainty through their unknown distributions. We now illustrate some of these points.

4.2 Parameter Uncertainties for each Dyke for Fixed Distributions

To illuminate the way the dynamical ranges of parameters, and their distributions, change the assessment of potential costs, we let each of p(A), p(B), M(A), and M(B) be uncertain by ±10% around the nominal values given above and, in this section, we take it that the distribution of each parameter is uniform and centered on the nominal values. We also allow the probability of the dykes being correlated to fluctuate around the 50% value by ±10%, i.e. to range from 45% to 55% also with a uniform distribution. Such ranges for the parameters are very much smaller than is customarily the case in actual situations, but they allow one to see most clearly the consequences of varying different assumptions as will become clear in the next two subsections.

Results of running Monte Carlo computations under the above framework are as follows. As shown in Figure 6.3, the mean value of the cost assessment takes on the value \$95.5MM with two contributions to its variance. The first contribution is due to the probability of correlation, which itself now has a range of uncertainty as shown in Figure 6.4, but with a mean value of \$328 MM^2, and a second component due to the uncertainty on the parameters entering the calculation of mean cost assessment which yields a variance contribution of \$22 MM^2. Thus the uncertainty on the mean cost assessment is dominated by the correlation probability term (which provides about 93% of the total variance), and which itself varies from about \$225MM^2 to \$552MM^2 (at 90% confidence) due to variations in the parameter values. The variance due to the uncertainty of parameters alone shows relative contributions, as exhibited in figure 6.5, dominated by the 10% uncertainty in the costs for dykes A and B, but the total uncertainty due to these uncertain parameters is only 7% of the total. Absolute dominance of the correlation probability uncertainty factor is paramount.

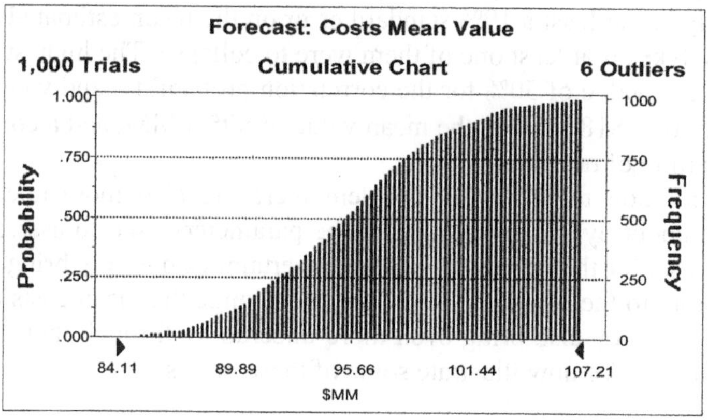

Fig. 6.3. Cumulative probability distribution for the mean cost assessment when all parameters are taken to be drawn from uniform distributions as described in text

It makes no commercial sense to try to improve the uncertainty on the parameter values until some attempt to narrow the uncertainty caused by the correlation probability is attempted. Thus, one major conclusion is that the choice of a correlation probability of 45-55 % is dominating all other concerns concerning the uncertainty of mean cost assessment and, as shown in figure 6.4, contributes a volatility to the mean assessment that ranges from about 17% to about 23 % (at 90% confidence).

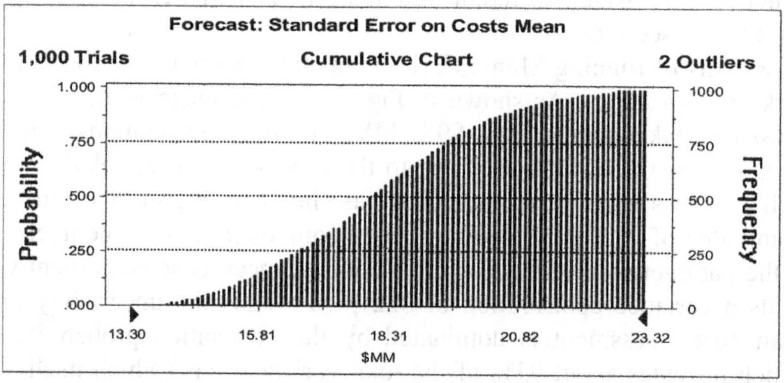

Fig. 6.4. Standard error on the mean cost assessment due to the uncertainty on whether the dykes are correlated or independent

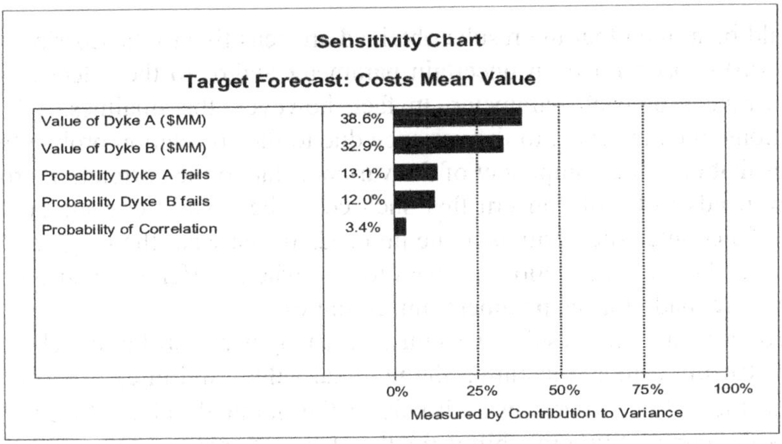

Fig. 6.5. Relative contributions to uncertainty in that fraction of the mean cost assessment that is due solely to parameter uncertainty (which contributes only about 7% of the total uncertainty with the other 93% being provided by the uncertainty in whether the two dykes are correlated or independent).

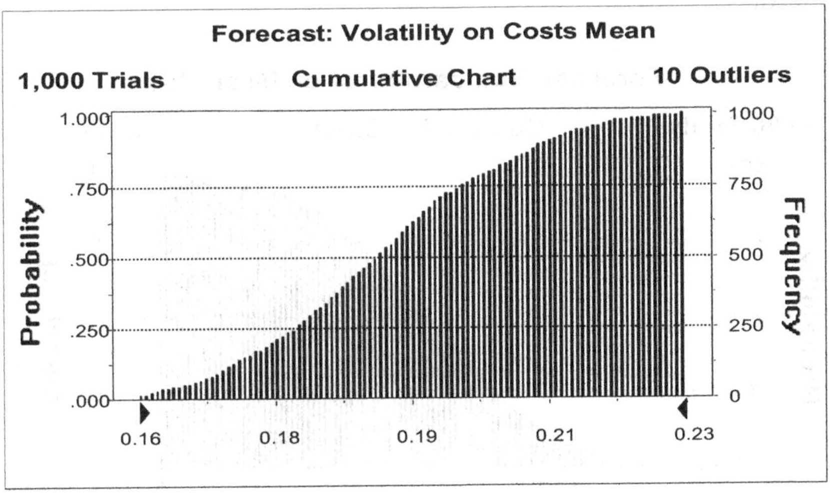

Fig. 6.6. Cumulative probability plot of the volatility of the mean cost assessment with uncertainty (nominal correlation probability of 50%) due mainly to the uncertainty of whether one has independent or correlated dykes for uniform probability distribution choices for the uncertain parameters

4.3 Non-Uniform Distribution Choices

It could be argued that the results obtained are sensitive to the distribution (uniform) chosen for each uncertain parameter and/or to the selected dynamic ranges of each parameter. In fact, however, the smallness of the variations in comparison to the variance due to the correlation probability, argues that only the component of the variance due to the correlation probability needs to be of concern that such could be a valid critique at this stage. To counter such criticism one proceeds to see what the range of the variance due to correlation probability is when different distribution choices are made for the parameter uncertainties.

To illustrate the results we consider triangular distribution choices symmetrically centered again on the nominal values and linearly reaching zero at the ± 10% values on each side of the nominal values. Figure 6.7 shows the cumulative probability for the standard error on the mean cost assessment with an average standard error of$18.1 MM and an uncertainty on this value of about $2MM. The corresponding mean cost assessment is itself then uncertain and, as shown in figure 6.8, has a volatility ranging from about 0.18 to about 0.22 centered on a mean cost assessment of about $95 MM.

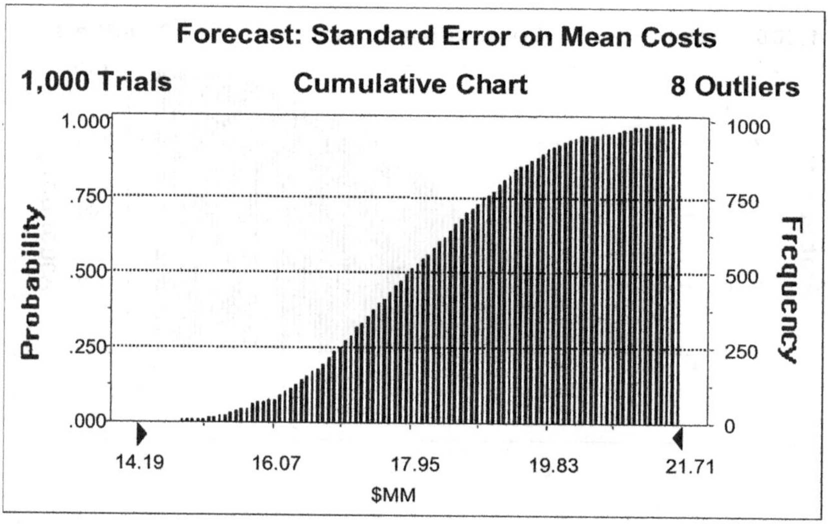

Fig. 6.7. Effective error on the mean cost assessment when the uncertain parameters are chosen from triangular distributions rather than uniform distributions but symmetrically centered on the nominal values given in text

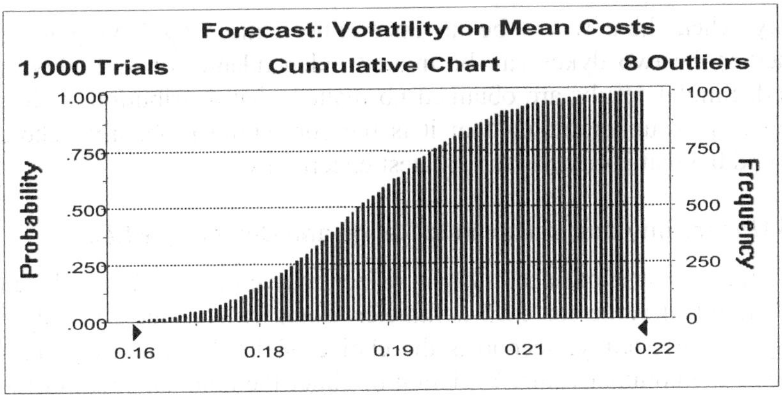

Fig. 6.8. Cumulative probability plot of the volatility of the mean cost assessment with uncertainty (nominal correlation probability of 50%) due mainly to the uncertainty of whether one has independent or correlated dykes for triangular probability distribution choices for the uncertain parameters

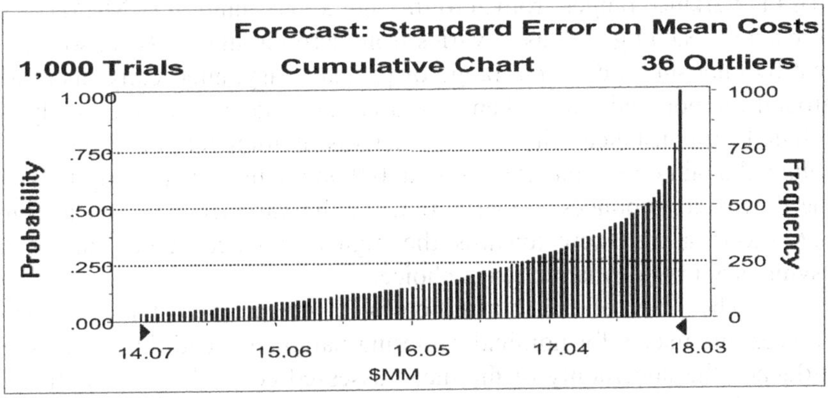

Fig.6.9. Asymmetric cumulative probability distribution of the error in the mean cost assessment due solely to uncertainty in whether the dykes are correlated or not when the correlation probability is chosen from a triangular distribution centered on a 50% probability of correlation but with an upside chance as high as 90% and a downside chance as low as 40%

These results are similar to those for the uniform distribution choice situation considered above, so that one can be confident in this case that it is not so much the choice of underlying distribution for the uncertain parameters that is the main cause of uncertainty but rather the fact that the correlation probability is not close to one of its extreme values of zero or unity when there would be no uncertainty caused by having to assess whether the two dykes can be treated independently or as highly correlated. Similar results are obtained no matter what distribution choices are made, again underpinning that it is the correlation probability choice of around 0.5 that is causing the greatest uncertainty.

4.4 Uncertainty due to Dynamic Range and Centering Effects

One further cause of uncertainty is related to two factors: first is the choice of a nominal value for each parameter around which one has a dynamic range of uncertainty; second is the choice of the dynamic range on each side of the nominal value. Variation of either the nominal value or the dynamic range, or both, also contributes uncertainty to the assessment of potential costs. Such is manifest just by considering the variance caused by the probability that the two dykes are correlated or not, for then the variance, as given by equation (6.5), shows a dependence proportional to $p_c(1-p_c)(CM-IM)^2$. A choice for p_c close to either zero or unity (as opposed to the nominal centering value 0.5 used in the illustrations above) indicates that the variance ranges from zero through a maximum of $(CM-IM)^2/4$ before again reaching zero as p_c varies from zero through unity. Also clear is that by choosing a dynamic range of p_c that is asymmetrically distributed around the nominal value, then so too is the variance asymmetrically distributed. For instance with a triangular distribution for the choice of p_c, centered again on 0.5 but extending to 0.9 on the upside and only to 0.4 on the downside, produces a distribution for the variance as given in figure 6.9, showing the skew towards the higher variance values due to the asymmetry in the dynamic range choice.

The point of this illustration is to indicate that it is the various choices one makes for nominal centering parameters and their ranges that influence the uncertainty of the mean assessed cost. The same is true for each and every parameter entering the calculation of assessment of potential costs.

5 Discussion and Conclusion

In any given situation decision makers have to determine whether or not to reinforce dykes with the hope of increasing strength, and so resistance against floodwater breakthrough, with potentially lowered damage costs, assuming the damage costs are more than the costs of remediation and strengthening of the dykes.

One of the major problems one faces is to decide whether one can treat such unbreached dykes as related in some way either to prior information from already broken dykes or to the hydrogeological concept in place that one surmises accounts for the already failed dykes (and the fraction of standing unbreached dykes as well!).

With a two dyke example, in which both dykes are still unbreached, this chapter has shown how one can go about estimating likely potential damage costs due to the chances at least one of the dykes will fail when one considers the dykes to be either independent or correlated. And this is precisely where the biggest uncertainty arises for one does not know ahead of actual dyke failure whether they are indeed independent or correlated. So just like any other aspect of uncertain assessments of events that have not yet happened, one needs to provide a probability that the two dykes are either independent or correlated. In turn this probability carries through to an uncertainty on the mean damage cost assessment even if all other parameters involved in the assessment are held at statistically sharp values.

What we have shown in this chapter is how to modify the argument advanced by MacKay (2003) in another context, to allow for this uncertainty caused by ignorance on the state of independence or correlation of the two dykes. In turn, one then has to consider whether the uncertainty on the damage cost assessment brought about by this ignorance is larger or smaller than the uncertainty brought about by estimates of the dyke parameters (potential costs and the probability each dyke will fail). Also one must determine whether the uncertainty on the dynamic range, nominal centering values, and also choices of distributions for each "fuzzy" parameter, have a greater or lesser influence on the uncertainty of the mean cost assessment than does the uncertainty brought about by not knowing with precision whether the dykes are correlated or not.

This interlocked suite of problems has to be addressed quantitatively and resolved prior to dyke reinforcement decisions, so that one knows which effects are most urgent to address if one wishes to narrow the range of uncertainty on the cost assessment should at least one of the dykes fail. As we showed by specific example, there is little point in trying to improve the uncertainty due to fuzziness of parameters related to each dyke if the dominant uncertainty is caused by indeterminacy on knowledge of whether the dykes are correlated or not.

It is these addenda to the basic argument that are crucial ingredients in providing information of relevance to decisions to reinforce, or not, dykes that potentially could fail under the next floodwater conditions.

Chapter 7

Biological Remediation of Environmentally Contaminated Water

Summary

Biological activity in a dam containing an organic depository (the Hufeisen See region) in East Germany cleanses organic contaminants, by a factor of about a million, from water seeping through the dam to an outside lake. This chapter sets up several possible steady-state models to account for the observations, depending on whether the biologically active components are tied to sedimentary particles in the dam material or are free to move with the seeping water. Allowance is made for the death rate of the biological components and also for their variable growth rates. Numerical examples are provided to illustrate the patterns of response in the different cases possible for seepage speeds that are either fast or slow compared to the biological cleansing rate. Comparison with determination of parameters and limitations on model possibilities, are not yet possible, nevertheless it is possible to determine dependences of parameters in each model in order to be in accord with the observations. Allowing for the known uncertainty in the measured data of contaminant concentration ratio, from the interior to exterior of the depository, allows an assessment to be made of the range of uncertainty of parameter relations. For the Hufeisen See depository it is shown that the uncertainty is relatively small in comparison to mean values of parameter relations. The overall implication is that the choice of appropriate model cannot yet be better constrained without new measurements of relative organic concentration through and in the dam material itself.

1 Introduction

In many areas of East Germany there are capped depositories of animal water organic waste that are separated from drinking water reservoirs by earthen dams. In several situations it is known that the seepage of water from the depository into the reservoir takes place through the dam. What is remarkable is that the seeping water is of extremely high purity in contrast to the connate waters in the organic depository. It has been suggested that this water cleansing is due to biological activity in the earthen dam as the depository water seeps through to the reservoir (Strauch 1996).

What is lacking at the present stage is a quantitative understanding of when such situations need to be remediated and when it is best to leave them alone to natural remediation. It is also not known to what extent the quality and heterogeneity of the dam permeability, together with the variable thickness of the dam, play roles of relative importance in controlling the seepage speed so that biological activity can continue to cleanse the waters. It is also not known what type of biological activity is the best for water cleansing nor is it known to what extent the combined system of dam plus biological activity is becoming saturated, so that at some time in the future the ability to cleanse the waters will fail.

While copious data exist on the present-day conditions of such systems (Glaesser 1994, Christoph 1995, Glaeser 1995, Strauch 1996, Maiss et al. 1998), there is little information on past activity and virtually no predictions of future expected behavior. It would be of tremendous value if one could develop quantitative models incorporating what little is known of the various processes and geometries, together with biological activity models, in order to ascertain the dominance of each component in controlling the system behavior. Two sorts of models are envisioned. First, a forward steady-state model would be of great benefit as an investigative device to aid in understanding the interaction of all processes taken together. Once the sensitivity and resolution are clearly addressed in this way then a second stage model can be constructed that will enable parameters and process rates to be determined (by use of inverse methods) to obtain the best agreement between observations and model behaviors.

In such a way one would then generate a procedure for addressing concerns of relative dominance determination for each process involved and also determine the relative sensitivity to the available data; thus one could determine what type of data and how much data are needed to improve resolution. More importantly, one could then determine if such data collection is worthwhile or whether one has enough to provide an accurate description of each such situation and its likely development in the future.

Once the basic understanding of process interactions is clear, it is then a relatively simple matter to construct numerical programs so that one can rapidly run through likely future scenarios and determine the most probable outcome based on the models and data to hand. Such a computer program would be of incalculable benefit in addressing many situations, each of which has uncertain and often unknown properties. At the very least one could transfer, by analogy, parameters determined from a better known situation to a lesser known situation and so recommend the sort of data collection required to enhance understanding and control of the more poorly known situation.

In this chapter, we examine the steady-state situation in a quantitative manner and also demonstrate, with the use of Excel programs and Crystal Ball®, how one can rapidly sort out sensitive versus non-sensitive parameters so that one can concentrate efforts to better constrain the dominant uncertainties. In addition, we give a first application of such models to the former brown coal open pit mine at North Bruckdorf, Germany, which is now the organic waste depository known as Halle-Kanena, and for which a voluminous report containing excellent and massive amounts of data has been compiled (Strauch 1996).

It seems appropriate in this chapter to show not only how such interactions can be evaluated but also how one develops mathematical representations and also the relevant numerical procedures.

2 Quantitative Steady-State Models in One Dimension

The simplest sort of biologically remediated problem to investigate is that depicted schematically in Figure 7.1, where Darcy flow of contaminated water from the organic waste depository is slowly permeating through a clay barrier, of width L and permeability K, separating the depository from a reservoir. In the barrier domain there are active anaerobic biological organisms that survive and even thrive by taking the organic contaminant products in the pervading waters and using them as a food source to grow, thereby removing such contaminants from the water. Thus the water that finally crosses the barrier region to the reservoir side is cleansed by such actions.

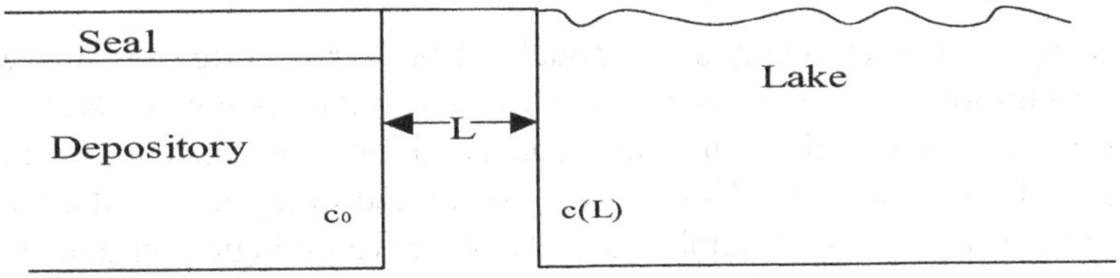

Fig. 7.1. Sketch of the one-dimensional problem geometry and conditions

There are two obvious extreme limits to this behavior. If the fluid flow that is transporting the contaminants is slow in comparison to the biological activity rate of contaminant removal, then the waters crossing the barrier will be completely cleansed of organic pollutants; on the other hand if the water flow is brisk in comparison to the biological activity rate then the flow passes quickly through the barrier and is still pollutant laden when it reaches the reservoir. In a more general situation the behavior must lie somewhere between these two extremes, and must also depend on the barrier thickness, the permeability, the biological activity rate, the fluid flow speed, and the lifetime of the active components of the biological system.

In addition, the behavior of the biological cleansing action must also depend on whether the active organisms are firmly tied to the solid particles of the barrier material or whether they are to be found in the interstitial pore waters. In the first case the organisms are not influenced by the flow of the water, be it large or small compared to the biological activity rate, while in the second case they can be swept along with the water. There must be a difference in the concentration of organic waste across the barrier in the two cases, and it may be that the difference in behavior can be used to determine, from observations, which sort of mechanism is the appropriate biological cleansing agent. We shall return to this point later after evaluating quantitatively the functional behaviors in both situations.

We do not consider direct absorption of organic contaminant by solid material in the barrier region in this chapter. It is certainly the case that such absorption must also be evaluated for its contribution to the total cleansing action, but our concern here is with the biological component alone. As we shall see, this biological effect has enough complicating factors in its own right without adding the extra complications of physical absorption and desorption.

2.1 Steady-State Model Construction

Start by setting up steady-state models of both transported organic waste concentration, c(z), with distance z across the barrier (which ends at $z = L$) starting at $z = 0$ with an in-going concentration c_0 as sketched in Figure 7.1, and also a model of biological growth and decay across the barrier, starting at $z = 0$ with a number density of active biological organisms N_0, and attaining the value N(z) at position z.

The steady-state variation of c(z) with position can be described by an equation of the form

$$v\partial c/\partial z = D\partial^2 c/\partial z^2 - k\,(N/N_0)\,c \qquad (7.1)$$

where v is the convective velocity of the fluid transporting the contaminant, and is usually described by a Darcy type of flow in the form $v = -KF/(\mu\rho)$ where μ is the fluid viscosity, ρ is the fluid density and where F is a driving pressure gradient that may be either hydrodynamic, capillary, ionic gradient, etc. For the purposes of this modeling we take v to be given but variable.

The first term on the right hand side of equation (7.1) describes diffusion (with fixed diffusion coefficient D) of the concentration during convection at velocity v through the barrier domain, while the last term describes the effect of biological concentration in removing contaminants through the rate factor k (with dimensions of inverse time) and in proportion to the number density, N, of active biological organisms (scaled relative to the surficial value N_0 at $z = 0$), and in proportion to the concentration, c, available to act as a food supply at position z in the barrier domain.

For the behavior of the active biological number density N with position z in the barrier domain one can write a somewhat similar steady-state equation in the form

$$(qv+v_{biol})\partial N/\partial z = D_{biol}\partial^2 N/\partial z^2 - N/T + \lambda N c/c_0 \tag{7.2}$$

which describes how the convective flux of the biological species is balanced by a biological diffusion process (the first term on the right hand side of equation (7.2)), a death rate of the organisms (described by the factor N/T on the right hand side where T is the death timescale) and growth term (described by the factor $\lambda Nc/c_0$ on the right hand side, and where λ is a scale constant with the dimensions of inverse time that describes the rate at which growth occurs as a consequence of the concentration food supply measured by c/c_0). The factor q is introduced so that one sets $q = 0$ for the case of biological activity tied to the solid particles of the barrier material, whereas for biological activity in the water-laden pores one has $q = 1$. In addition, the velocity v_{biol} measures the propensity of the active biological component to head towards the richest supply of food and hence is proportional (for simple models) to the normalized concentration gradient $\partial(c/c_0)/\partial z$, and will be written

$$v_{biol} = \Lambda c_0^{-1}\partial c/\partial z \tag{7.3}$$

throughout this chapter, where Λ is positive constant with the dimensions of length^2/(time) in order that the dimensions match on both sides of equation (7.3). With this behavior for the biological velocity one can write equation (7.2) in the form

$$(qv + \Lambda\, c_0^{-1}\partial c/\partial z)\partial N/\partial z = D_{biol}\partial^2 N/\partial z^2 + \lambda N c/c_0 - N/T \qquad (7.4)$$

The coupled non-linear equations (7.1) and (7.4) are to be solved subject to the conditions $N = N_0$ and $c = c_0$ on $z = 0$, and with the ingoing flux of contaminant on $z = 0$ specified, and also the outgoing flux specified on $z = L$, the outgoing side of the barrier. In the case where the diffusion coefficients for biological activity, D_{biol}, and contaminant diffusion, D, are negligible in relation to other terms determining steady-state balance conditions, it is sufficient to specify c_0 and N_0 on $z = 0$ in order to determine the complete behavior of the contaminant concentration and also the biological number density at all locations in $0<z<L$.

To investigate the rich variety of solution characteristics possible under different conditions we examine here several patterns related to the size of various terms in the two equations (7.1) and (7.4). These cases are considered next.

2.2 Special Cases of Steady-state Model Behaviors in the Absence of Diffusion

2.2.1 No diffusion, solid attachment, no death rate.

The parameters in equations (7.1) and (7.4) for these conditions are $q=0$, $1/T = 0$, $D = 0$, and $D_{biol} = 0$. In this case one has the simplified but still highly non-linear coupled equations

$$v\partial c/\partial z = -k\,(N/N_0)\, c \qquad (7.5)$$

and

$$(\Lambda c_0^{-1}\partial c/\partial z)\partial N/\partial z = \lambda N c/c_0 \qquad (7.6)$$

which have to be solved subject to the boundary conditions $N = N_0$ and $c = c_0$ on $z = 0$. The general solution of the equations is not difficult to obtain because by dividing equation (7.6) by equation (7.5) one obtains

$$\partial N/\partial z = -\,(v\lambda/k\Lambda)\, N_0 \qquad (7.7)$$

Equation (7.7) has the solution

$$N = N_0\,(1 - z/z^*) \quad \text{in } 0<z<z^* \qquad (7.8)$$

where $z^* = 1\,/\,(v\lambda/k\Lambda)$, while in $z>z^*$ consideration of equations (7.5) and (7.6) yields the solution $N = 0$.

In coincidence with the solution (7.8), the solution of equation (7.5) is then given by

$$c = c_0 \exp(-(kz/v)[1-z/(2z^*)]) \quad \text{in } z<z^* , \tag{7.9a}$$

and

$$c= c_0 \exp(-kz^*/(2v)) < c_0 \qquad \text{in } z>z^* . \tag{7.9b}$$

Because the barrier is of width L, interest centers on whether L is larger or smaller than z^*. If $L>z^*$ then N=0 in the domain $L>z>z^*$ and the concentration of organic pollutants is fixed in the same domain at the value given by equation (7.9b). On the other hand, if $L<z^*$, then there is a finite value of biological activity at z = L given by $N(L) = N_0\,(1-L/z^*)$ and the concentration of pollutants is also higher than given by equation (7.9b) and has the value $c(L) = c_0 \exp(-(kL/v)[1-L/(2z^*)])$ on z = L. For special numerical values of parameters, sketches of the different behaviors in both cases are presented in Figure 7.2 to illustrate the functional behaviors with distance z through the barrier domain.

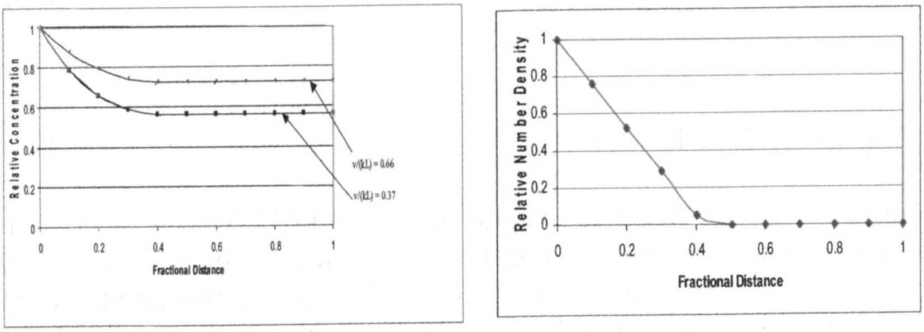

Fig. 7.2. (a) Relative concentration through the barrier domain for two values of v/(kL) showing the decreased concentration as the transport velocity slows; (b) Relative number density through the barrier corresponding to the concentration shown in part (a) for sediment-tied biological activity with no death rate

2.2.2 No diffusion, solid attachment, finite death rate.

In this case we retain $D_{biol} = 0$, D = 0, q = 0 but allow the death rate term N/T to be finite in comparison to the terms already investigated in the previous subsection. The corresponding equations to be solved are

$$v\partial c/\partial z = -k\,(N/N_0)\,c \tag{7.10a}$$

together with

$$(\Lambda\partial c/\partial z)\partial N/\partial z = \lambda N\,c - N\,c_0/T \tag{7.10b}$$

again under the boundary conditions $N=N_0$ and $c=c_0$ on z=0.

One can note that the right hand side of equation (7.10b) changes sign at a critical concentration, c^*, given by $c^* = c_0/(\lambda T)$. At the critical concentration value one has $\partial N/\partial z = 0$, but $\partial^2 N/\partial z^2 = N(\lambda/\Lambda) \geq 0$. So if c starts above c^* (i.e. if $\lambda T > 1$) and decreases towards a value c^*, then it can never cross c^*. Thus the only solution available is N=0 on $c=c^*$. For solutions that start with a concentration $c<c^*$ (i.e. if $\lambda T<1$) such a constraint is not present, but they have $\partial N/\partial z>0$ everywhere so that the biologically active number density increases with distance away from the barrier entrance at z =0. For values of $c \gg c^*$ equation (7.10b) reverts to the case of almost infinite lifetime for the biological activity equation, as in the example above, while for $c \ll c^*$, the death rate of the biological component is much larger than the supply of nutrient ($\lambda T \ll 1$). In that case equation (7.10b) can be written approximately as

$$(\Lambda \partial c/\partial z)\partial N/\partial z = -N\, c_0/T, \text{ in } c<c^* \tag{7.11}$$

Equations (7.10a) and (7.11) then have the solutions

$$N^2 = N_0^2 + 2[(N_0 v)^2/(k^2 T\Lambda)](\,c_0/c-1) \tag{7.12a}$$

and

$$\int_{(c/c_0)^{1/2}}^{1} ds/[(1-A)s^2+A]^{1/2} = kz/(2v) \tag{7.12b}$$

where $A = 2[v^2/(k^2 T\Lambda)]$. The integral in equation (7.12b) can be done in closed form, the functional form of the result depending on whether A(>0) is greater than, equal to, or less than unity. We return to this point later in the chapter using the results from Appendix 1. Note also that one must have $\lambda T \ll 1$in order that the solution given though equations (7.12a) and (7.12b) be valid.

The general solution (not making the assumption c is either large or small compared to c^*) to the coupled non-linear equations (7.10a) and (7.10b) can also be effected as follows. Write $\partial N/\partial z = \partial N/\partial c . \partial c/\partial z$ in equation (7.10b) and then use equation (7.10a) to replace $\partial c/\partial z$ in favor of-$k(N/N_0)c/v$. The result is the equation

$$\partial N^2/\partial c = 2v^2 N_0^2/(\Lambda k^2 c^2)[\lambda c - c_0/T] \tag{7.13}$$

Equation (7.13) integrates immediately to yield

$$N^2 = N_0^2\,[1+2v^2/(\Lambda k^2)\{\lambda \ln(c/c_0) + (1/T)(c_0/c-1)\}] \equiv N(c)^2 \tag{7.14}$$

and, when the solution (7.14) is used in equation (7.10a), it then follows that one can write the solution for c(z) in the form

$$\int_{c}^{c_0} dc'/[c' N(c')] = kz/v \tag{7.15}$$

Thus in this case, too, one can obtain a general solution in terms of quadratures so that numerical representations can be given easily. Some such simple cases are shown in figure 7.3 to illustrate the change in behavior of c(z) due to the presence of the finite lifetime effect compared to the case of infinite lifetime (1/T=0) examined above.

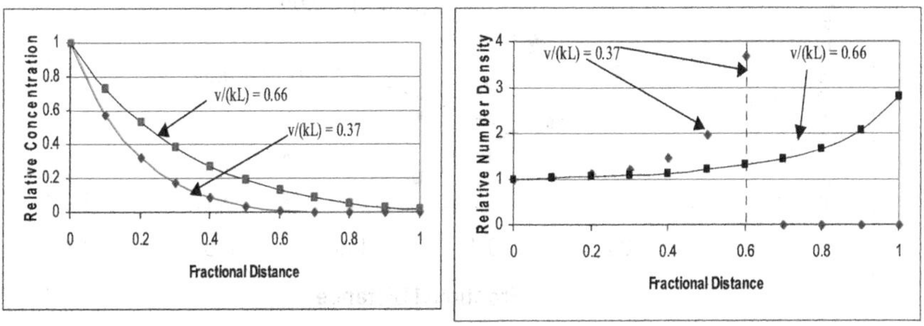

Fig. 7.3. (a) Relative concentration through the barrier for the same cases as for figure 7.2 but with a short lifetime for the sediment-tied biological activity with a short lifetime (high death rate); (b) Relative number density through the barrier region for the two values of v/(kL). Note the integrable singularity at about z/L = 0.6 as discussed in text

2.2.3. No diffusion, fluid attachment, no death rate.

In this case the assumption is that the biological agents are present in the fluid-laden pore spaces and so are subject to the influence of the flowing water transporting the organic contaminant. Then one has the parameter values q = 1, 1/T =0, D_{biol} = 0, D = 0. The corresponding coupled non-linear equations for contaminant and biological activity then take the form

$$v\partial c/\partial z = -k\,(N/N_0)\,c \qquad (7.16a)$$

and

$$(v + \Lambda c_0^{-1}\partial c/\partial z)\partial N/\partial z = \lambda N c/c_0 \qquad (7.16b)$$

The only case of interest here is when $vc_0 \gg |\Lambda\partial c/\partial z|$ because otherwise the case devolves back to that of example (i). And when $vc_0 \gg |\Lambda\partial c/\partial z|$ then one has

$$N = N_0\,(1+\lambda/k - (\lambda/k)c/c_0), \qquad (7.17a)$$

together with

$$c(z) = c_0\,(1+\lambda/k)\exp[-(k/v)z(1+\lambda/k)]/[1+(\lambda/k)\{1+\exp(-(k/v)z(1+\lambda/k))\}] \qquad (7.17b)$$

And this solution is valid as long as $vc_0 \gg |\Lambda\partial c/\partial z|$. When $vc_0 \ll |\Lambda\partial c/\partial z|$ then one can use the solution given in example (i) in that regime of z values. Figure 7.4 shows several representative behaviors for the solutions for contaminant concentration and biological activity density depicted by equations (7.17a) and (7.17b).

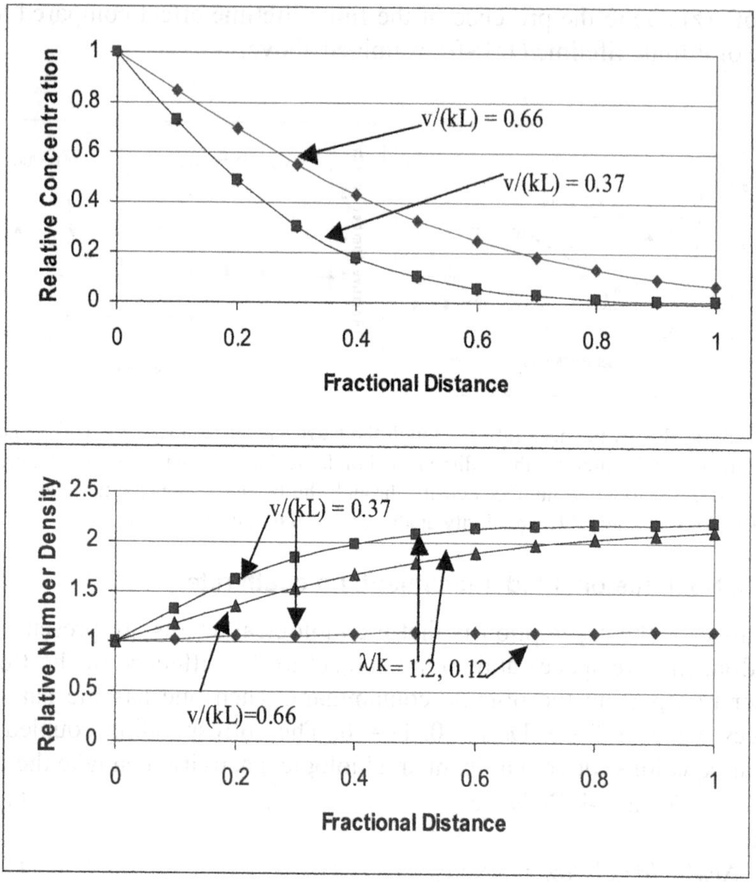

Fig. 7.4. (a) Relative concentration through the barrier region for fluid-attached biological activity for the same two values of v/(kL) as the previous examples but with the growth rate to feeding rate factor λ/k kept fixed at 1.2; (b) Relative number density through the barrier region for infinite lifetime (1/T =0) showing the sensitivity to changes in the growth rate to feeding factor rate (λ/k = 1.2 and 0.12) but in all cases indicating an increase in the biological component as distance through the barrier region increases away from the influx point of organic contaminant concentration

2.2.4 No diffusion, fluid attachment, finite death rate

In this situation the difference compared to example (iii) arises from the extra presence of the term –N/T in the biological density equation. Thus one is asked to solve the coupled pair of equations

$$v\partial c/\partial z = -k\,(N/N_0)\,c \qquad (7.18a)$$

and

$$(v + \Lambda c_0^{-1}\partial c/\partial z)\partial N/\partial z = \lambda Nc/c_0 - N/T \qquad (7.18b)$$

Once again, the only case of novel interest is that with $vc_0 >> |\Lambda\partial c/\partial z|$ because otherwise one can just use the solution from equations (7.10a) and (7.10b) under the condition that $vc_0 << |\Lambda\partial c/\partial z|$. Under the condition that $vc_0 >> |\Lambda\partial c/\partial z|$ equation (7.18b) reduces to

$$v\partial N/\partial z = \lambda Nc/c_0 - N/T \qquad (7.18c)$$

The coupled non-linear equation pair (7.18a) and (7.18c) can be solved simply (by division of one equation by the other as previously) to yield

$$N(c) = N_0[1+(1/kT)\{\ln(c/c_0)+\lambda T(1-c/c_0)\}] \qquad (7.19a)$$

together with

$$\int_{c/c_0}^{1} u^{-1}du/[1+(1/kT)\{\ln u+\lambda T(1-u)\}] = kz/v \qquad (7.19b)$$

Again the integral in equation (7.19b) can be done in closed form yielding different functional behaviors depending on the various parameters in the integral. Illustrations are given in Appendix 2. Note also that, as the concentration decreases, equation (7.19a) indicates than N(c) eventually crosses zero and, for lower concentrations than the critical value, c_{crit} (given by setting the right hand side of equation (7.19a) to zero) the appropriate solutions of equations (7.18a) and (7.18c) are N=0 and $c=c_{crit}$. The determination of whether c(z) ever reaches c_{crit} ($<c_0$) in the barrier region is obtained from the evaluation of the integral on the left hand side of equation (7.19b) in relation to the value kL/v. If the integral evaluated between $1>c/c_0>c_{crit}/c_0$ yields a value greater than kL/v then c never approaches c_{crit} in $0<z<L$, else c crosses c_{crit} at some finite z value less than L.

This sequence completes all the interesting cases that can arise in the absence of the diffusion terms. The presence of diffusion radically changes the structure of the basic equations from coupled non-linear, but first order, differential equations to second order equations. The complexity of the solutions is then just that much greater whenever diffusion is not ignorable. This point is made by considering first the general situation, described by equations (7.1) and (7.4), and performing some simple transformations on the coupled equations, and second by considering dominance of various terms, just as was done for the equations in the absence of the diffusive terms.

2.3 Diffusive Models in the Steady-state Limit

When a diffusion term is necessary in either the biological activity equation or the contaminant concentration equation, the implication is that such a diffusion term must be large compared to at least one of the other terms on the right hand sides of equations (7.1) or (7.4), else there would be no need to include such diffusive behavior.

Now if the diffusion of contaminant concentration were to exceed the loss of contaminant by biological activity (represented by the term $-k(N/N_0)c$ on the right hand side of equation (7.1)) then there would be no "cleansing" of the flow-through by biological activity but rather just a spreading out of the contaminant in position through the barrier domain. In such a situation the contaminant concentration would be given by

$$c(z) = c_0 + (vc_0 - F_0)(\exp(vz/D) - 1) \qquad (7.20)$$

where the contaminant flux at location $z = 0$ is $vc_0 - D\partial c/\partial z \equiv F_0$

The concentration of contaminant at the output face on $z = L$ would then be just

$$c(L) = c_0 + [vc_0 - F_0](\exp(vL/D) - 1) \qquad (7.21)$$

and is completely uninfluenced by the biological action- something that is not of interest in terms of the biological cleansing of Darcy flow contaminant transport being remediated by ongoing biological activity.

Accordingly, in terms of the underlying framework of the influence of biological activity one can ignore the situation where diffusion dominates in influencing the contaminant transport picture and so concentrate solely on the situation where biological activity compensates the net flow. Then write

$$v\partial c/\partial z = -k\,(N/N_0)\,c \qquad (7.22)$$

as the relevant transport equation for the organic contaminant component. However, it is not so clear that the biological component can be so simply and similarly treated. The point here is that the transport equation for the biological component, including diffusion, is

$$(qv + \Lambda c_0^{-1}\partial c/\partial z)\partial N/\partial z = D_{biol}\partial^2 N/\partial z^2 + \lambda Nc/c_0 - N/T \qquad (7.23)$$

with $q = 1$ or $q = 0$, depending on whether one is considering fluid ($q = 1$) or sediment-tied ($q = 0$) biological remediation components.

But in either event, the right hand side of equation (7.23) then allows one to compare the diffusive term with either, or both, of the feeding term ($\lambda Nc/c_0$) or the death rate term ($-N/T$). Inclusion of the biological diffusion term implies it is larger than at least one of the other factors on the right hand side of equation (7.23), but there is no reason that it must be larger than both simultaneously. Thus two cases exist if one considers diffusion inclusion: either (i) $N/T \ll \lambda Nc/c_0$, but the diffusion term is not necessarily small compared to $\lambda Nc/c_0$; or (ii) $N/T >> \lambda Nc/c_0$, but the diffusion term is not necessarily small compared to N/T. And for each of these two cases one has to distinguish whether one is considering sediment-tied biological activity or fluid dominated biological activity.

2.3.1 Fluid Dominated Biological Activity

In this situation ($q = 1$) one is again interested only in the case where $vc_0 >> \Lambda|\partial c/\partial z|$ because otherwise it is as though the fluid convection plays no role in modifying the biological remediation (i.e. as though the biological activity were sediment-tied).

Then one is interested in the solutions to

$$v\,\partial N/\partial z = D_{biol}\partial^2 N/\partial z^2 + \lambda Nc/c_0 - N/T \qquad (7.24)$$

coupled to equation (7.22) under the constraint $vc_0 >> |\Lambda\partial c/\partial z|$. Simplification of the coupled equations can be achieved by using equation (7.22) in the right hand side of equation (7.24) to obtain

$$v\,\partial N/\partial z = D_{biol}\partial^2 N/\partial z^2 - (vN_0\lambda/(kc_0))\,\partial c/\partial z + (vN_0/(kT))\,c^{-1}\partial c/\partial z \qquad (7.25)$$

Equation (7.25) has an exact first integral given by

$$vN = D_{biol}\partial N/\partial z - (vN_0\lambda/k)\,(c/c_0 - 1) + (vN_0/(kT))\ln(c/c_0) + A \qquad (7.26)$$

where A is a constant of integration to be determined by imposition of the boundary conditions.

Some general reduction of equation (7.26) can be achieved by replacing $\partial N/\partial z$ by $\partial N/\partial c . \partial c/\partial z = -(\partial N/\partial c)\,Nc\,(k/(vN_0))$ when equation (7.26) can be written

$$vN + D_{biol}\,Nc\,(k/(vN_0))\partial N/\partial c = -(vN_0\lambda/k)(c/c_0-1) + (vN_0/(kT))\ln(c/c_0) + A \qquad (7.27)$$

Equation (27) can be reduced to the form

$$\partial N/\partial x = [H(x)-N]/N \qquad (7.28)$$

using $\ln(c/c_0)\,[D_{biol}\,(k/(v^2N_0))]^{-1} = x$ as a new independent variable and where

$H(x) = -(N_0\lambda/k)\,(c/c_0 - 1) + (N_0/(kT))\ln(c/c_0) + A$, with c given explicitly as a function of x. From equation (7.22) the relation between position, z, in the barrier medium and the variable x is then given by

$$\int_x^1 dx'/N(x') = kz\{(N_0 v)\,[D_{biol}\,(k/(v^2N_0))]\}^{-1} \qquad (7.29)$$

Exact analytic solutions to equation (7.28) are not available for arbitrary H(x) but note that approximations can be obtained easily if H(x) is either very small or very large in magnitude compared to N(x). If $|H(x)| >> N$ then equation (7.28) yields the approximate solution

$$N^2 = N_0^2 + 2\int^{x} H(x')dx' \qquad (7.30)$$

while if $|H(x)| << N$ then

$$N = N_0 - x \qquad (7.31)$$

Thus, one can piece together the fundamental behavior of the diffusion dominated biological activity under fluid transfer conditions for the organic contaminant and with the biological activity being fluid controlled. In addition, the structure of equations (7.28) and (7.30) allows very simple numerical procedures to be used with high numerical stability. This advantage arises largely as a direct consequence of being able to write down an exact first integral (in closed form) to the second order diffusion equation for the fluid controlled biological activity, thereby reducing the general problem to two coupled, highly nonlinear, but only first order equations.

2.3.2Sediment-tied Biological Activity

When the biological component is tied to the sediment particles ($q = 0$) then the relevant equations to consider are

$$\Lambda c_0^{-1} \partial c/\partial z \partial N/\partial z = D_{biol} \partial^2 N/\partial z^2 + \lambda N c/c_0 - N/T \qquad (7.32)$$

and

$$v \partial c/\partial z = -k\,(N/N_0)\,c \qquad (7.33)$$

In this case it is more difficult to find any exact integrals and recourse must be had to numerical schemes. However, if the left hand side of equation (7.32) is very small, so that diffusion dominates the trend of the sediment-tied biological component to seek regions of highest organic concentration, then one can approximate equation (7.32) by

$$D_{biol} \partial^2 N/\partial z^2 + \lambda N c/c_0 - N/T \approx 0 \qquad (7.34)$$

And equation (7.34) does have an exact first integral given by

$$D_{biol} \partial N/\partial z - (vN_0\lambda/k)\,(c/c_0 - 1) + (vN_0/(kT))\ln(c/c_0) = B \qquad (7.35)$$

where B is a constant of integration to be determined by the boundary conditions. Again replacing $\partial N/\partial z$ by $\partial N/\partial c . \partial c/\partial z = -(\partial N/\partial c)\, Nc\, (k/(vN_0))$ in equation (7.35) yields

$$D_{biol}\,(\partial N/\partial c)\, Nc\, (k/(vN_0)) = B + (vN_0\lambda/k)\,(c/c_0 - 1) - (vN_0/(kT))\ln(c/c_0) \qquad (7.36)$$

which can be integrated in closed form to give

$$N^2 = N_0^2 - 2vN_0/(k\,D_{biol})[\,(B - vN_0\lambda/k)\ln(c/c_0) + (vN_0\lambda/k)(c/c_0 - 1) - 1/2\,(vN_0/(kT))[\ln(c/c_0)]^2\,] \qquad (7.37)$$

Equation (7.37) expresses N(c) as an explicit function of the concentration c, so that one then has the connection between location in the barrier and concentration given through the quadrature

$$\int_{c}^{1} dc'/[c'N(c')] = kz/[vN_0] \qquad (7.38)$$

Other situations are best investigated numerically.

This completes the mathematical listing of steady-state situations for this chapter. The next section of the chapter explores some of the patterns of behavior that can be obtained depending on individual conditions and anticipated parameter ranges.

3 Illustrative Patterns of Behavior

The four basic non-diffusive steady-state solution patterns are illustrated in this section of the chapter for various parameter values entering the equations in order to show the basic curves of response. Even in the case where diffusion of both the contamination and the biological activity can be neglected, there is still a large number of parameters that control the basic patterns of variation with position in the barrier region. The relative strengths and nonlinear combinations of parameters is what determine the solution behavior for each case. The basic five parameters are v, Λ, k, λ and T. These five, in various combinations, control the solution structure. In addition, there is the width, L, of the barrier in relation to particular limiting lengths, such as z_*, that determines the crossover from one solution type to another.

A complete investigation of all parameter ranges in relation to all structural dependences of solutions for the relative concentration, $c(z)/c_0$, and the biological relative number density, $N(z)/N_0$,as functions of position z through the barrier region, would be a formidable undertaking and one that would defeat the purpose of this chapter, which is to show how the spatial variations of particular solution classes relate to the influence of parameter combinations special to each class. More general investigations can then be undertaken appropriate to particular applications once the general patterns of development are available.

Accordingly we have taken the four non-diffusive classes of mathematical solutions given in the previous section and for each have constructed an Excel spreadsheet program. Each of the four classes above has been numerically modeled to show the ebb and sway of patterns of behavior for a few variations of the parameters so that an appreciation can readily be obtained of what happens in each case. We consider the numerical illustrations, in turn, using the same subsection nomenclature as for the mathematical development section.

3.1 No diffusion, solid attachment, no death rate.

Figures 7.2a and 7.2b show, respectively, the variation of relative concentration and relative biological number density corresponding to equations (7.8) and (7.9). Note that because the relative number density is sensitive to only the one parameter z_*, while the concentration is sensitive to both z_* and to the ratio v/k, for different values of v/k one obtains precisely the same steady-state relative number density variation with fractional position z/L through the barrier region. What differs, for a fixed z_* value, is the level at which the relative concentration c/c_0 is constant once the fractional position crosses the value z_*/L, at which N=0. Small values of v/k correspond to more rapid depletion of the concentration compared to larger values, as exhibited. The reason for this effect is that the rate of concentration loss is proportional to the rate factor k so that, for a given flow speed v, a higher value of k corresponds to a more rapid diminution of the concentration, as exhibited in figure 7.2a. If z_* is increased (corresponding to a larger value of the biological speed of motion, v_{biol}, towards the richest supply of organic contaminant on z = 0) then the effect is even more pronounced over a larger spatial scale because the relative concentration then takes on the value $\exp(-kz_*/(2v))$, which decreases as z_* increases, and the number density is more slowly decreasing for increasing z_*, giving the biological organisms a larger range of distances in the barrier region to deplete the contaminant concentration.

3.2 No diffusion, solid attachment, finite death rate.

Adding a finite lifetime T to the basic behavior described in the previous section changes completely the basic pattern of behavior, particularly at small lifetimes in relation to the rate at which biological enhancement of number density can take place as a consequence of consuming the organic contaminant, as measured through the rate parameter λ, so that a fundamentally different steady-state behavior ensues for $\lambda T<1$ compared to $\lambda T>1$. The latter case is basically similar in structure to the previous situation because there is little difference between a large but finite lifetime relative to an infinite lifetime. So here we concentrate on the small lifetime situations to show the structural differences in behavior compared to the previous subsection. For $\lambda T<1$ one can again vary the spatial scale v/k with remarkably different behaviors resulting. For the same two values of v/k as used previously, figures 7.3a and 7.3b record the variation with fractional position in the barrier region of relative concentration and relative number density, respectively.

Figure 7.3a shows that now there is no steady-state fixed concentration reached in the barrier region; either the contaminant concentration reaches the zero value or it steadily declines throughout the region, Precisely which of the two situations occurs depends sensitively on the value of v/k because it controls how the variation of biological active number density behaves. As shown on figure 7.3b, for small values of v/k the number density rises without bound (relative to its value on $z = 0$), eventually becoming infinitely large in the barrier region. What is happening here is that the solid-attached organisms are doing their best to head in the direction towards $z = 0$, where the richest concentration of contaminant is, but die so fast on their way there that the only steady state available is an infinite number of organisms somewhere deeper in the barrier region.

For larger values of v/k, the speed of flow of the contaminant concentration is large enough to overcome the need of the biological organisms to reproduce massively (effectively infinitely fast) to maintain a stable population, food is brought past their sediment –controlled locations at a rapid enough rate to keep them supplied despite their high death rate. This fact is seen in figure 7.3b (for the curve marked $v/k = 0.66$) where, while the number density rises as fractional distance increases through the barrier region, it nevertheless stays finite. And also one then has a small but finite concentration at the barrier exit. Note that the singular cases where the relative concentration rises to infinity are of the so-called integrable variety, in the sense that as z approaches the critical value z_{crit} where $N \rightarrow \infty$, the behavior of N is proportional to $(z_{crit}-z)^{-1/2}$ which yields a finite total count of biological activity per square cm because the integral of $(z_{crit}-z)^{-1/2}$ over z yields $(z_{crit}-z)^{1/2}$ which tends to zero as $z \rightarrow z_{crit}$.

One could argue that the infinite number density cases are non-physical and that remaining processes, excluded in this simple investigation, need to be included (a statement with which we agree); either time-dependence or diffusion or both for example. While true, nevertheless the point is made most sharply with these extreme illustrations that it is the variation of rapidity of the flow (as measured by v/k) in relation to the short lifetime (as measured by $\lambda T<1$) that is responsible for these anomalous results; and that is precisely the central theme we wished to illustrate.

3.3 No diffusion, fluid attachment, no death rate.

Patterns of behavior change markedly from the two previous cases when the biological activity flows with the concentration and is not sediment-tied. Figures 7.4a and 7.4b show the variation with fractional distance through the barrier for relative concentration and relative number density, respectively, for three values of parameters, the first two are precisely as for the previous two situations and the third is for λ/k reduced from 1.2 to 0.12, so that the growth of biologically active material (measured by the rate constant λ) is slowed relative to the processing rate (measured by the rate constant k) of the contaminant concentration by the biological matter. Thus, feeding on the contaminant does not produce as much new biological material as previous cases provided so that the gradient of relative number density decreases through the barrier region.

Keeping the processing rate constant, k, at the same values as previously means that the contaminant concentration variation with lateral position through the barrier domain keeps the same behaviors as in the previous two subsections, as indicated by the shapes of decline of the relative concentration curves through the barrier domain, given in figure 4a for different parameter values. However, because of the slower growth of the biological component as λ/k decreases, in the steady state situations it then follows that a lower relative number density ensues. This sort of lowered increase in relative number density through the barrier domain is exhibited in figure 7.4b for various parameters.

3.4 No diffusion, fluid attachment, finite death rate

This final numerical example compares the effects of a finite lifetime on the patterns of response. Figure 7.5a shows the relative concentration behavior through the barrier domain at different choices for both v/k, λ/k, and λT.

The three curves exhibited show that the sensitivity is more dominated by the variation of λ/k ranging from 1.2 to 0.12 than it is to the variation of λT, which ranges from 0.99 to 0.09. But in all cases the combined effect of a lower growth rate of the biological component (as measured through the lowering of the ratio λ/k) and a short lifetime (as measured by the lowering of the parameter λT) is to enhance the rate at which the contaminant concentration is decreased with increasing position through the barrier region. The relative number density through the barrier for different choices of the parameters is exhibited in figure 7.5b, where it can be seen that a smaller lifetime, T, together with a smaller growth rate of the biological component both act to lower the relative number density significantly relative to longer lived and faster growth situations, as was to be anticipated.

These four groups of illustrative patterns of behavior provide some idea of the richness of possible behaviors depending on the relative strengths of parameters and also on whether one considers the biologically active component to be tied to the sediment particles or flowing with the concentration through the pore space.

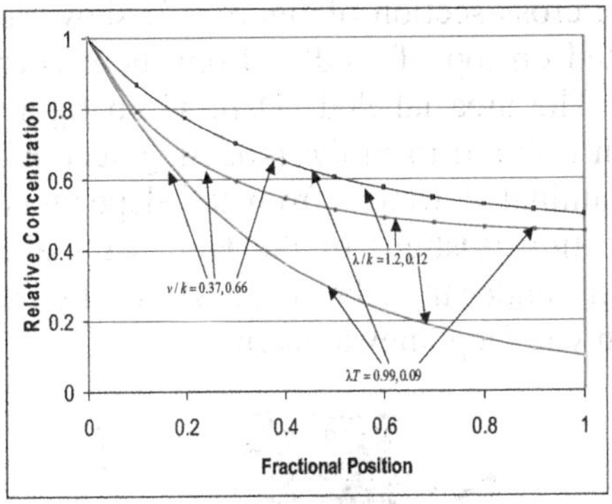

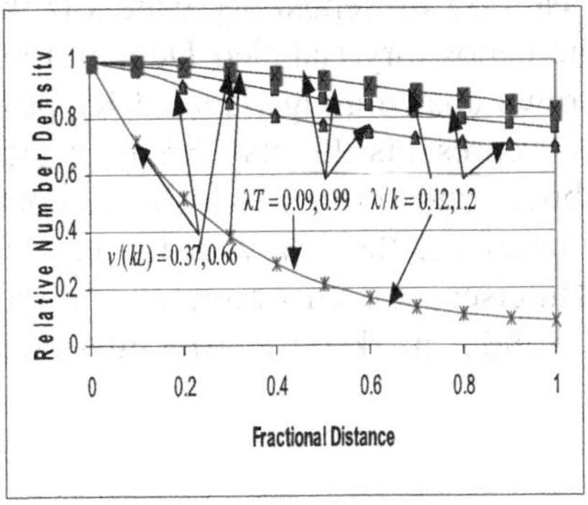

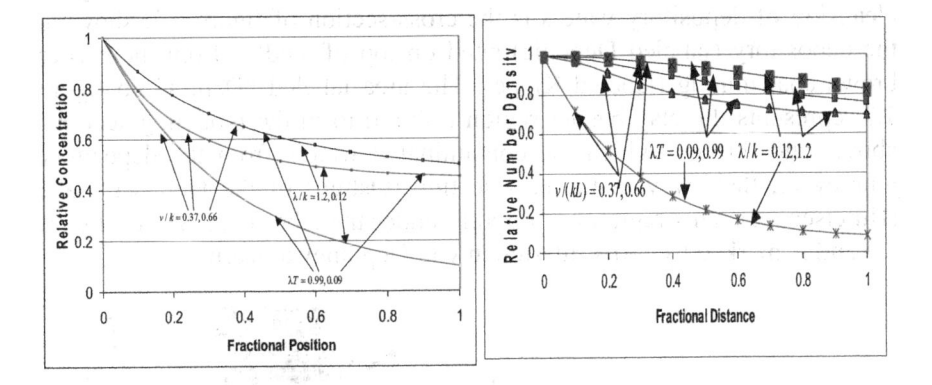

Fig. 7.5. (a) Relative concentration through the barrier for fluid-attached biological activity for finite, short ($\lambda T<1$) lifetime conditions. Three curves indicate the dominant variation for different values of v/(kL), different ratios of λ/k, and different values of λT; (b) Relative number density variations through the barrier region for the parameter values shown in figure 7.5 (a), showing the decrease of the number density as fractional distance through the barrier increases

In addition, one should not forget there are also the effects of diffusion of both the concentration and also the biologically active component that can influence the patterns of behavior together with the effects of physical absorption and desorption. But inclusion of all such processes would make for a very long chapter indeed, and one that would defeat the purpose of illustrating as simply as possible the sorts of response than can be coaxed out of the system.

4 Comparison with Observations at the Organic Depository, Halle-Kanena.

A plan view of the organic waste depository at Halle-Kanena is given in figure 7.6, showing the positions of measurement sites and also the outline of the depository in relation to the surrounding lake (Hufeisensee). Much has been written on the Halle-Kanena depository and its underlying sediment and structural support; the best compendium is the report edited by Strauch (1996), to which report we refer the interested reader for detailed information on site locations, measurements made, and inferences drawn there from over the course of more than a decade. Additional detailed investigations of various components of the depository are to be found in Glaesser (1994), Christoph (1995), Glaeser (1995), and Maiss et al.(1998).

Perhaps of greatest interest in connection with the biological cleansing of depository waters is the cross-section of figure 7.7, showing the depository (labeled Deponie) sited on top of landfill from the earlier brown-coal mining (labeled Kippe). The area labeled "Damm" on figure 7.7 represents the clay barrier region referred to in the modeling sections above, and through which the contaminated waters from the depository journey on their way to the east section (Ostteil) of the Hufeisen Lake (Hufeisensee). This region lies directly under the label ENE/SW on figure 7.7 and is marked by "Austrittsstellen von Deponiewaessern".

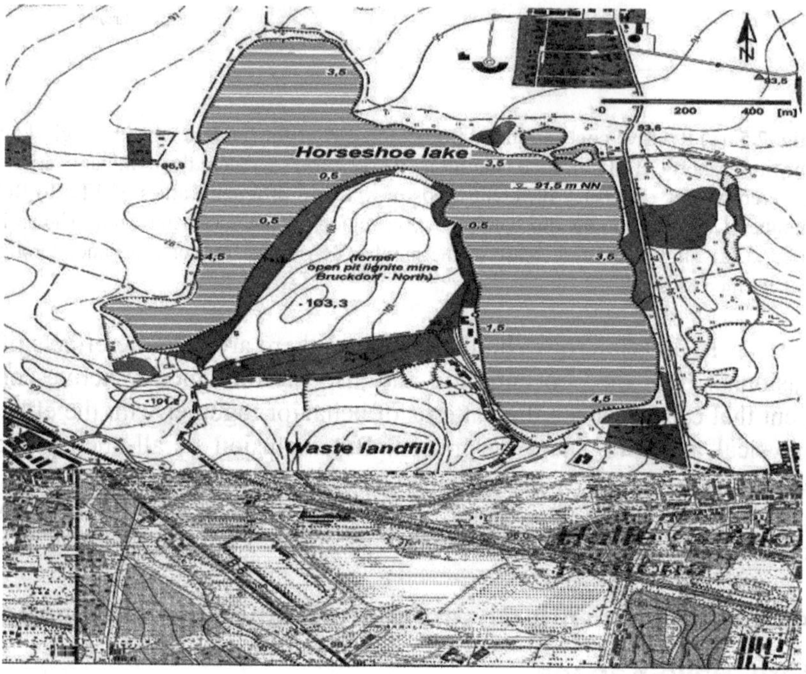

Fig. 7.6. Plan view of the depository Halle-Kanena and the Hufeisen Lake made by the open pit mining. The measurement positions for the groundwater are marked by open circles, the position of a depth profile and measurements in the depository is marked by a triangle, while the outline of the depository is marked by the dotted line. The dashed lines represent walkable pathways across the depository. The arrow in the upper right hand corner, marked "GW-ANSTROM" ,marks the direction of groundwater flow into the Hufeisen Lake

gion with extremely low mineralization (from the water surface down to a depth marker around 75m using the vertical scale on figure 7.7). Below that depth marker, there is a sharp transition to a domain of highly mineralized water, extending to almost the bottom of the water and below which a region with significant H_2S represents the basal waters. Residual brown coal (labeled Braunkohle) on the SW side has its top formation contact coincident with the transition from low mineralization to high mineralization, as shown on figure 7.7.

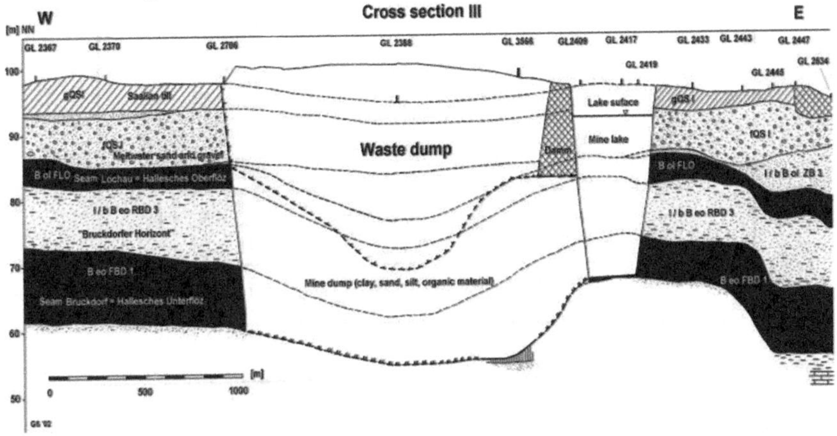

Fig. 7.7. Cross-section through the depository showing the relative dimensions of the depository and the exit points of the water from the depository. See text for further details

Sketched in figure 7.8a are the concentrations of PAH (polyaromatic hydrocarbons;in German, PAK-Polyaromatisch Kohlenwasserstoff)) measured at various sites inside the depository (labeled S1-S4) and also in the waters at the exit flux points (labeled W1-W5) shown in figure 7.7. To be noted are two major facts: first, the reduction in relative concentration of PAH contaminants in passing through the barrier region is about a factor of a million; second, irrespective of where in the depository or where in the waters the sampling points are located, the concentration in the depository is effectively independent of measurement location, and likewise the concentration in the exit waters is also independent of measurement location. It is these two facts that allow use of a one –dimensional model to address the concerns of biological remediation. The point here is that while one may not know for certain the actual flow path of the PAH contaminants through the barrier region (which is manifestly not a one-dimensional entity as can be seen from figure 7.6), the fact is that there are no major differences in the concentrations with respect to measurement sites. Thus, a one-dimensional model behavior captures the essence of the contaminant flow without the need for more complex, but more rigorous, multi-dimensional models. The available data are consistent with dominance of a one-dimensional behavior and will be modeled as such here.

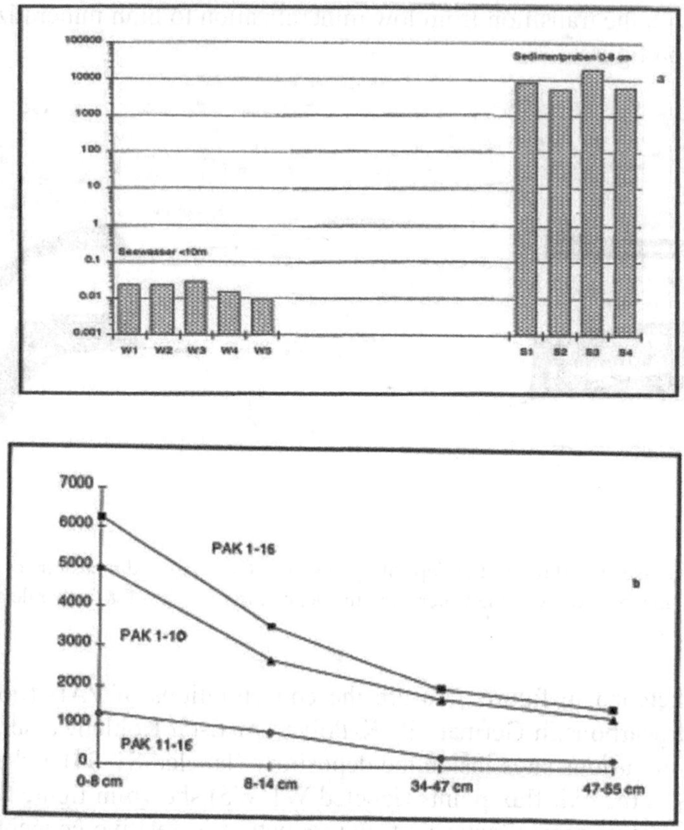

Fig. 7.8. (a). Concentration of PAH in the lake water at depths less than 10 meters and also in the depository at depths between 0-8cm. Note the relative drop in concentration by a factor of about a million as the waters exit the depository; (b) Depth and formation dependence of the PAH contamination for a sediment probe in the depository in the east basin part of the Hufeisen Lake. The overall differences (marked PAH 1-10, PAH 11-16) refer to different components of the PAH contamination, while the curve marked PAH 1-16 refers to the sum of all components

One might be concerned that the PAH contaminant is variable spatially with depth rather than with lateral position. Figure 7.8b shows the depth dependence of different PAH components for one sediment probe in the eastern section of the depository. It can be seen that the dominance is over a depth range of some 50 cm or less, so that the majority of the PAH contaminant is flowing in a unit small compared to the barrier height (about 10 m based on figure 7.7), and so fluctuations due to such effects can be ignored in this first analysis of model and observation comparisons.

Effectively, then, what one has as a measurement is the ratio $c(L)/c_0 \approx 10^{-6}$ but with an unknown value for L, but presumably L lies between the geometric width of the barrier and the length of the barrier. Equally unknown are the flow speed v of the PAH contaminants through the barrier, the parameters k, T, λ, Λ describing the rates of biological growth and death in relation to contaminant flow, and the question of whether or not the biological activity is tied to the sediment particles or flows with the contaminant flux. It is therefore clear that a single measure, such as $c(L)/c_0$, is not going to allow determination of all such parameters, many more measurements would be needed to unscramble all of the parameter values. However, what is possible to achieve is to determine combinations of parameters for different models that will allow satisfaction of the observed measure. We illustrate this approach with just two simple examples to illustrate behaviors and uncertainties.

4.1 Model 1. No diffusion, solid attachment, no death rate.

The match of this model to the observations can be most easily be achieved by rewriting equations (7.9a) and (7.9b) in the form

$$(kL/v) = \ln R/(1-L/(2z^*), \text{ in } L/z^*<1 \qquad (7.39a)$$

$$(kL/v) = 2\ln R(L/z^*), \text{ in } L/z^*>1. \qquad (7.39b)$$

where $R=c(L)/c_0$. It is then simple to plot the variation of kL/v versus L/z* for particular values of R (in our example 10^{-6}) and so obtain the acceptable values.

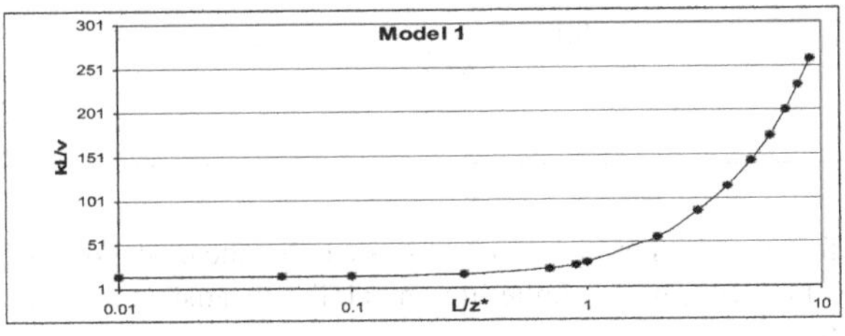

Fig. 7.9. Curve for the values of kL/v versus L/z* such that the model of no diffusion

Figure 7.9 shows a plot of the behavior of equations (7.39), showing that a broad range of acceptable solution parameters exists. When it is remembered that $z^* = k\Lambda/(v\lambda)$, it follows that a high flow velocity for the contaminant (and so a low value of kL/v) can be compensated for by a low value of L/z* (as shown on figure 7.9) corresponding to a high value of z* and therefore a high value of Λ, so that one has a high speed of biological components tied to the sediment particles in the direction of the entry point (z=0). For each choice of kL/v, there is always a value of L/z* that will allow satisfaction of the observed ratio R.

4.2 Model 2. No diffusion, fluid attachment, no death rate

This case has been chosen so that a direct comparison can be made between the situation with sediment-tied biological activity and fluid attached biological activity.

In this situation one writes equation (7.17b) in the form

$$kL/v = (-\ln R - \ln(1+\lambda/k) + \ln[1+\lambda/k(1-R)])/(1+\lambda/k) \quad (7.40)$$

so one sees immediately that for any choice of R there are two independent positive parameter combinations: kL/v and λ/k. These two parameter groups can be varied according to equation (40) and maintain a constant value of R ($=c(L)/c_0$).

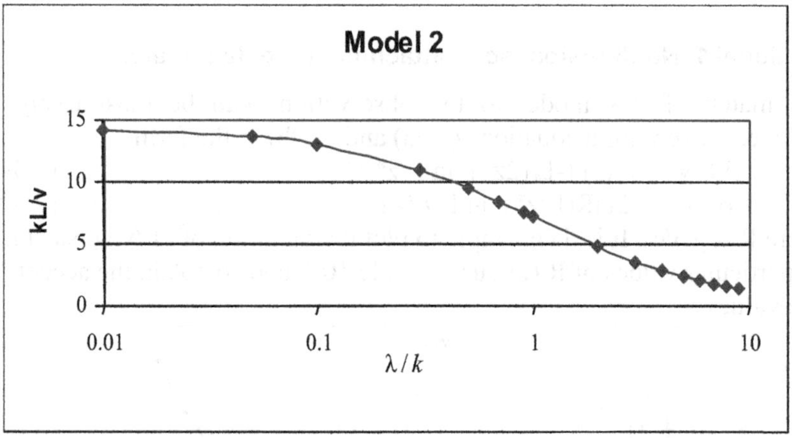

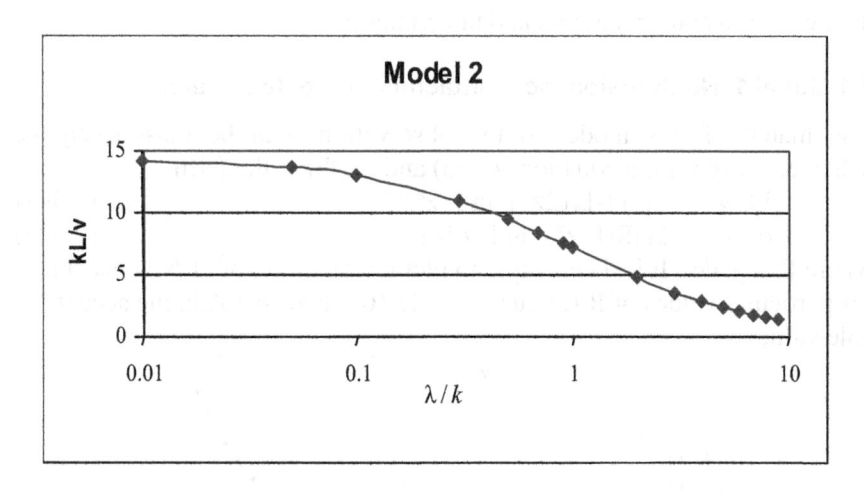

Fig. 7.10. Curve for the values of kL/v versus λ/k such that the model of no diffusion, fluid attached, biological activity, with infinite lifetime (1/T =0) satisfies the observed efflux to influx ratio $R = 10^{-6}$

Figure 7.10 shows the variation of kL/v versus λ/k for R= 10^{-6} as previously. Note that the behavior of kL/v is now very different than for the sediment-attached situation. The main reason for the differences is due to the biological growth rate factor λ in relation to the rate factor k determining the rate at which the contaminant is removed by the biological activity. For a high speed of transport, v, through the system (and so a low value of kL/v) one requires a rapid growth rate of the biological component (also transported with the fluid at speed v) in order that the concentration be reduced as observed. For low values of flow transport speeds there is ample time for the biological component to reduce the contaminant concentration and so a low value of λ/k suffices to maintain the observed concentration reduction.

The remaining two situations described in the text can also be investigated to determine the range of the relevant parameters that will allow agreement of the model outputs with the observations, but the point we wished to make has been accomplished: there are broad ranges of parameter values for each model that will allow agreement with observations. And the fact that the observations are limited to just one ratio means that the parameters are not uniquely determined. Illustration of these two points was the main aim of this comparison section of the chapter.

5 Uncertainty and Sensitivity

The discussion throughout this chapter so far has been deterministic in terms of model developments, observed concentration depletion of PAH contamination, and determination of parameter ranges that allow models to match observed values. However, several factors indicate that one must move beyond this limited viewpoint. First, the depository at Halle-Kanena is by no means one-dimensional; the ingress and egress positions for fluids are not determinable by single plane lines (at z=0 and z =L respectively). Second, the contamination concentration, both in the depository and in the surrounding waters, is not precisely the same at all measurement sites (see figure 7.8a where variations at the egress measurement positions are from $(1\text{-}3)\text{x}10^{-2}$ in relation to the variation of $(0.7\text{-}2.0)\ \text{x}10^{4}$ in the depository measurements for a total dynamic range of $(2.0\text{-}0.23)\text{x}10^{6}$).Third, the concentration ,at least at one measurement site in the depository (see figure 7.8b), is variable with depth and also dependent on the specific PAH components. Taken together, all these factors imply that there is not an unique relation between model parameters to describe exactly the concentration reduction variation.

There are two very different procedures to address this uncertainty concern: either one constructs an incredibly detailed multi-dimensional model and attempts once more to determine its parameters to encompass the variation of the observations; or one uses stochastic methods in attempts to bracket the range of uncertainty of parameter values and, at the same time, determine which parameters are causing the biggest uncertainty. In this way one obtains some idea of where to focus efforts in any attempt to narrow down the uncertainty. It is this second approach that we exhibit here.

For each choice of the ratio R ($=c(L)/c_0$) one could calculate a curve of kL/v versus either L/z* (for Model 1) or λ/k (for Model 2). But because R is uncertain and can vary in the range $(2.0\text{-}0.23) \times 10^{-6}$ it follows that there will be a slew of curves for all the particular values of R chosen. In addition, it is not known what observational distribution to ascribe to the value R because four internal depository measurements and four external values are hardly sufficient to determine the true underlying distribution. So two factors play a role in assessing the uncertainty on the parameters: the range of R values and the distribution from which they are culled.

To illustrate how one can handle this uncertainty, here we consider Model 2 only and use Crystal Ball® to determine the range of uncertainty on kL/v as λ/k varies for two distribution choices for R, uniform and triangular. Many other choices of distribution could be made, and likely should be in order to address the total suite of possibilities, but it will suffice for our purposes to consider just two. We need to determine the extent to which it is either the choice of underlying distribution that is dominating the uncertainty on kL/v or whether it is the range of R that is more dominant. If the range is dominant then the results of sensitivity are basically uninfluenced by the distribution choice made, while if the distribution choice dominates the uncertainty then it would not be so important to narrow the range of uncertainty of R but rather to do a better job on determining the distribution of values. The two cases of triangular and uniform distributions serve the purpose of illustrating this point nicely.

5.1 Uniform Distribution for R

Inspection of equation (7.39) for Model 1 shows that in both the cases of $L/z^*<1$ and $L/z^*>1$ the dependence of kL/v is proportional to lnR for any choice of L/z*. Thus the statistics of lnR are mirrored directly in the statistics of kL/v. Likewise, inspection of equation (7.40) shows that the same dependence operates almost exactly because, for $R<<1$, the term $\ln\{[1+\lambda/k(1-R)]/(1+\lambda/k)\}$ is always very close to zero for all values of λ/k. For a uniform choice of the distribution for R, in the range $(0.23\text{-}2)\times10^{-6}$, we have then used Crystal Ball® to generate the distribution of lnR. This distribution can then be used to generate curves of kL/v versus

L/z* for Model 1, and of kL/v versus λ/k for Model 2 at any chosen statistical measure. Shown in figure 7.11a are the corresponding curves taken at one standard deviation above and below the mean value of R. To be noticed is the very close parallel of the curves, suggesting that there is very little uncertainty on the relevant measures of uncertainty in the sense that there is not much departure from the mean curve at the one standard error values for the uniform distribution choice.

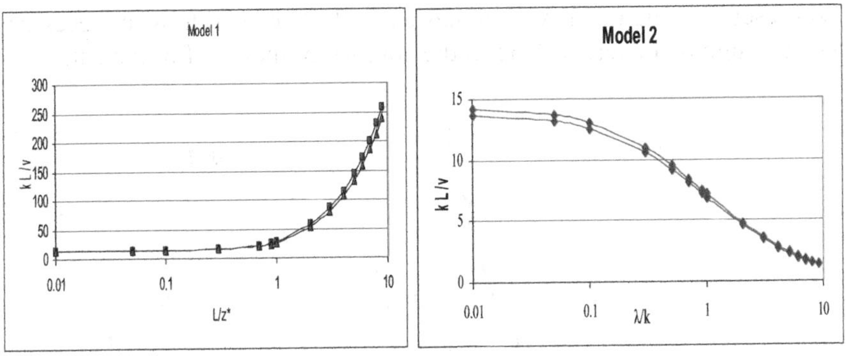

Fig. 7.11. (a) Curves of kL/v versus L/z* drawn for the model of figure 7.9 but with the curves at one standard error distance for lnR around its mean value when $R/10^{-6}$ varies in the range 0.23-2.0 using a uniform probability distribution for R; (b) as for part (a) but for the model of figure 7.10

5.2. Triangular Distribution for R

Changing the distribution choice for R changes the resultant behaviors for kL/v versus L/z* and λ/k at the one standard error level. We have chosen a triangular distribution that is centered at the mid-point of the range for $R/10^{-6}$ in 0.23-2.0 and that tails off to zero at the two extremes of 0.23 and 2.0 respectively. Because there is now a greater chance that random values for R chosen in the permitted range will be centered closer to the most likely value halfway through the range, rather than being spread uniformly as in the previous example, one anticipates that the curves for kL/v at the mean and at plus and minus one standard error should be more tightly grouped, and this is indeed the case as can be seen for both Model 1 and Model 2 behaviors in figures (7.12a) and (7.12b), respectively.

The dominant result of these simple statistical tests is to show that there is not too much uncertainty in the patterns of behavior for parameter determination based solely on the mean values discussed in Section IV of the chapter. The dominant uncertainty is a small percentage of the mean value curves, indicating that there is still a very large range of parameters that will satisfy the observed data, even when uncertainty on the data is allowed for. Basically the message is being conveyed that the data available (viz. ratio of concentration in the waters surrounding the depository to the concentration measured in the depository itself) are not sufficient to allow unequivocal choices to be made for parameter ranges and values. While we have illustrated this stochastic point for only two of the four steady state models in the absence of diffusion, the same lack of determinism is also present for the other two models as well. It seemed to us unnecessary to go through that exercise here in the absence of more definitive data.

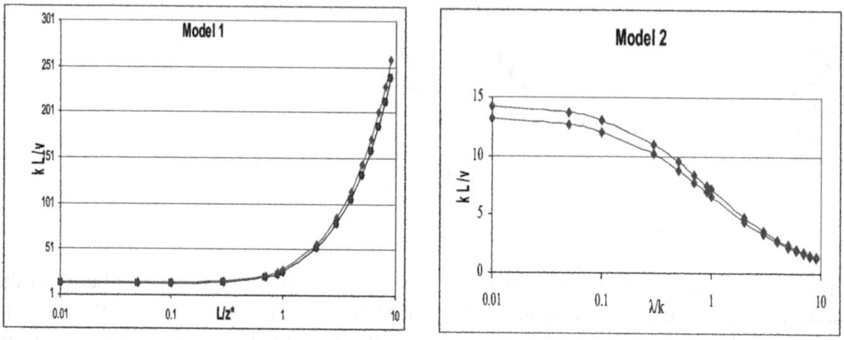

Fig. 7.12. (a) As for figure 7.11(a) but when the distribution of R is chosen to be triangular centered on the mid-point value of the range and with zero probability at the ends of the range. Note the much narrower uncertainty than for figure 7.11(a); (b) as for part (a) but for the model of figure 7.10. Note that the triangular distribution uncertainty is narrowed for R relative to that for the uniform probability

6 Discussion and Conclusion

The ultimate goal of biological remediation models of organically contaminated waters in depositories is to assess the dominant parameters entering such models so that one can obtain a better idea of the risk factor associated with the depositories in terms of environmental safety. On the road to achieving that goal there are a variety of problems that need to be resolved. The simplest models in such a framework are the one-dimensional steady-state situations addressed in this chapter. As has been shown, on their own such models have a large number of parameters depending on what is thought to be the appropriate behavior of the biological components feeding on the organic contaminant concentration. If one cannot resolve the parameters in such simple models based on available data, then resolution of more complex multi-dimensional time dependent models (with even more parameter values to be determined) is that much less effective.

For the case of the Halle-Kanena depository it would appear that many more measurements are needed over time, and also within the barrier separating the depository from the Hufeisen Lake, before one can separate different biological models that provide the same output for relative concentration of organic PAH. It would also seem that there is, as yet, very little resolution possible on the ranges of parameters allowed because, even for perfect data, there are very broad acceptable, correlated, ranges of allowed parameters. For less statistically sharp data, the stochastic analysis presented here shows how one goes about addressing such concerns, although the limited data available plus the massive reduction in the organic concentration (by a factor of a million or so) from inside to outside the depository suggest that any uncertainty of about a factor 3-10 on either side of the mean value has little to contribute to helping resolve any better the unknown parameters- the dominant behavior is more than adequately captured by the mean value with only slight uncertainty. Nevertheless, the models generated and investigated here are then available to be used with a more complete set of measured information, or in other settings where the contamination may show greater degrees of variation than seems to be available for the Halle-Kanena depository.

Appendix 1. Solutions of Equation (7.12b).

The structure of the solutions to equation (7.12b) depends on the numerical value of the positive constant A. Three cases need to be discussed: $A = 1$; $A < 1$; and $A > 1$.

The case A =1.

For $A = 1$, the solution to equation (7.12b) can be written down by inspection and yields

$$c/c_0 = (1 - kz/(2v))^2 \tag{A1}$$

provided that $kz/(2v) < 1$ and that $kL/(2v) < 1$. In $kz/(2v) > 1$ the solution is $c=0$. The corresponding value of N as a function of position z in the barrier domain is then written down by inserting the solution (A1) into equation (7.12a).

The case A<1.

In this case use of the substitution $s = (A/(1-A))^{1/2}\tan\theta$ allows the integral in equation (7.12b) to be evaluated in closed form yielding the result

$$c/c_0 = (A/(1-A))[\tan\{2\tan^{-1}((\tan b-\tanh x)/(1-\tan b\tanh x))\}]^2 \qquad \text{(A2)}$$

where $b = 0.5\tan^{-1}[((1-A)/A)^{1/2}]$, and $x = kz(1-A)^{1/2}/(2v)$. And equation (A2) is appropriate as long as $0<c/c_0<1$. Use of equation (A2) in equation (7.12b) also then produces an expression for the z dependence of the biological number density.

The case A>1.

Slightly more care needs to be exercised in this case because the integral in equation (7.12b) then has a square root singularity at $s = (A/(A-1))^{1/2} \equiv s^*>1$, which fortunately however cannot lie in the range of integration whenever $c<c_0$. The substitution $s = s^*\sinh\varphi$ allows the integral in equation (7.12b) to be completed in closed form providing the analytical expression

$$c/c_0 = (A/(A-1))[\sinh\{\sinh^{-1}(1/s^*)-kz(A-1)^{1/2}/(2v)\}]^2 \qquad \text{(A3)}$$

And again this solution can be used in equation (7.12a) to provide the spatial dependence of the biological number density with position through the barrier region. Once more the solution is appropriate only when $0<c<c_0$ and outside of this regime one reverts to $c = 0$ as the relevant value.

Appendix 2. Solutions of Equation (7.19b).

The integral on the left hand side of equation (7.19b) does not permit a general analytic closed form expression. But for $\lambda T<<1$ or $\lambda T>>1$ analytic approximations are possible for all values of kT. Consider each case.

The case $\lambda T<<1$.

In this situation the term lnu in the integrand far outweighs the term $\lambda T(1-u)$ so that one can write the accurate approximation

$$\int_{c/c_0}^{1} u^{-1}du/[1+(1/kT)\ln u] = kz/v \qquad \text{(B1)}$$

which can be simply integrated in closed form to yield

$$c/c_0 = \exp[-kT(1-\exp(-z/(vT))] \qquad \text{(B2)}$$

The case $\lambda T>>1$.

In this case the term $\lambda T(1-u)$ dominates over lnu except whenever $c/c_0<<\exp(-\lambda T)$, but then c is so small that the error one makes is negligible. In that case one can write the accurate approximation

$$\int_{c/c_0}^{1} u^{-1}du/[1+(\lambda/k)(1-u)] = kz/v \qquad \text{(B3)}$$

which can be integrated simply in closed form to yield

$$c/c_0 = (1+\lambda/k)\exp(-(1+\lambda/k)kz/v)/[1+(\lambda/k)\exp(-(1+\lambda/k)kz/v)] \qquad \text{(B4)}$$

Chapter 8

Heavy Metal Contamination Removal by Bacterial Activity in Seeping Depositories

Summary

Heavy metal removal from a toxic waste depository in East Germany occurs through the action of bacterial scavenging as contaminant-laden waters transit a bounding dam. Measurements inside and outside the toxic domain indicate this removal can amount to over 90% of particular heavy metals from the exiting waters. This chapter examines the steady-state behavior of interconnected bacterial number density dependence on the heavy metal concentration, removal of the bacteria by heavy metal poisoning, and simultaneous removal of the heavy metals by bacterial scavenging. Adjustment of the birth and death rates of bacteria, together with a propensity for the bacteria to seek out and scavenge the richest zones of heavy metal contamination in the waters seeping through the dam, are the main intertwined components that allow heavy metal removal.

There is also a critical concentration such that bacteria cannot scavenge in waters with heavy metals above the critical concentration; the closeness of the input concentration to the critical value at the toxic side of the bounding dam plays a significant role in the spatial distribution of heavy metal removal and also in bacterial number density concentration. Measurements both inside and outside the toxic depository allow one to bracket combinations of some of the biological parameters to account for the observations, but there are no uniquely determined values; such would require measurements in the dam materials. For a variety of health risk reasons such dam probes are not likely to be allowed in the near future so that the determinations are limited.

1 Introduction

One of the more significant environmental problems in East Germany is heavy metal contamination in toxic waste depositories. These heavy metal contaminants (Cu, Zn, Cd, Cr, and Ni in particular) present an environmental hazard whenever they find ways to escape from the containing depositories. Some of the more obvious processes for such "leakage" are leaching by rainwater, fracturing of a previously sealed depository due to changes in stress conditions, ground water infiltration with associated subsurface leaching of heavy metal contaminants, and direct flow through the containment dams.

This chapter reports on the presence of anaerobic bacterial activity in a depository that markedly cleans seepage waters, laden with heavy minerals, that pass through a containment boundary. More simply put, "scavenging" by the bacteria precipitates heavy metals onto the sedimentary particles of the containment dam so that the residual waters are notably cleansed of heavy metals. The bacteria apparently develop an affinity for heavy metal scavenging, and pay the price by shortening their lifetimes. As a consequence, there must then exist a critical concentration of heavy metals such that higher concentrations will not permit a thorough cleansing of contaminated waters by bacterial action. Equally, with the bacteria being rooted on sedimentary particles, the bulk speed of motion of a spatial distribution of bacteria will be such that they seek out the richest sources of contaminant concentration, so that the distribution will progressively evolve in the direction of highest contaminant concentration. There is, then, a competition between growth and death of the bacterial activity versus contaminant convected flow. Heavy metals scavenged by the bacteria are deposited locally on the sedimentary substrate at bacterial death, so that a build-up of heavy metals takes place on the sediments with concomitant removal of the heavy metals in the through-flowing waters.

This chapter describes how this competition between contaminant flow and bacterial scavenging operates quantitatively, with illustrations from an East German case history. It is, of course, also possible that physical and/or chemical scavenging play roles in heavy metal removal as well as bacterial activity. Indeed, one would anticipate that all three processes would be in play at the same time. While the various strengths of each of the three processes as heavy metal scavengers are not know (and presumably depend on sediment mineralogy and groundwater chemistry), what is known is that in the East German example to be discussed later, there are several orders of magnitude change in bacterial concentration between waters inside the containment area and outside. Such a significant change cannot be due to seasonally related effects, implying that some other major activity is taking place to produce this drastic concentration shift. One could consider just chemical and/ or physical remediation processes on their own, of course, but then one would be left with no explanation at all for the observed orders of magnitude change in bacterial concentrations.

For this reason at least, we have investigated what bacterial activity alone can do to remediate heavy metal contamination. First we develop the quantitative model dependence including illustrative synthetic situations so that one has some background to use as a template when evaluating the data from the East German example. Second, we then go through the East German data and measurements to show how they lead one to consider that the biological remediation model can be used to account for the data. Third we show how the East German data and the quantitative model can be interactively combined to attempt to determine the biological parameters in the model using the data both inside and outside the toxic waste depository.

2 Quantitative Development

Consider the one-dimensional situation depicted in Figure 8.1 where waters contaminated with heavy metals enter a depository-bounding dam (at position $z = 0$) with a concentration c_0 of such metals. The waters are considered to flow at a constant speed v and to transport the heavy metals at the same speed.

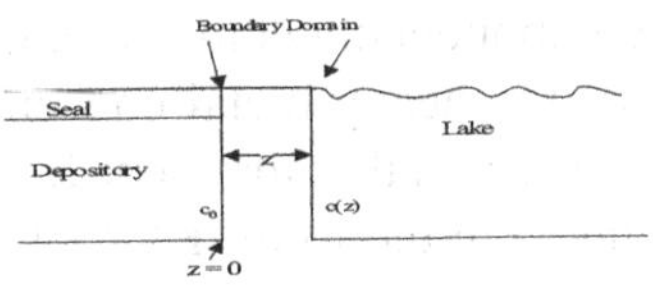

Fig.8.1. Sketch of the one-dimensional situation between depository waters contaminated by heavy metals (to the left) and external waters (to the right) separated by a bounding dam region

For a bacterial distribution number density N(z) with value N_0 at location $z = 0$, the bacterial distribution with position z in the barrier domain is made up of three components: a "convective" spatial gradient term $w\partial N/\partial z$, where w is the bulk transport speed of the bacterial distribution , which, in the case of the bacteria seeking the rich domains of heavy metal concentration, is given by $w = (L/c_0)\, w_0 \partial c/\partial z$ with L a scale length, and w_0 a constant speed chosen so that the physical dimensions are maintained;

the second component is the birth rate component of the bacterial distribution, taken in the form $kN(z)$, with k the birth rate constant (dimensions $time^{-1}$); the third component is the death rate of the bacterial distribution made up of two pieces: a natural death rate with a time scale $1/\tau$ and an enhanced component to the death rate driven by the bacterial consumption of heavy metal that can be written in the form $\lambda c/c_0$ where λ is also a rate constant (units $time^{-1}$) describing the death component. The total equation describing the steady-state transport of the bacterial number density in the barrier domain can be written

$$(L/c_0)\, w_0(\partial c/\partial z)\, \partial N/\partial z = kN - N(1/\tau + \lambda c/c_0). \quad (8.1)$$

Note that in the absence of a richness gradient speed ($w_0 = 0$) for the bacteria and also in the absence of an accelerated death rate ($\lambda = 0$), one would have a balance between birth and natural death rates with $k = 1/\tau$. This natural balance is broken due to the enhancement of the death rate and also due to the richness-seeking gradient, so that the distribution of the spatial bacterial number density will be altered as a consequence.

For the heavy metal concentration variation through the barrier domain, there are several components: a convective transport term $v\partial c/\partial z$, representing the transport of the heavy metal concentration with the connate waters; a loss of the heavy metal component from the flowing waters due to bacterial addiction, taken in the form $-K(c/c_0)(1-c/c_*)N(z)/N_0$, where N_0 is the bacterial number density at the influx boundary at $z = 0$, K is a rate constant with the dimensions of $time^{-1}$, and c_* is the critical concentration at which the bacteria are poisoned by the adsorption of heavy metals (for concentrations $c > c_*$ the bacteria are no longer capable of removing heavy metals from the flowing waters and so the loss due to bacterial addiction is then set to zero); a third component is a diffusive variation of the heavy metal concentration due to variations in the permeability of the barrier material that influence the heavy metals and can be written in the form $D\partial^2 c/\partial z^2$ so that the equation describing steady-state water transported concentration of heavy metals in the barrier domain can be written as

$$v\partial c/\partial z = D\partial^2 c/\partial z^2 - K(c/c_0)(1-c/c_*)N(z)/N_0\,, \quad \text{in } c < c_* \quad (8.2)$$

For $c > c_*$ the heavy metal removal by bacteria is not possible and the concentration would remain at the input value c_0 ($> c_*$) in the absence of other removal mechanisms. We consider here only the situations where bacterial activity can remove heavy metals from the flow seeping through the barrier containment region (i.e. those situations where $c < c_*$).

General quadrature solution to the nonlinearly coupled equations (8.1) and (8.2) is possible in the absence of the diffusive term in equation (8.2), so that effects of bacterial removal are then the dominant concern. This solution procedure is now described.

In the absence of diffusion equation (8.2) reduces to

$$v\partial c/\partial z = -K(c/c_0)(1-c/c_*)N(z)/N_0 \tag{8.3}$$

Then write equation (8.1) first in the form

$$[(L/c_0)\,(w_0/v)K(-c/c_0)(1-c/c_*)N(z)/N_0\,]\partial N/\partial z = kN-N(1/\tau+\lambda c/c_0) \tag{8.4}$$

where $\partial c/\partial z$ in equation (8.2) has been replaced by the right hand side of equation (8.3). Then rewrite $\partial N/\partial z$ as $(\partial N/\partial c)\partial c/\partial z$ and again use equation (8.3) to eliminate $\partial c/\partial z$, yielding the resulting equation for N as a function of c in the form

$$(L/c_0)w_0[K(c/c_0)v^{-1}(1-c/c_*)N(z)/N_0\,]^2(\partial N/\partial c) = kN-N(1/\tau+\lambda c/c_0) \tag{8.5}$$

Equation (8.5) simplifies to

$$a\,(N(z)/N_0)\,c_0\partial(N/\,N_0)/\partial c = (k-1/\tau-\lambda c/c_0)\,(c/c_0)^{-2}(1-c/c_*)^{-2} \tag{8.6}$$

where $a = Lw_0(K/v\ c_0)^2$.For simplicity set $n = N(z)/N_0$, and $y = c/c_0$, with $n = 1$ on $y = 1$, when equation (8.6) can be written more transparently as

$$an\partial n/\partial y = y^{-2}\,(1-yc_0/c_*)^{-2}(k-1/\tau-\lambda y)\,. \tag{8.7}$$

The general solution to equation (8.7) can be given immediately as

$$(a/2)(n^2-1) = m(k-1/\tau)[2\ln y-2\ln\{(1-ym)/(1-m)\}-1/(ym)+1/m+1/(1-ym)-1/(1-m)]-\lambda[\ln(y)-2\ln\{(1-ym)/(1-m)\}+1/(1-ym)-1/(1-m) \tag{8.8}$$

where $m = c_0/c_* < 1$.

Quadrature of equation (8.3) then follows directly with

$$Z \equiv (Kz/vc_0) = \int_y^1 dy'/[y'(1-my')n(y')] \tag{8.9}$$

It is then evident that the structural dependence of the normalized bacterial number density n(y) on normalized heavy metal concentration y ($=c/c_0$) is key to understanding how the heavy metal concentration varies with distance through the barrier domain.

2.1 Non-Adaptive Bacterial Properties

For instance, suppose the bacteria do not adapt their biological properties to the presence of the heavy metal concentration so that they maintain a steady-state balance between birth and death rates, with $k = 1/\tau$. The equation connecting bacterial number density and contaminant concentration then reduces to

$$n^2 = 1 - (2\lambda/a)[\ln(y) - 2\ln\{(1-ym)/(1-m)\} + 1/(1-ym) - 1/(1-m)] \qquad (8.10)$$

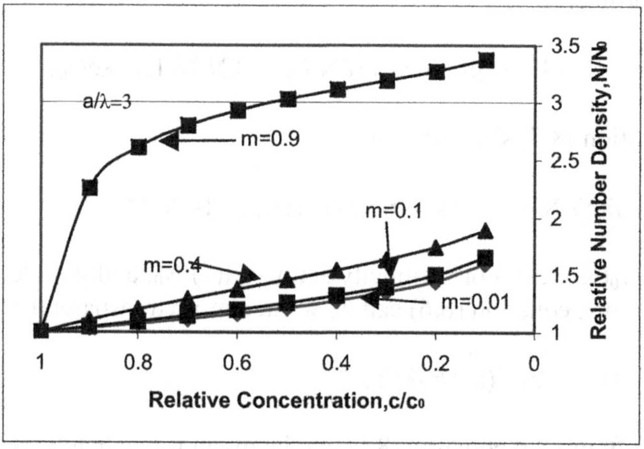

Fig. 8.2. Dependence of the normalized bacterial number density on concentration for the case of non-adaptive biological parameters as given through equation (8.10); curves represent various values of the ratio of initial concentration at the dam boundary to critical concentration

A sketch of the variation of n ($= N/N_0$) with decreasing relative concentration, y ($= c/c_0$), is shown in figure 8.2 for different values of m<1 but with $a/\lambda = 3$. Note that n systematically increases as the concentration of heavy metals decreases, reflecting to some extent the decrease in the death rate of the bacteria with lowered heavy metal concentration and also the fact that as c_0 approaches c_* it requires a larger population of bacteria to effect the same level of heavy metal removal.

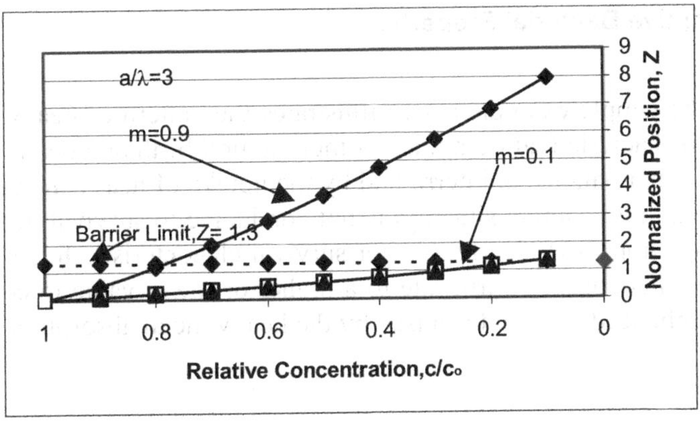

Fig. 8.3. Dependence of the normalized heavy metal concentration as a function of distance through the dam, as given through equation (8.9), for the case of non-adaptive biological parameters and for various ratios of the initial concentration at the dam boundary to the critical concentration

For the same parameter values, from equation (8.9) one can also compute the variation of heavy metal concentration with increasing distance z into the barrier domain, sketched in Figure 8.3 for the normalized barrier position Z (= Kz/vc_0) and for both m = 0.1 and m = 0.9 to show how the variation of concentration is sensitive to the ratio (= m) of the original concentration c_0 (at z = 0) to the critical concentration c_* For m = 0.9 note that the population of bacteria rises faster at large distances from the input flux point (located at z = 0), reflecting the lessened efficiency of the bacteria in removing the heavy metals and so requiring a larger population density to achieve removal.

The total removal of the heavy metal concentration through the barrier region can be estimated by considering the steady-state input heavy metal contaminant flux vc_0 and the water-borne flux vc(x) at the exit position z = x of the barrier domain. The difference measures the amount deposited between position z=0 and z = x per second as $A(x)=v(c_0-c(x))$ $gmcm^{-2}$ sec^{-1}. For instance note that at a fixed normalized distance of Z =1.3, figure 8.3 shows that the relative concentration is reduced to 80% of the initial input concentration for m=0.9, and is reduced to 10% for m=0.1.Thus the efficiency of removal is sensitive to how close the original input concentration is to the critical limit c_*, with only 20% heavy metal flux removal in the case m=0.9, and 90% removal in the case m = 0.1 for Z =1.3 marking the normalized barrier limit.

2.2 Adaptive Bacterial Properties

While the simple example above illustrates the pattern of heavy metal removal for bacteria that do not adapt their natural birth and death rates, and so maintain a balance uninterrupted by the uptake of heavy metals, in reality such a non-adaptive situation is not conducive to survival of the bacterial population. It is clearly a better survival characteristic for the bacteria if they can adapt their birth rate to a higher or lower value to compensate for the enhanced death rate caused by the heavy metal absorption.

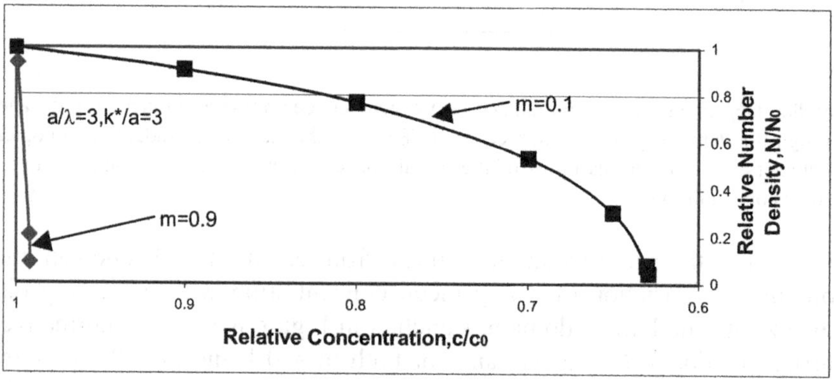

Fig. 8.4. Dependence of the normalized bacterial number density on concentration for the case of adaptive biological parameters for the bacteria with a higher birth rate; curves represent various values of the ratio of initial concentration at the dam boundary to critical concentration

To illustrate how such a survival procedure operates and how it influences the heavy metal concentration, in equation (8.8) set $k-1/\tau=k_*$ ($\neq 0$) to reflect this adaptive behavior. The pattern of behavior of the normalized number density given in equation (8.8) with relative concentration variation now depends not only on how close m is to the critical value of m=1, as well as on the ratio a/λ, but also on the ratio of k_*/a. Thus there is a competition between the changed birth rate of the bacteria, as measured by the terms factored by k_* in equation (8.8), and the death rate caused solely by the bacterial affinity for heavy metals, as measured by the terms factored by λ in equation (8.8). To provide a direct comparison between the previous case of non-adaptive bacterial properties and the present situation we consider the same ratio of $a/\lambda=3$, and also the two situations $m=0.1$, and $m=0.9$. For the ratio k_*/a, we consider a value large in magnitude compared to λ/a so that the influence of adaptive characteristics is easily seen. To this end we have chosen $k^*/a=3$, compared to $\lambda/a=1/3$, for the first illustration and $k^*/a= -3$ for the second illustration.

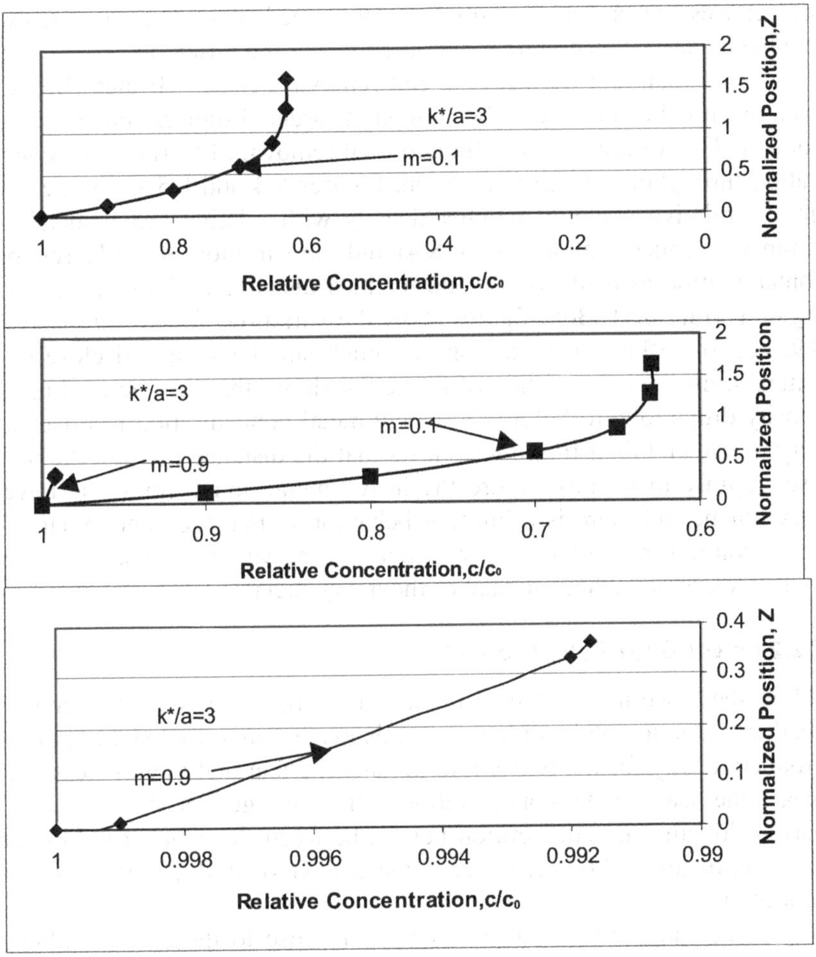

Fig. 8.5. Dependence of the normalized heavy metal concentration as a function of distance through the dam, as given through equation (8.9), for the case of adaptive biological parameters for the bacteria with a higher birth rate, and for various ratios of the initial concentration at the dam boundary to the critical concentration. The three groups of curves are described fully in text

2.2.1Higher Birth Rate, k*/a=3.

While it might be thought that a higher birth rate is conducive to a greater removal of heavy metal contamination, in fact just the opposite occurs. The reason is not difficult to see because a higher birth rate causes a greater removal of contaminant near the influx point z =0; however such a rapid removal in turn causes the concentration gradient to sharpen, which then causes the bacterial number density speed of motion to increase so that the bacteria concentrate more quickly near the influx point.

To maintain the heavy metal removal near z = 0 then forces the bacterial number density to be limited at larger distances into the barrier domain. The upshot is a very limited total removal of heavy metal concentration throughout the barrier region. Figures 8.4 and 8.5 show the variation of relative bacterial number density with relative heavy metal contamination concentration (figure 8.4) and the variation of the heavy metal contamination as a function of normalized distance Z through the barrier region (figure 8.5). Both figures show the variations for m = 0.1 and m = 0.9, i.e. for influx concentrations c_0 much smaller than and close to the critical limit of c_*. Inspection of figure 8.4 shows that the bacterial number density drops to zero before the heavy metal concentration has decreased very much, and then figure 8.5 shows that the distance through the barrier rises rapidly to infinity before the heavy metal concentration is lowered very much; and there is a limiting behavior so that the contaminant concentration never becomes lower than a limiting value because there are no more bacteria to provide uptake of the heavy metals.

2.2.2 Lower Birth Rate, k_*/a = -3.

While the lowering of the birth rate in competition with the enhanced death rate due to uptake of heavy metals may seem at first sight not to be a productive way for the bacteria to mutate, the fact that there is less uptake means the heavy metal concentration will have a gentler slope through the barrier domain; the competition here is between the ratio of the effective death terms divided by the lowered bacterial velocity due to the lower concentration gradient.

For an initial influx concentration close to the critical value, the ability of the bacteria to absorb heavy metals is lowered and so the concentration gradient with increasing position through the barrier is also lower. Accordingly, the relative number density of bacteria as the concentration decreases should be higher than for an initial concentration much less than the critical value. This expectation is substantiated as shown in Figure 8.6 for the two cases of m = 0.9 and m = 0.1, respectively. In addition, Figure

8.7 shows that the relative concentration of heavy metals is maintained to a larger distance through the barrier domain for influx concentrations much smaller than critical value than for higher initial concentrations, representing the change in the bacterial bulk velocity as a result of the decreased contaminant spatial gradient.

3 East German Case History: Observations at the Organic Depository, Halle-Kanena.

A plan view of the organic waste depository at Halle-Kanena, east of the city of Halle (Saale) is given in figure 8.8. Marked on the figure are the positions of measurement sites and also the outline of the depository in relation to the surrounding Horseshoe Lake.

Much has been written on the Halle-Kanena depository and its underlying sediment and structural support; the best compendium is the report edited by Strauch (1996), to which report we refer the interested reader for detailed information on site locations, measurements made, and inferences drawn over the course of more than a decade or two. Additional detailed investigations of various components of the depository are to be found in Glaesser (1994), Christoph (1995), Glaeser (1995), and Maiss et al. (1998).

Perhaps of greatest interest in connection with the biological cleansing of depository waters is the cross-section of figure 8.9, showing the depository sited on top of landfill from the earlier lignite coal mining The area labeled "Dam" on figure 8.9 represents the clay barrier region referred to in the modeling sections above, and through which the contaminated waters from the depository journey on their way to the east section of the Horseshoe Lake This region lies directly under the label ENE/SW on figure 8.9.

For the shallow waters on the eastern side of the Horseshoe Lake there is a region with medium mineralization of 2300 μS/cm from the water surface down to a depth marker around 75m using the vertical scale on figure 8.9. Below that depth marker, there is a sharp transition to a domain of highly mineralized water, extending to the bottom of the water, and for which a region with significant H_2S represents the basal waters. Residual lignite coal on the NE side has its top formation contact coincident with the transition from low mineralization to high mineralization, as shown on figure 8.9.

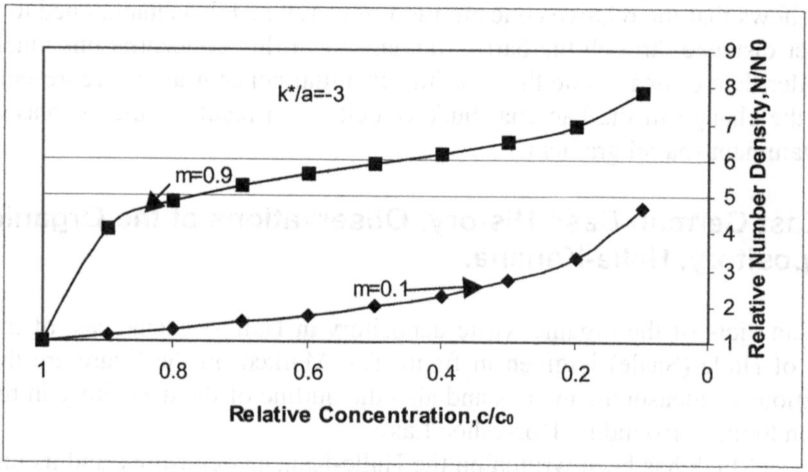

Fig. 8.6. Dependence of the normalized bacterial number density on concentration for the case of adaptive biological parameters for the bacteria with a lower birth rate; curves represent various values of the ratio of initial concentration at the dam boundary to critical concentration

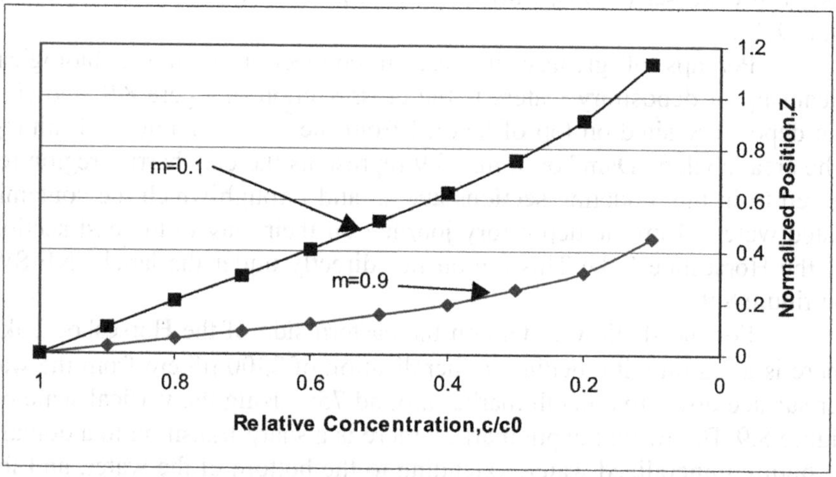

Fig. 8.7. Dependence of the normalized heavy metal concentration as a function of distance through the dam, for the case of adaptive biological parameters with a lowered bacterial birth rate, and for various ratios of the initial concentration at the dam boundary to the critical concentration

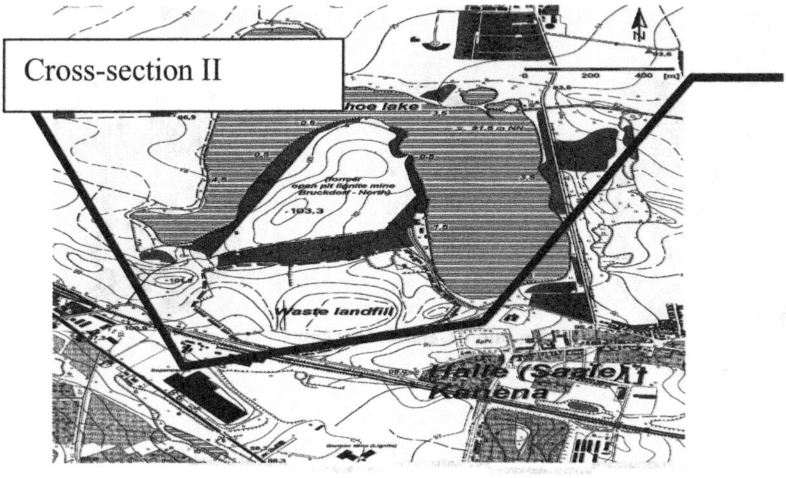

Fig.8.8. Plan view of the organic waste depository at Halle-Kanena, known as Horseshoe Lake

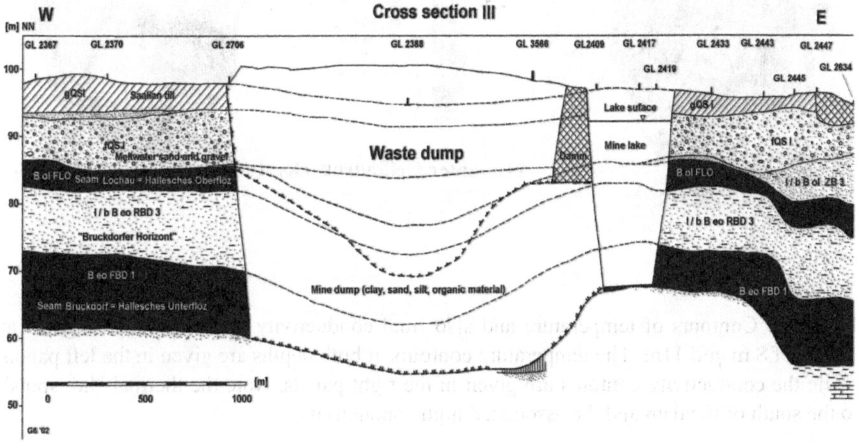

Fig. 8.9. Schematic cross-section (from NW to NE) through the Horseshoe Lake organic waste depository (see text for detailed explanation)

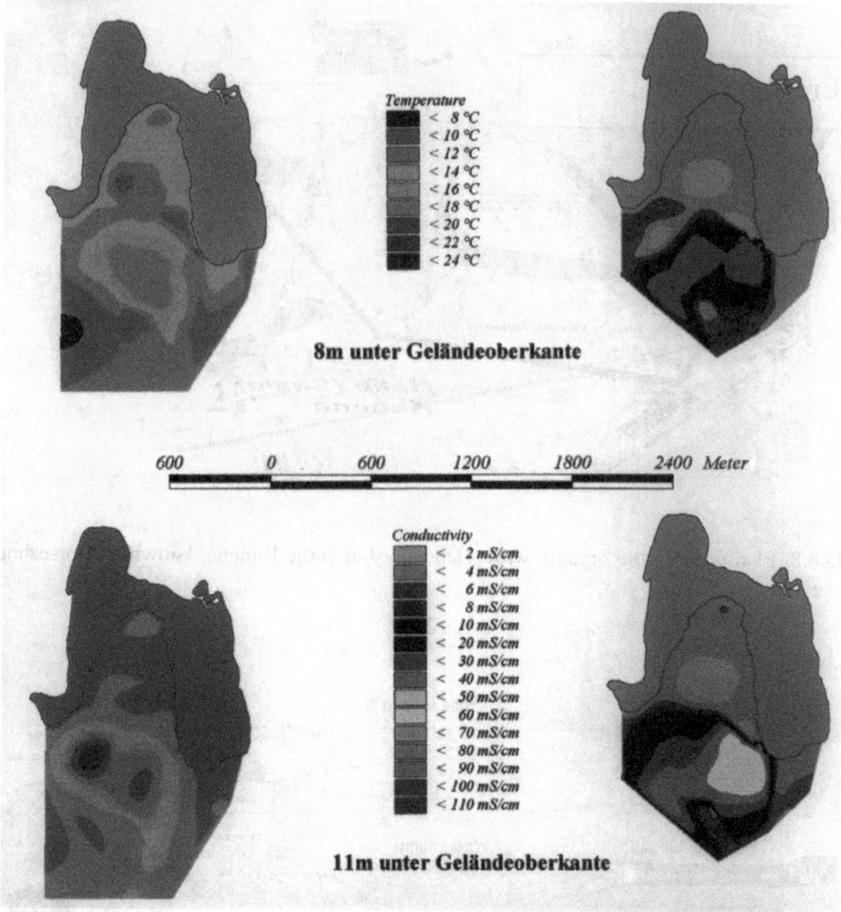

Fig. 8.10. Contours of temperature and also fluid conductivity in the area investigated at depths of 8 m and 11m. The temperature contours at both depths are given in the left panels while the conductivity contours are given in the right panels. Note the thermal "hot spots" to the south of the dam and the associated high conductivity

In addition, subsurface temperature measurements and fluid conductivity measurements have been made both in the toxic region to the north of the dam and in the region to the south of the dam, as shown in figures 8.10 and 8.11 for depth ranges 8 and 11 m, and 15 and 20 m, respectively. The panels on the left side of each of figures 8.10 and 8.11 show the temperature contours, while those on the right show the

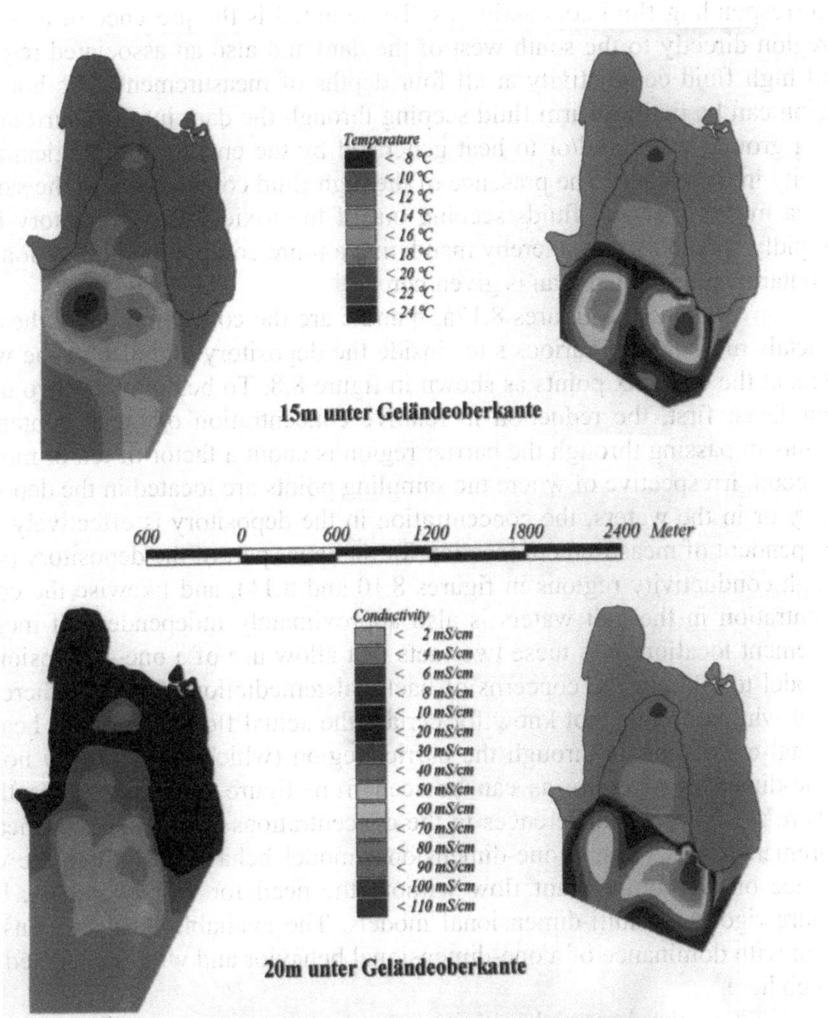

Fig. 8.11. Contours of temperature and also fluid conductivity in the area investigated at depths of 15 m and 20 m. The temperature contours at both depths are given in the left panels while the conductivity contours are given in the right panels. Note the thermal "hot spots" to the south of the dam and the associated high conductivity

corresponding fluid conductivities. To be noted is the presence of a "hot" region directly to the south west of the dam and also an associated region of high fluid conductivity at all four depths of measurement. The hot region can be tied to warm fluid seeping through the dam into the surrounding ground water and/or to heat generated by the enhanced biological activity in the waters. The presence of the high fluid conductivity in the same area means that any fluids seeping out of the toxic waste depository are rapidly spread further, thereby mandating a more complete investigation of contaminant removal than is given here.

Sketched in figures 8.12a, b and c are the concentrations of heavy metals measured at various sites inside the depository and also in the waters at the exit flux points as shown in figure 8.8. To be noted are two major facts: first, the reduction in relative concentration of metal contaminants in passing through the barrier region is about a factor of ten or more; second, irrespective of where the sampling points are located in the depository or in the waters, the concentration in the depository is effectively independent of measurement location for the inner part of the depository (see high conductivity regions in figures 8.10 and 8.11), and likewise the concentration in the exit waters is also approximately independent of measurement location. It is these two facts that allow use of a one–dimensional model to address the concerns of bacterial remediation. The point here is that while one may not know for certain the actual flow path of the heavy metal contaminants through the barrier region (which is manifestly not a one-dimensional entity as can be seen from figure 8.8), the fact is that there are no major differences in the concentrations with respect to measurement sites. Thus, a one-dimensional model behavior captures the essence of the contaminant flow without the need for more complex, but more rigorous, multi-dimensional models. The available data are consistent with dominance of a one-dimensional behavior and will be modeled as such here.

For the bacterial activity, measurements were made over the course of a year in 1993. Shown on figure 8.13 are the counts of bacterial activity in the toxic area waters just to the north of the dam and also in the waters to the south of the dam. While the counts of the bacteria vary from summer to winter at both locations, as expected, the points to note are that the average bacterial counts in the waters north of the dam are at least three orders of magnitude less than those in the waters to the south of the dam. The suggestion here is that the high concentration of heavy metals in the toxic depository north of the dam are effectively limiting the bacterial activity due to a more rapid death rate of the bacteria due to heavy metal uptake.

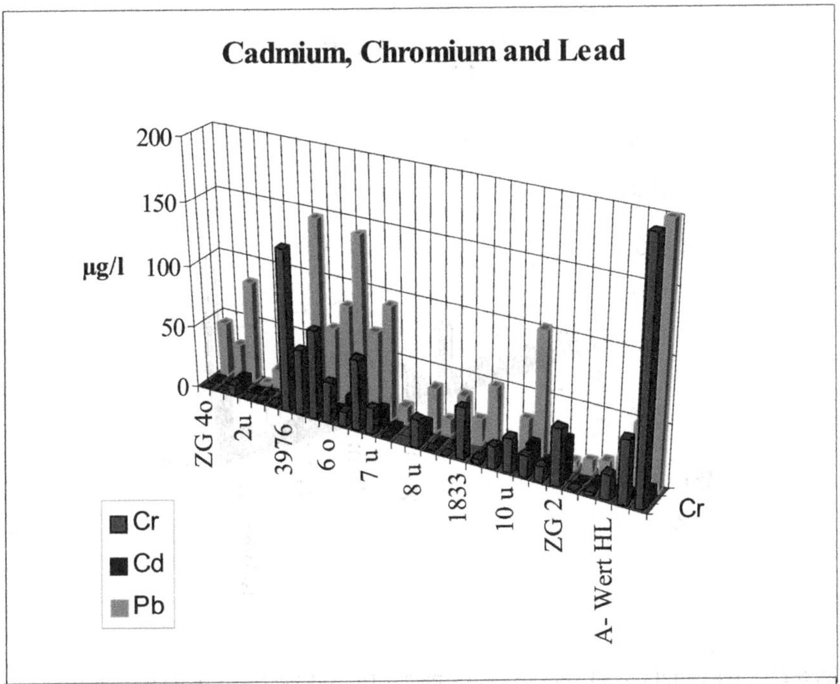
Cadmium, Chromium and Lead
200
150
µg/l 100
50
0
ZG 4o
2u
3976
6 o
7 u
8 u
1833
10 u
ZG 2
A- Wert HL
Cr
Cr
Cd
Pb

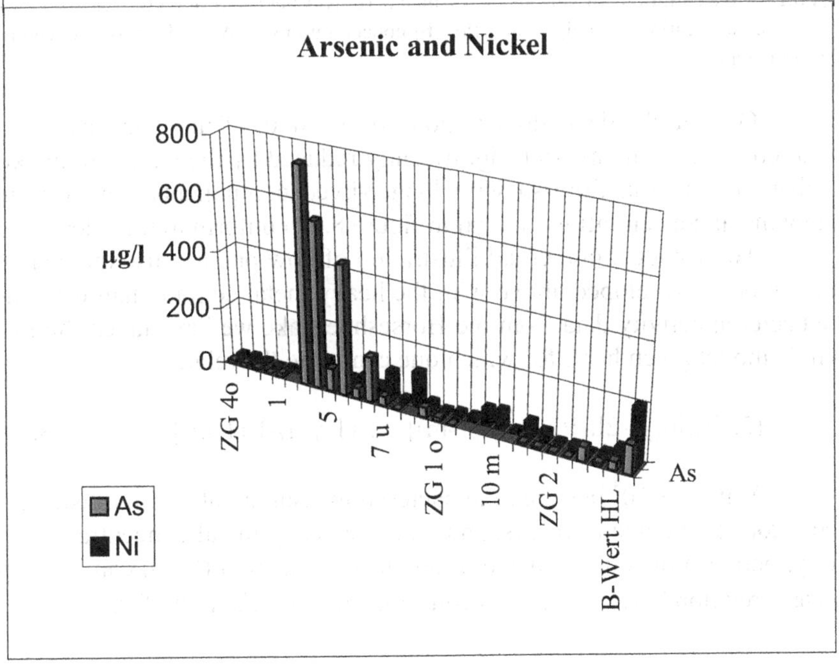
Arsenic and Nickel
800
600
µg/l 400
200
0
ZG 4o
1
5
7 u
ZG 1 o
10 m
ZG 2
B-Wert HL
As
As
Ni

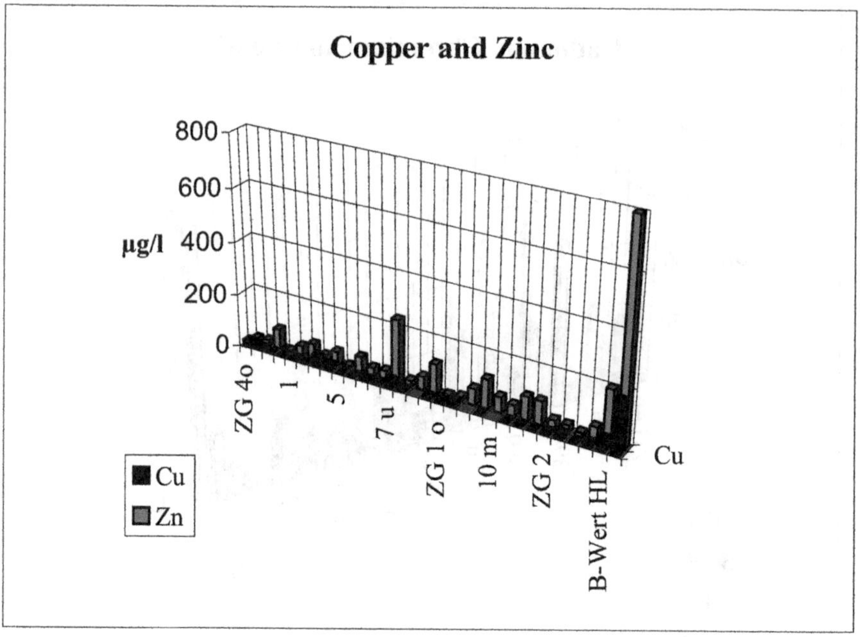

Fig. 8.12. Concentrations of various heavy metals (plus Arsenic) through the waste depository: (a) concentrations of Cd, Cr and Pb; (b) concentrations of As and Ni; (c) concentrations of Cu and Z

Outside the depository region, to the south of the dam, the heavy metal concentration has been significantly reduced by the bacterial uptake, so that the bacteria flourish with longer-lived components. The various heavy metal concentrations shown in figure 8.12 conform to this idea.

To provide a numerical illustration of how one can use the quantitative models developed above with the heavy metal contaminant data and the bacterial activity data from the Horseshoe Lake we have taken the first simple model given here, for which one can write the relation

$$n^2 = 1 - (2\lambda/a)[\ln(y) - 2\ln\{(1-ym)/(1-m)\} + 1/(1-ym) - 1/(1-m)] \tag{8.10}$$

With $n \approx 10^3$ between the bacterial measurements from outside the depository to the inside values, and with the heavy metal concentration ratio, y, between outside to inside of about 0.1, equation (8.10) can be written as a relation between the two constants m and $2\lambda/a$ in the form

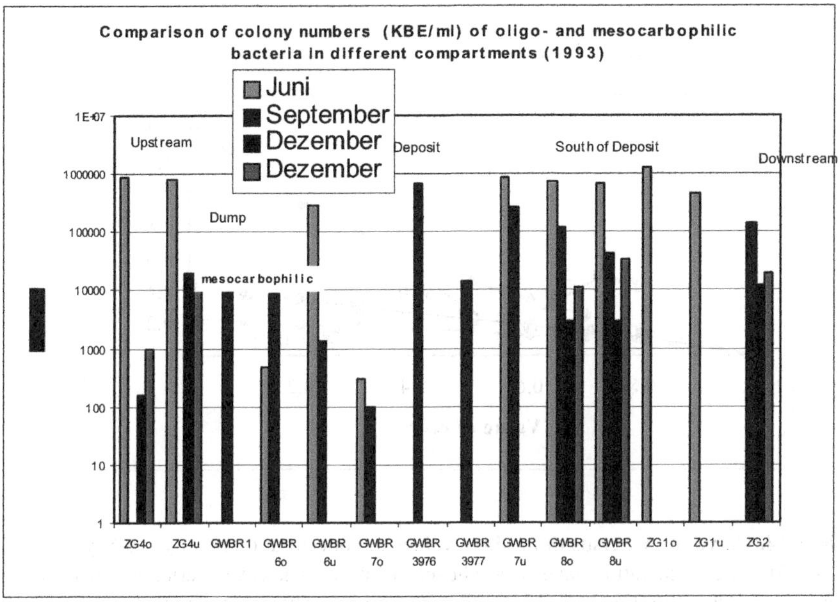

Fig. 8.13. Counts of bacterial activity in the toxic area waters just to the north of the dam and also in the waters to the south of the dam. While the counts of the bacteria vary from summer to winter at both locations, as expected, the points to note are that the average bacterial counts in the toxic waters north of the dam are at least 3 orders of magnitude less than those in the waters to the south of the dam

$$2\lambda/a = (1-n^2)/[\ln(y)-2\ln\{(1-ym)/(1-m)\} + 1/(1-ym)-1/(1-m)] \tag{8.11}$$

Write $(2\lambda/a)/(1-n^2) = -Y$, as a single grouping of parameters, equation (8.11) delivers a simple relation connecting Y and m for different measurements of the concentration ratio, y. Figure 8.14 shows the curves of Y versus m for three values of the heavy metal concentration ratio, y = 0.5, y = 0.1,and y = 0.01.

The three ratios have been chosen to more than cover the variation in measured concentration ratios. What is clear from figure 8.14 is that Y is a relatively stable parameter, varying only from around zero (at critical concentration m = 1) to around 1.5 at m = 0, as y varies by more than two orders of magnitude. Thus it is possible to comment that even for such a simple model two factors are apparent: first, such models can account for the known observations in a general sense with associated estimates of the

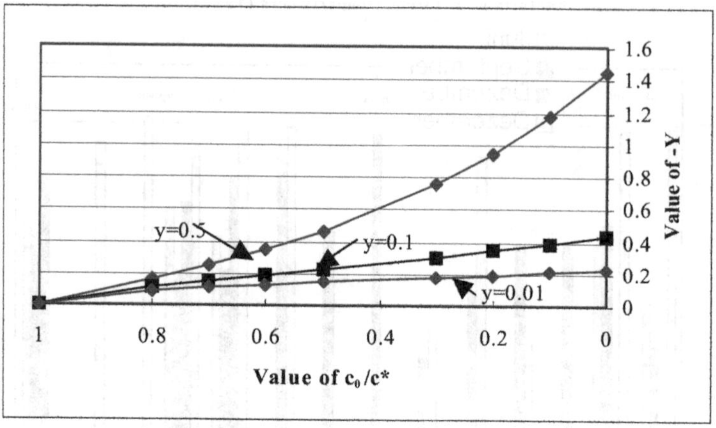

Fig. 8.14. Curves of Y versus m for three values of concentration ratio, y = 0.5, y = 0.1,and y = 0.01. The three ratios have been chosen to more than cover variations in measured heavy metal concentration ratios. Note that Y is a relatively stable parameter, varying only from around zero (at critical concentration m = 1) to around 1.5 at m = 0, as y varies by more than two orders of magnitude

biological and heavy metal concentration parameters also being available; second, it is not possible to uniquely determine which groupings and values of parameters are describing the observations. This second point is clear because of the allowed variation of the two parameters in the simple model above that indicates a range of both parameters will do an adequate job of matching the data variations. This second point also indicates that measurements of the heavy metal concentration in the dam region itself (in both the through-flowing waters and also on the associated sedimentary materials comprising the dam) are needed to refine any better the parameter values.

However, there are very strong reasons for NOT wishing to disturb the dam materials. One does not know to what extent such data collection would exacerbate any structural instability regimes of the dam, with the potential for dam leakage or collapse that would provide an even worse environmental situation than already exists in the area, which is densely populated. One does not know what other contaminants are present in the dam materials that could be impacted adversely by such probing, with potential concomitant release. One does not know to what extent the milieu conditions supporting the bacterial population would be influenced by such probing, leading possibly to total death of all the bacteria by oxygen poisoning, and so undiminished escape of heavy metal contaminants in the through-flowing waters. For all these reasons there is considerable resistance from the State authorities to any such probe activities in the containing dam and, to date, no such measurements have been made in a systematic manner.

It would seem that any sharper focus on parameter values must be delayed to some distant future time.

4 Discussion and Conclusion

The presence of heavy metal contaminants in toxic waste depositories is a significant environmental danger. What we have attempted to show in this chapter is that one particular natural biological remedial activity due to anaerobic bacteria can play a substantial role in limiting the heavy metal contamination that seeps through a containing dam in the connate waters of the toxic depository. We have shown that it is possible to use the observations of heavy metal variations from the waters in the depository and also in the regions external to the depository to at least bracket the ranges of parameters that are involved in bacterial scavenging of the heavy metals.

What is not so easy to determine are more or less unique values for the parameters involved without probing into the dam materials, something that is not necessarily a very good idea because of the potential for causing even larger environmental hazards, as detailed above. It is also not clear to what extent the biological bacterial cleansing activity will change in the future, nor is it clear to what extent physical and/or chemical scavenging also play roles of significance. But, at least for the Horseshoe Lake depository, it is possible to say that bacterial scavenging can indeed remove heavy metals from the through-flowing waters, and that one can determine allowed ranges of parameters associated with such removal. Whether the same can be done for other toxic depositories in East Germany we do not know because detailed data are not so readily available as they are for Horseshoe Lake. We would welcome such investigations in attempts to understand better which of the various models developed here can be more sharply delineated, and we would welcome further measurements that could more sharply determine parameter ranges for each individual case.

Chapter 9

Quantitative Risks of Death and Sickness from Toxic Contamination: General Population

Summary

This chapter shows how to estimate the likely death and sickness rates due to release of toxic material when the probability of a spill, the amount spilled, the amount transported by air or water away from the primary spill site, and the types of material spilled have uncertainties. In addition, using risk methods that include quantitative procedures for determining the relative contribution, relative importance, and relative sensitivity of the uncertainties, which of the uncertain parameters is causing the greatest degree of uncertainty in the death and sickness rates can also be identified. This affords the reduction of uncertainty. In this way, focusing on limiting the parameters that are dominant without spending effort and money on parameters that contribute little lessening of uncertainty of death and sickness rates is possible. In addition, synergistic effects are considered for two toxic materials combining to produce either a more virulent or benign toxic effect than the original materials. For such synergistic involvement, death and sickness rates are also quantitatively described with the dominance ordering of uncertain parameters involved in the synergism. The numerical illustrations exemplify the method for assessing the risk and uncertainty and also illustrate the use of uncertainty to more precisely determine what is necessary to ameliorate the risk to the maximum extent possible.

1 Introduction

One of the more significant problems in environmental science concerns the spillage of one or more toxic materials and the assessment of attendant probable toxicity to humans. Spillage may occur accidentally or with intent and may occur on land, in the atmosphere, rivers, or oceans. Intentional spillage may transpire to avoid proper disposal of environmental contaminants, as an act of war or terrorist activity, or from improper destruction of toxic contaminants. Even so, spillage of toxic materials has the potential to cause sickness and death to portions of humanity. For example, Oliver (2003) has written, "The most recent medical study available suggests that 19% of all persons exposed to beryllium will come down with a beryllium-related illness. Since AT&T Nassau minerals processed beryllium oxide and beryllium alloys and employed 1,150 people, what was the damage? This does not include 3,400 additional family members who were potentially exposed to the beryllium dust on employees' work clothes when they came home. It also does not include the community at large, exposed to the beryllium dust emitted from the AT&T Nassau metals shredder." Presumably, one also needs to consider the concentration of beryllium in defining the 19% who will become sick, and of deciding if the 19% has any uncertainty itself due to other factors to which people are exposed. Nevertheless, a continuing danger from beryllium–related sickness is present and this toxicity effect needs further awareness.

In terms of toxic contamination due to unintended radioactive release of material, perhaps the three most readily identified spills have been due to: atmospheric Sr release from the Windscale, UK reactor; the Three Mile Island release in the U.S.; and, the Chernobyl disaster in the Ukraine. The long-term sickness and death rates of the local population and also future unborn generations must be assessed for such catastrophes.

In short, the unanticipated and unwanted accidental release of biological, chemical, and nuclear contaminants has major repercussions for humanity's health. Planned contaminant releases, either by commission or by obvious neglect or omission of adequate safety controls, have even more profound repercussions. The 1984 Bhopal, India disaster, the Superfund Cleanup of many contaminated sites in the U.S. after many decades of neglect, or the explicit 1991 gassing of Iraqi people by Saddam Hussein's regime are notable examples. Many more events have occurred, are likely to occur, or have the potential to occur without tighter controls to attempt to limit the risks.

In addition, the toxic effects of materials released into a given milieu can be considered harmless, but that material may interact synergistically with both chemically and/or biologically derived materials already present, either as earlier spillage or occurring naturally, to provide a synergistic result that is highly toxic. Conversely, materials considered toxic may combine synergistically to produce a resultant material much less harmful or even benign relative to the individual materials.

Also, spilled materials do not necessarily remain at the spill location. Transport by air currents, water by river currents, ocean currents, ingestion by fish or other aquatic creatures that are not sessile, or groundwater transport are all obvious modes of movement of spilled materials. The areas affected by such transport of toxic materials and the corresponding areas where accumulations of intrinsically or synergistically derived toxic materials occur are then also hazardous to humanity.

Three classes of toxic hazards to humanity are presented: 1) a concentration of toxic materials above a terminal activity concentration (TAC) level sufficient to cause death; 2) a concentration of toxic materials below the TAC level but higher than a critical sickness concentration (CSC) sufficient to cause long-term sickness but not death; and, 3) a concentration below the CSC defined as developing either no sickness or only minor, recoverable, sickness.

Major difficulties in attempting to provide quantitative assessments of toxic hazards due to spillage are due to a variety of unknown or poorly known factors as well as uncertainty of parameters entering estimating procedures. The major purpose here is showing how to investigate the probability of death or sickness based on uncertain information and, perhaps more importantly, how to determine which of the poorly known parameters is dominant in causing the greatest uncertainty in the estimates of death and sickness. This procedure enhances the focus on attempts to narrow ranges of uncertainty caused by the dominant parameters absent the inordinate effort to be more precise on other parameters, whose sharpened determination would do little to alleviate the uncertainty because the parameters causing the largest uncertainty were not investigated.

The procedure is based on a Microsoft Excel program (e.g., Appendix D) that is extremely fast requiring under 0.05 sec to operate. The Excel code alone provides a deterministic assessment of direct death rates and sickness rates as well as rates from synergistic effects. Input parameters are spill amount, transport fractions by air and water, area coverage for each transport mode, TAC, CSC, and probability of spill for each of 10 materials. In order to allow for variations in the input values because the various parameters are not well known, the Excel code interfaces with the Monte Carlo program, Crystal Ball®, that allows any and all parameters to be varied in any ranges. In order to construct the probable distribution of outputs (in this case death and sickness rates for different conditions and transport modes) a variety of choices is also incorporated for each of the underlying distributions for the parameters from which Crystal Ball® will choose.

Using Crystal Ball® produces several advantages: it interfaces seamlessly with any Excel program providing a framework from which to perform Monte Carlo calculations; takes the same operating time to perform a fixed number of Monte Carlo computations irrespective of the number of input variables chosen or their underlying distributions (i.e., if it takes 10 sec to perform 1000 Monte Carlo runs with just one uncertain input variable, then 5, 10, 100 or any other number of input variables will take the same time) ; the various outputs chosen to examine for their dependence on the ranges of the input variables not only are represented by a probability distribution but Crystal Ball® also provides a representation of the Relative Contribution (RC) of each of the uncertain input variables to the variance of each output being investigated. Thus, an immediate graphical sense of which input variable is causing the greatest relative uncertainty in the output of interest is available-a concern when attempting to assess which input variable to investigate further for sharper determination of death and sickness rates. In addition, Crystal Ball® allows varying the underlying distribution chosen for each input variable (from which it draws representative values to use in the Monte Carlo computations).

Thus, the Monte Carlo calculations can be rapidly iterated to determine whether changing the underlying distribution choices produces significant changes in the estimated variance of any and all outputs chosen- part of the so-called Relative Sensitivity (RS) argument. The dynamic range of each input variable can be changed just as quickly to determine whether the dynamic range is also causing a significant shift in the same variances-the other part of the RS argument.

Finally, Crystal Ball® also produces statistical descriptions of the causes of output variability. Two simple quantitative measures of absolute uncertainty on output values of interest are easily constructed. Volatility (Warren 1978, 1979, 1981, 1983a, 1983b, 1988), often provided by the value $v = (P(90)-P(10))/P(50)$, measures how the uncertainty range of an output is related to a centrally chosen value (here P(10), P(50) and P(90) are cumulative percentage values of obtaining a particular output worth or less). A volatility much less than unity ($v \ll 1$) indicates very little uncertainty on the output variable whereas a volatility much larger than unity ($v \gg 1$) indicates considerable uncertainty. Based on this volatility measure, the RC of each input parameter uncertainty to the range of an output of interest are then important when dealing with a high volatility situation for the output. A second measure often used to express the absolute uncertainty of an output is the equivalent log-normal coefficient, μ, determined by $\mu^2 = \ln(1+\sigma^2/E^2)$ where σ^2 is the variance on the mean value, E, of an output of interest (see Appendix A). Both the volatility and the lognormal coefficient provide similar measures of absolute uncertainty, so volatility is treated solely as a quantitative measure. Elsewhere, Thomsen and Lerche (1997) have shown how to use the equivalent lognormal coefficient in handling uncertainty due to migration loss and retention loss of generated hydrocarbons. The combination of RC, RS, and RI allows the determination of where and when to improve estimates of inputs in order to sharpen knowledge of outputs.

2 Toxic Model Construction

The basic program for handling toxic spill and transport for up to ten independent materials is shown in Appendix D. Effectively, the program is composed of four parts. First, the TAC and CSC are specified for each of the ten components, so that a measure of the direct toxicity properties for each material is available. In the program version exhibited in Appendix D, the TAC and CSC units are gm/m^2 although any relevant units can be used (e.g., spore count/unit volume, dpm/gm/cc, etc.) because a direct comparison of expected concentrations of toxic materials against the TAC and CSC values is made. Thus, the units are required to be the same for each but can be modified as appropriate. Once TAC and CSC properties of each material are specified, the spill characteristics are determined. The program requires an estimate of the total amount of each material spilled, the probability of a spill for each of the ten components, the fraction of spilled materials remaining in place at the spill site, the area of the spill site covered by toxic materials, and the transported fraction of spilled material (i.e., the difference between total spilled amount, considered as unity, and the fraction left in place).

The third element of the program requires that estimates of the percentages of the transported toxic materials that are transported by water or air and the corresponding area coverage of each by the transport processes. With this information for each material component, the concentration (in gm/m^2) due to air or water transport at the contaminated sites can be calculated. With the values for TAC and CSC in the same units, the direct sickness and/or death rates per 100,000 population from each contaminant at the in-place spill location, the area affected by contaminant due to water transport, and the area contaminated by air transport can be calculated. The sickness and death rates can be computed allowing for the probability of a spill (the so-called likely sickness and likely death rates) as well as by ignoring the spill probabilities (i.e., assuming a spill has occurred and the direct estimates of sickness and death rates are computed).

The fourth component of the program deals with synergistic effects. For simplicity, only two materials at a time can provide a synergistic effect. The synergism is assumed proportional to the concentrations of the two materials so the material with the lowest concentration determines the synergism (other models of synergism are also possible and are easily programmed). The synergism may result as providing either a more toxic or less toxic situation than the direct toxicity of each material alone. Accordingly, the TAC and CSC for the synergistic mix, which can be very different than the corresponding values for the individual materials, must also be specified. Assuming synergism does occur, both the direct sickness and death rates and the likely sickness and death rates can be computed allowing for the probabilities of a spill of each of the two synergistic components.

3 Numerical Illustrations of Deterministic Toxic Sickness and Death Rates

In order to frame the calculations for death and sickness rates including

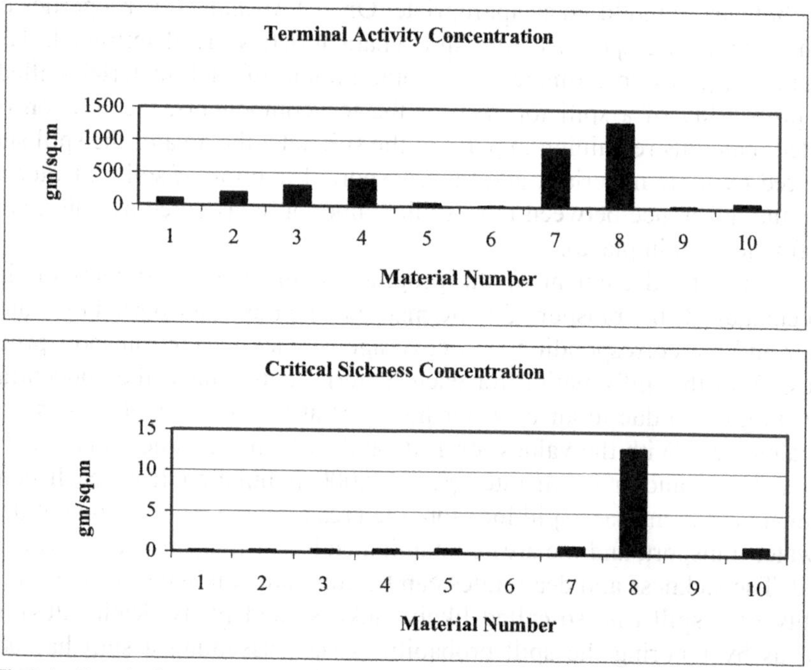

Fig.9.1. Terminal Activity Concentrations (TAC) and Critical Sickness Concentrations (CSC) for 10 materials

uncertainty on input parameters, pictorial representations of the various components of the basic program calculated using the input values given in Appendix D are presented.

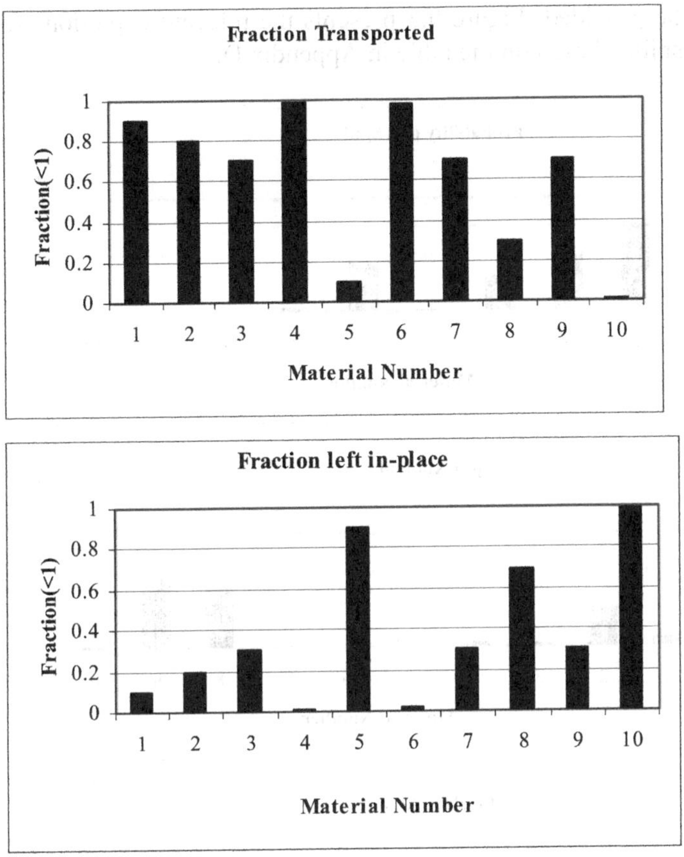

Fig. 9.2. Fractions of 10 materials transported away from the spill site and the fractions left in place

First, Figure 9.1 provides a representation of the TAC and CSC values for the 10 toxic materials listed to exhibit which materials have the greatest toxicity. Figure 9.2 shows the representation of each material fraction that is transported (irrespective of whether that transport is by air or water) and also the fraction remaining at the original spill site. Figure 9.3 illustrates the spill probability for each material, the amount spilled, and the area covered at the spill point by each contaminant (the so-called in-place area). After specifying each material fraction transported from the site, Figure 9.4 shows the percentage of materials transported by air and water together with the estimated contaminated areal extent for each transport mode. The concentrations of each material at the in-place spill site after air and water transport can be provided; Figure 9.5 presents the relevant depictions for each material spilled based on the table in Appendix D.

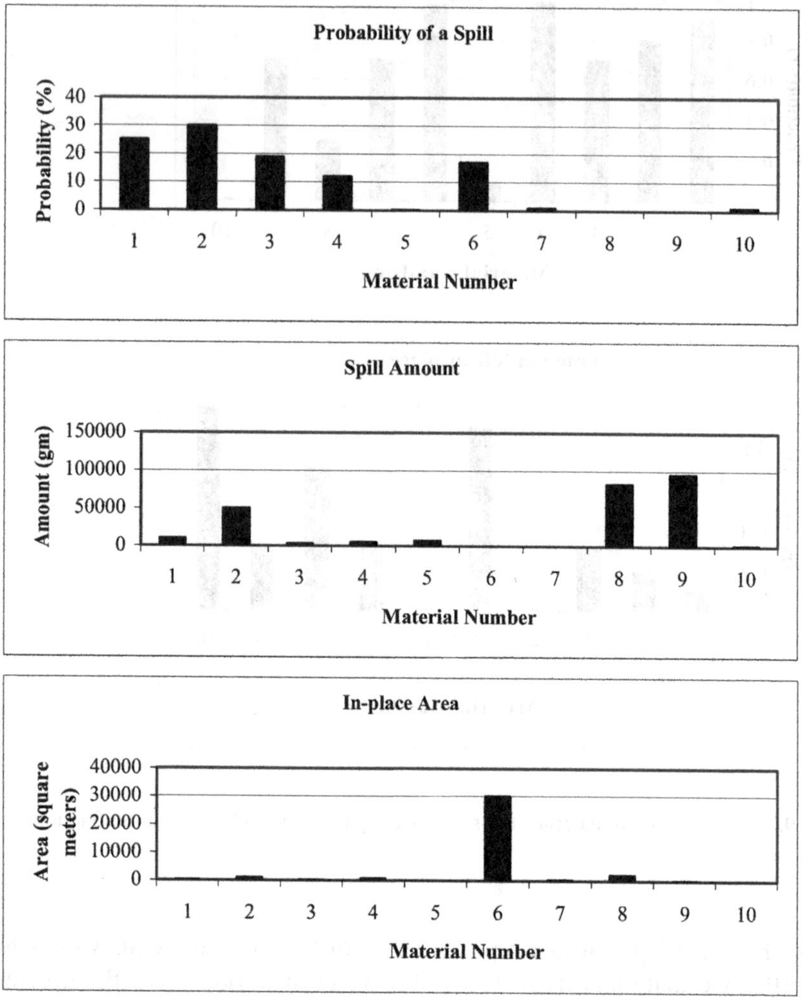

Fig. 9.3. Probabilities of spills for 10 materials, the amounts spilled and the in-place areas covered by the original spilled materials

Because the TAC and CSC for each material are given, determining whether a given concentration (no matter whether in-place, airborne or waterborne) exceeds the TAC and causes death; whether the relevant concentrations lie above the CSC but below the TAC and cause populations sickness; or, whether the concentrations lie below the CSC and do not

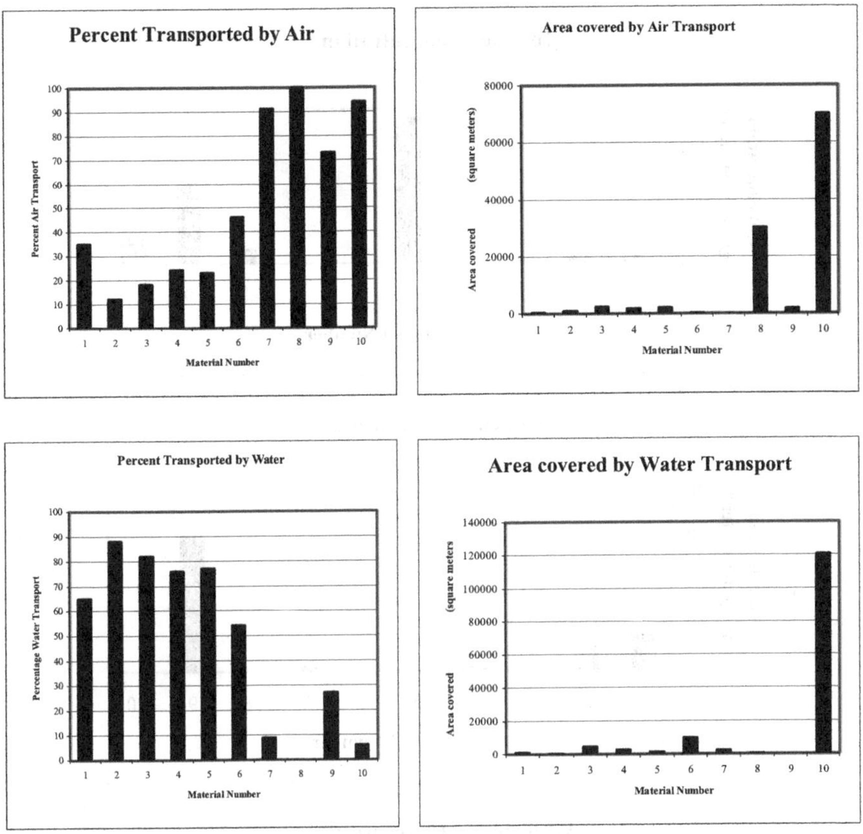

Fig. 9.4. The percentage of materials transported by air and the area covered by their later deposition together with the percentage transported by water and the area covered by the water transport. The percentages are for the fractions transported shown in Figure 9.2

harm the population is not difficult. Figure 9.6 shows the death rates per 100,000 population for in-place, airborne, and waterborne concentrations; while Figure 9.7 shows the corresponding sickness rates for each transport mode.

When allowance is made for synergistic effects of two combined materials when both TAC and CSC for the synergistic resultant can differ significantly from the corresponding values for the individual materials, the death and sickness rates per 100,000 population ignoring the spill probability for each of the two synergistic materials can be depicted (Figure 9.8). Also, the likely death and sickness rates when due allowance is

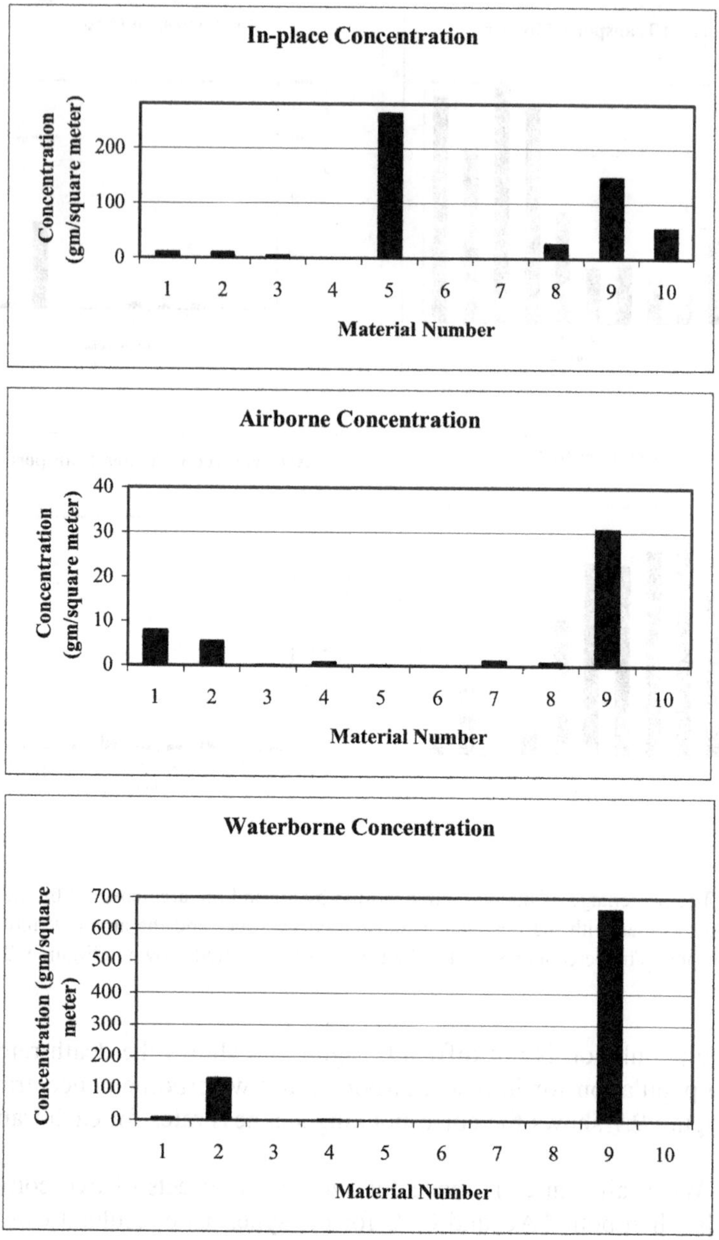

Fig. 9.5. Plots of the in-place, airborne, and waterborne concentrations of 10 materials

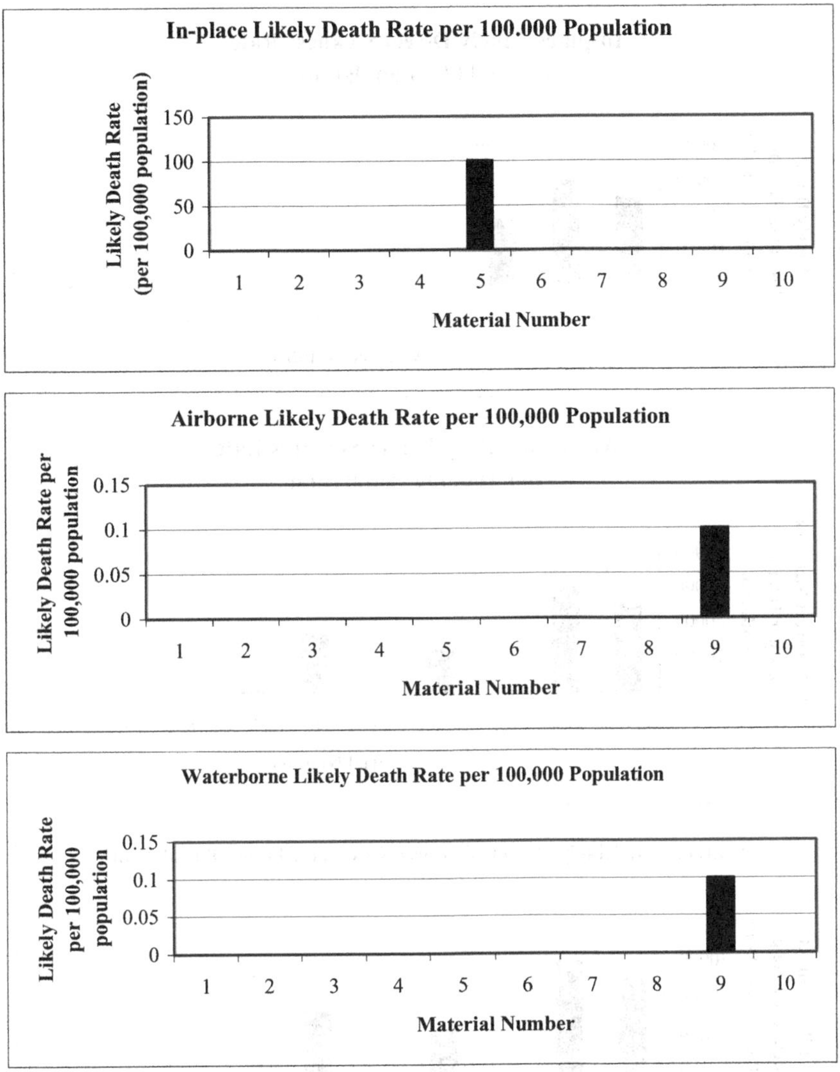

Fig. 9.6. The in-place, airborne, and waterborne likely death rates per 100,000 population as a result of spillage for 10 materials. The rates include the probability of a spill occurring

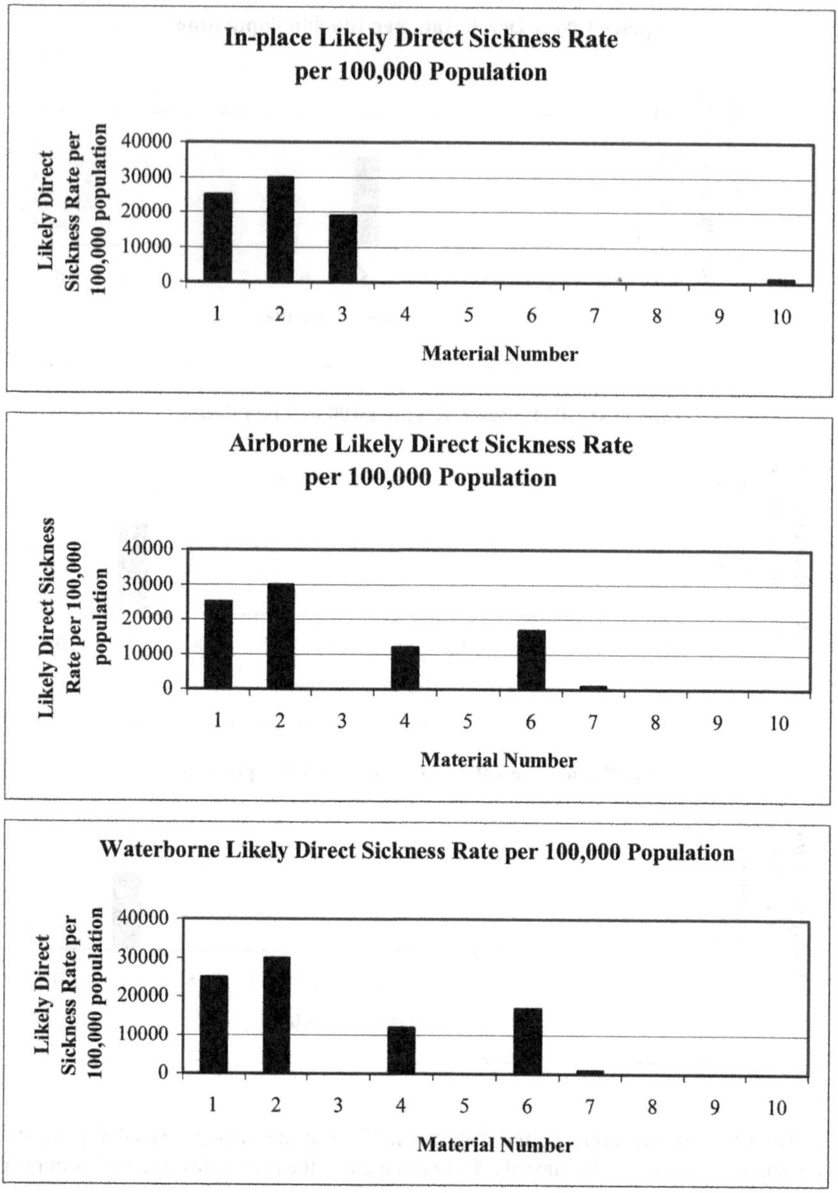

Fig. 9.7. The in-place, airborne, and waterborne likely sickness rates per 100,000 population as a result of spillage for 10 materials. The rates include the probability of a spill occurring

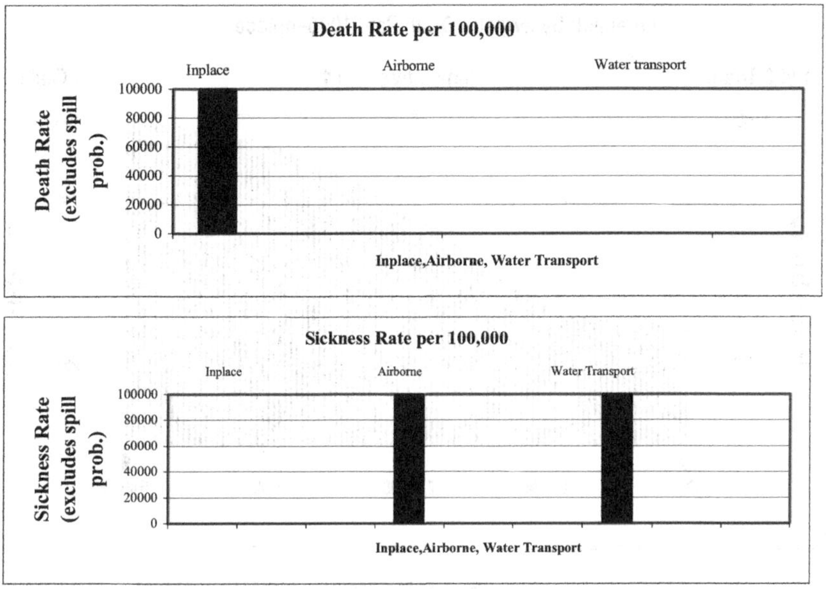

Fig. 9.8. Death and sickness rates per 100,000 population (excluding spill probability) due to in-place, airborne, and waterborne effects for synergism parameters in Appendix D

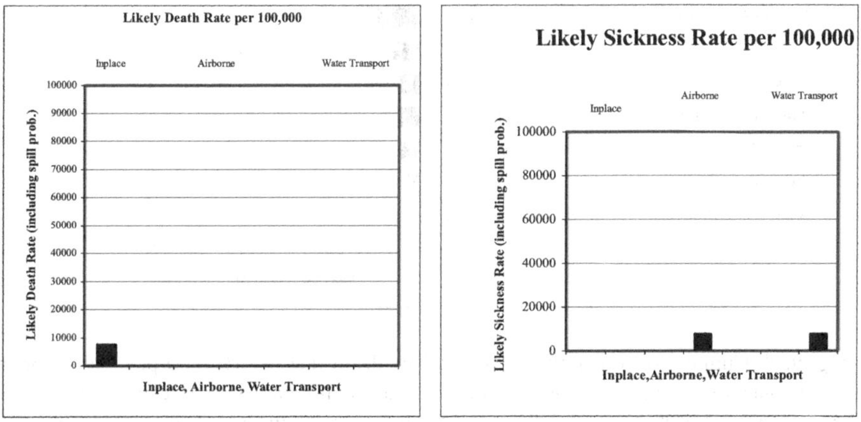

Fig.9.9. Same as Figure 9.8 but including the spill probability as shown in Figure 9.3

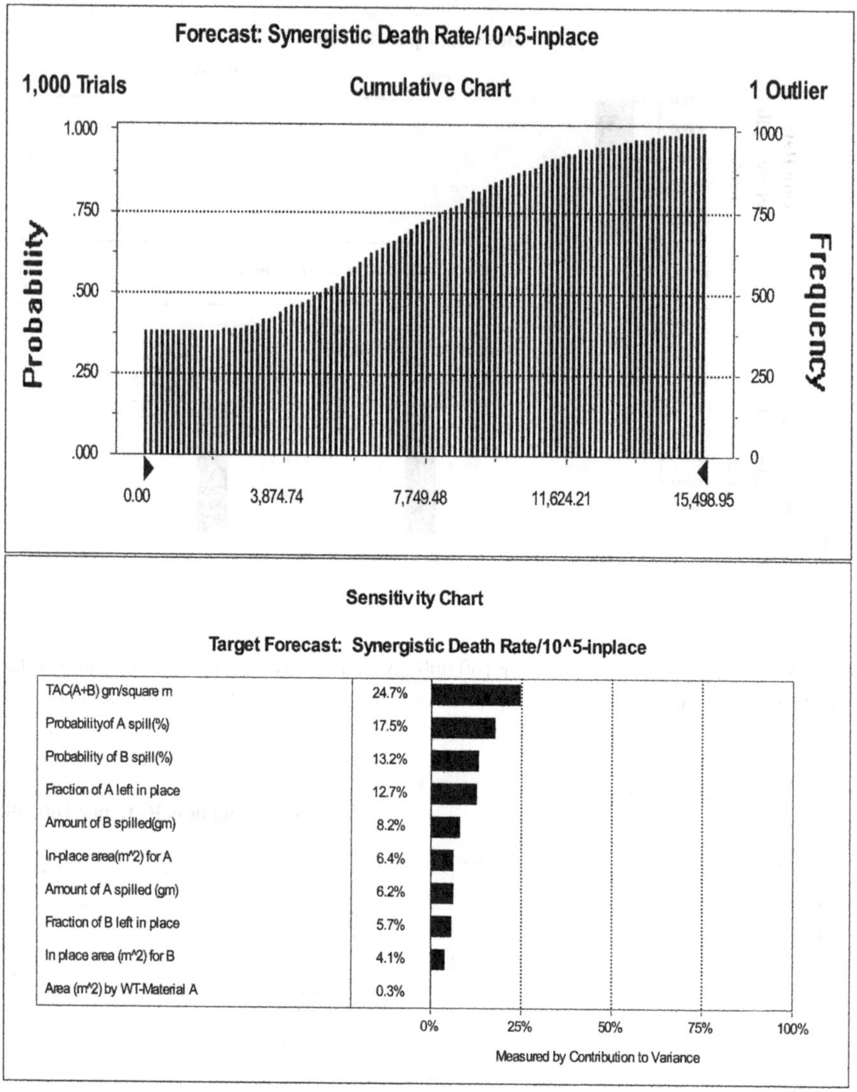

Fig. 9.10. Cumulative probability of synergistic death rate per 100,000 population for in-place contaminant fractions allowing for uncertainty in parameter values as mentioned in text; relative contribution of the uncertain parameters to the variance of the synergistic death rate. These results relate to uniform distribution choices for the uncertain parameters

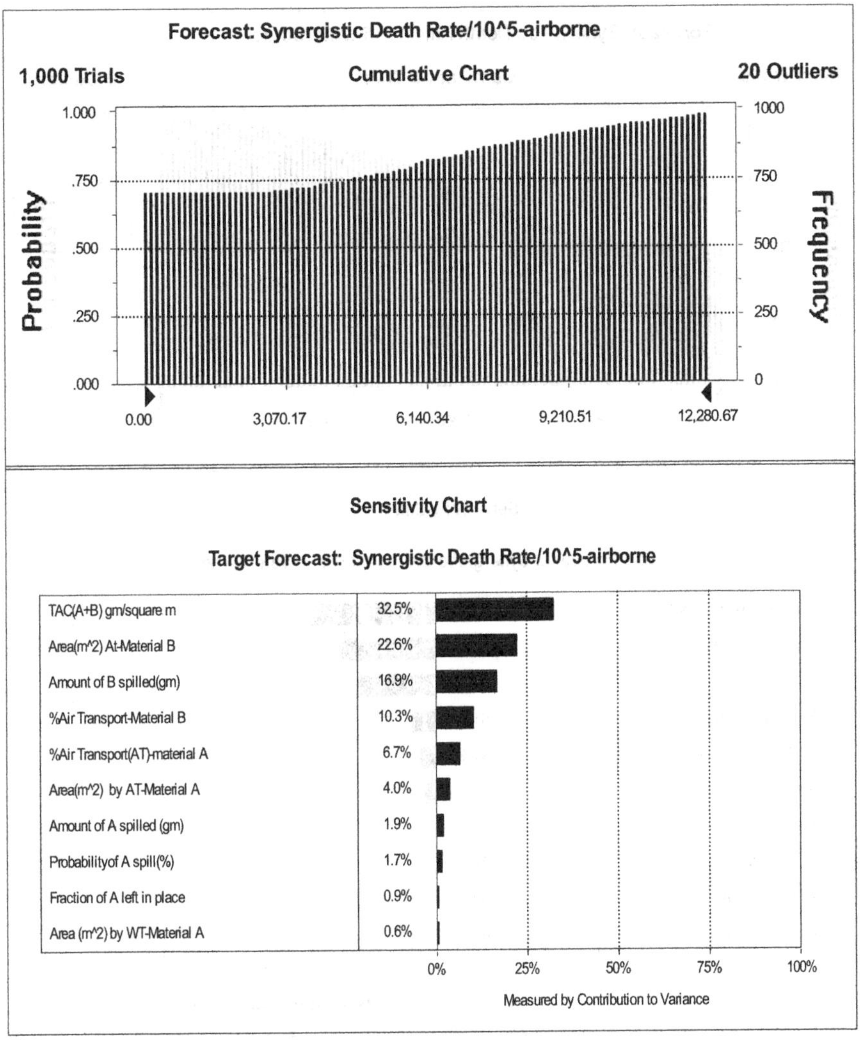

Fig. 9.11. Cumulative probability of synergistic death rate per 100,000 population for airborne contaminant fractions allowing for uncertainty in parameter values as mentioned in text; relative contribution of the uncertain parameters to the variance of the synergistic death rate. These results relate to uniform distribution choices for the uncertain parameters

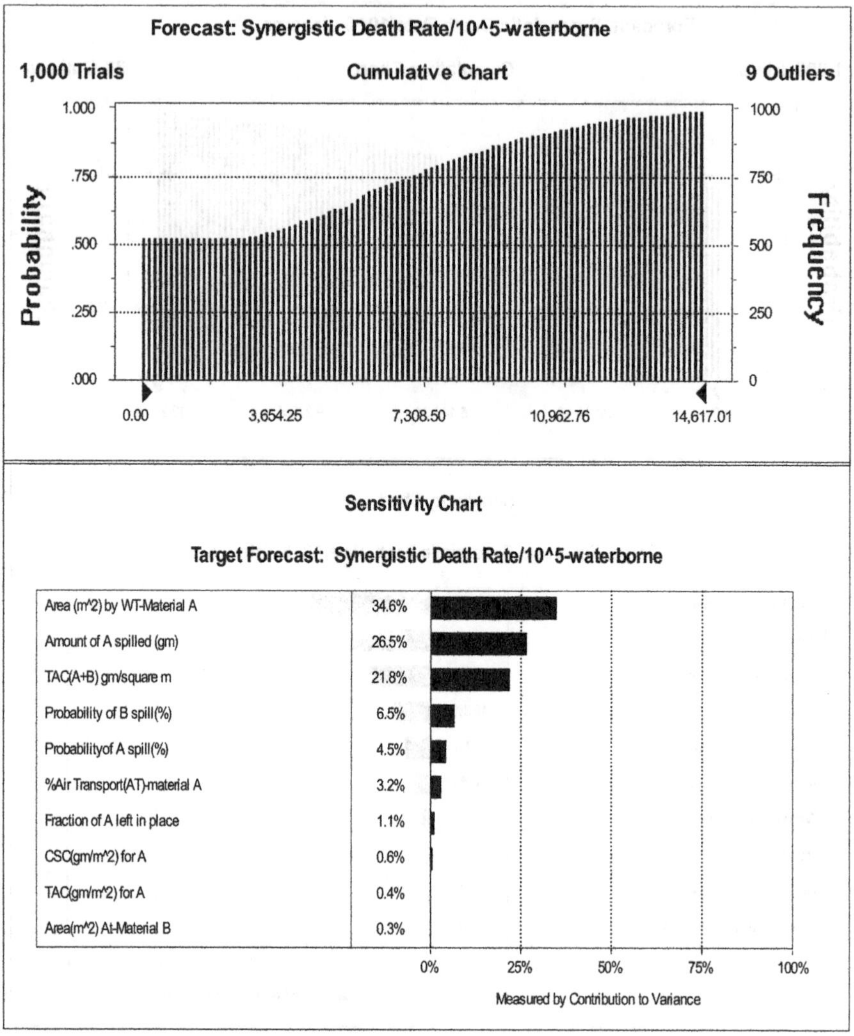

Fig. 9.12. Cumulative probability of synergistic death rate per 100,000 population for waterborne contaminant fractions allowing for uncertainty in parameter values as mentioned in text; relative contribution of the uncertain parameters to the variance of the synergistic death rate. These results relate to uniform distribution choices for the uncertain parameters

made for spill probabilities of both components can be estimated (Figure 9.9).

In short, the numerical program allows extremely rapid calculation of both direct and synergistic sickness and death rates for spills of up to ten materials. The program also allows computation of different death and sickness rates due to spilled material fractions that either remain at the spill site or are transported by air or water to different locales. In addition, the program allows synergistic effects for in-place, airborne and water transport of toxic materials.

However, the method is not adequate with this set of deterministic death and sickness estimates for a variety of reasons discussed below.

4 Uncertainty and Risking of Toxic Hazard Death and Sickness Rates

In the precise deterministic examples of the previous section several assumptions were made that are of less than unassailable veracity. Perhaps the major assumption is that TAC and CSC are the same values for all members of society for a given toxin; however, empirical observations show that individuals' health response to toxins vary markedly. Some will die at very low toxin concentrations while others will feel virtually no effect. Thus, the TAC and CSC values should have some degree of spread allowed for both individual materials and for synergistically derived toxins to mirror this pattern of behavior.

In addition, until a sufficient number of spills occur, the estimates of spill probability for a given material are either based on theoretical guesswork a very small number of occurrences, or innocuous materials spills under similar conditions to those that occur if toxic material were to be transported. In any event, spill probability is surely an uncertain parameter.

Each material type amounts that could be spilled are equally uncertain, as well as the in-place area that would be contaminated. Additionally, the fractions of materials transported, the percentage transported by air or water, and the secondarily contaminated areas as a consequence of such transport are uncertain. Presumably, this uncertainty in parameter values estimates leads to considerable differences of position corresponding to the lethal or sub-lethal conditions of a given event. For instance, in connection with destruction of Iraqi nerve gases on 4 March and 10 March 1991 at Khamisiyah during the first Gulf war, Gamboa (2003) reported that computer models used to estimate fallout were flawed and "The models were created with inaccurate data, and the height of the plume resulting from the 1991 weapons explosion was underestimated". What is unknown is whether more or less troops were exposed to nerve gas risk than the Pentagon's initial estimate of 100,000 soldiers. Presumably, in this instance the amount of toxic material released and its airborne coverage area are both uncertain, although the spill probability is 100% due to deliberate action to attempt to destroy the nerve gases. In addition, Showstack (2003) reported on a 2 June 2003 U. S. House of Representatives hearing where two reports (National Academy of Sciences and U. S. General Accounting Office) pointed out current plume model inadequacies for contaminant release - including inadequate quantification of plume height. This provides strong corroborative support that risking procedures remain the most viable option for addressing such unknowns and their impacts on death and sickness rates.

What is clear, however, is that many parameters are uncertain in making death and sickness estimates. The purpose of this section is to show how to use risking procedures based on Monte Carlo simulations to determine not only which parameters need improvement before any others (RC) but whether doing this is worthwhile (RI). The RS problem will also be addressed.

For simplicity, only materials A and B listed in the tables of Appendix D are considered. All other materials are deemed to have zero spill probabilities and so do not participate in the assessment.

For each situation examined the uncertain input parameters are taken to have a range of ±50% around centrally chosen values. For instance, the amount of material A spilled is taken to be uncertain but lie between 10000 ± 5000 gm. In addition, the spill probability for material A, as well as the in-place area, fractional transport, percent transported by air, TAC, CSC, airborne contaminant area, waterborne contaminant area, synergistic TAC and CSC, and the same quantities for material B, are all assumed to be ±50% uncertain around their nominal values recorded in Appendix D. But because the specified range for each variable involved does not indicate how the uncertain values are distributed, part of the RS problem remains. Also, no reason is presented for the choice of ±50% except as a pedagogical device. The question arises as to how the outputs of interest depend on both the type of distribution chosen for each variable and also any variation in the range limits for each variable as well as whether it is the uncertainty in the ranges or the distributions that contributes most to the output uncertainty? This total RS problem is addressed later in this section.

With the ranges above specified, the first concern is how the uncertainties influence the death and sickness rates of the various in-place, airborne, and waterborne components of the spilled materials. Note from Appendix D that the synergistic death rate TAC is much smaller than the TAC of either materials A or B taken on their own (even when allowance is made for the ±50% uncertainty) implying an aggressive synergism for enhancing the death rate. Accordingly, the first illustration concentrates only on the synergistic death and sickness rates.

4.1 Uniform Distributions, Aggressive Synergism

Because no priori knowledge exists for the distribution of the uncertain input parameters in the estimates of synergistic death and sickness rates, an assumption can be made that each parameter can be chosen from a uniform distribution population centered on the nominal value given in Appendix D and ranging uniformly to the ±50% values around that central value. The Monte Carlo program Crystal Ball® was then applied to the basic death and sickness program of Appendix D and 1000 runs were done, each time the values of the basic variables were chosen at random from the uniform distributions. The Monte Carlo simulations were also run with less than 1000 runs to ensure that by 1000 runs the compendium of numerical results had stabilized and so fairly represented the output statistics.

Figures 9.10 through 9.15 show the corresponding *cumulative* probabilities for death (Figures 10 through12) and sickness rates (Figures 9.13 through 9.15) per 100,000 population for the synergistic components for in-place, airborne, and waterborne effects with the basic uncertain parameters listed above.

Starting with Figure 9.10, a nearly 37% probability that no deaths will result from the in-place synergism of materials A and B, 50% chance that fewer than 5000 deaths per 100,000 population will result, and less than approximately 13,000 deaths at 90% probability. In comparison, the corresponding results for airborne and waterborne synergistic death rates indicate approximately 70% chance that no deaths will result from airborne contamination and approximately 52% chance that no deaths will result from waterborne synergistic contamination. There are, equally, less than approximately 4500 deaths likely at 75% chance for the airborne contaminant, and less than approximately 7300 deaths at the 75% chance for the waterborne contaminant; while at 90% chance there are likely to be less than approximately 12000 deaths due to airborne effects and less than approximately 12500 due to waterborne contamination.

The dominant factors causing the spread in the cumulative probabilities for each of the three death rate figures are also shown on Figures 9.10 through 9.12. Note from Figure 9.10 that the dominant two factors contributing to the uncertainty for the in-place death rate are the TAC for the synergistic mix of A and B (providing almost 25% of the total uncertainty) and the spill probability for material A (providing almost 18% of the total uncertainty). The remaining factors each cause less than approximately 13% of the uncertainty. For the airborne death rate, the dominant two factors contributing to the death rate uncertainty are the synergistic TAC (33% contribution) and the area uncertainty for airborne transport of material B (23% contribution); while for waterborne transport the two dominant contributors are the area contaminated by waterborne transport of material A (almost 35%), and the amount of material A that is spilled (27%) with the TAC for synergism a not so distant third contributor at almost 22%. Thus each transport mode provides different dominance of the various uncertain components to the cumulative probability of death.

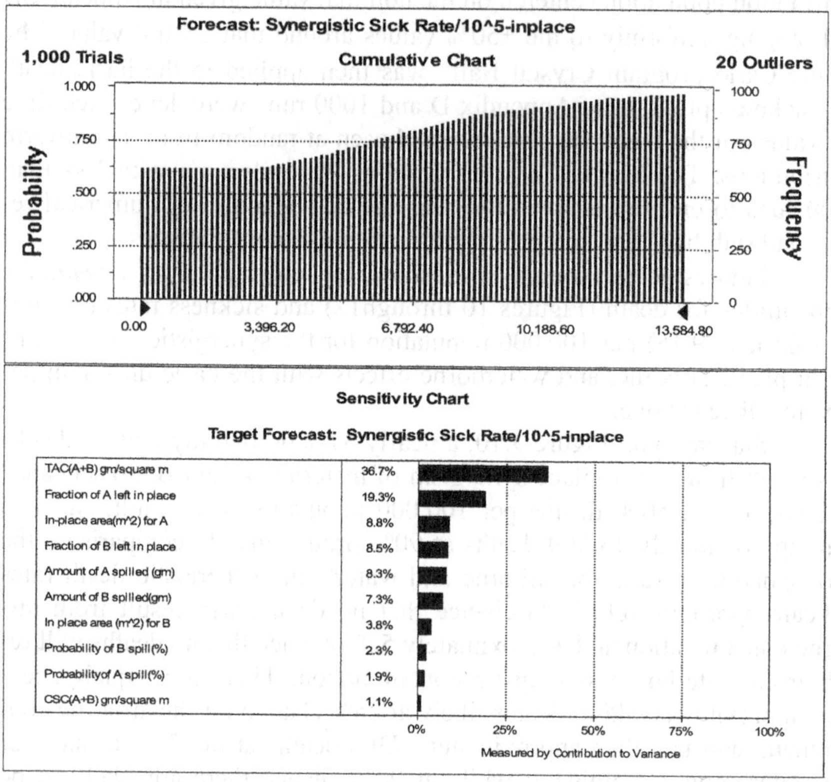

Fig. 9.13. Cumulative probability of synergistic sickness rate per 100,000 population for in-place contaminant fractions allowing for uncertainty in parameter values as mentioned in text; relative contribution of the uncertain parameters to the variance of the synergistic death rate. These results relate to uniform distribution choices for the uncertain parameters

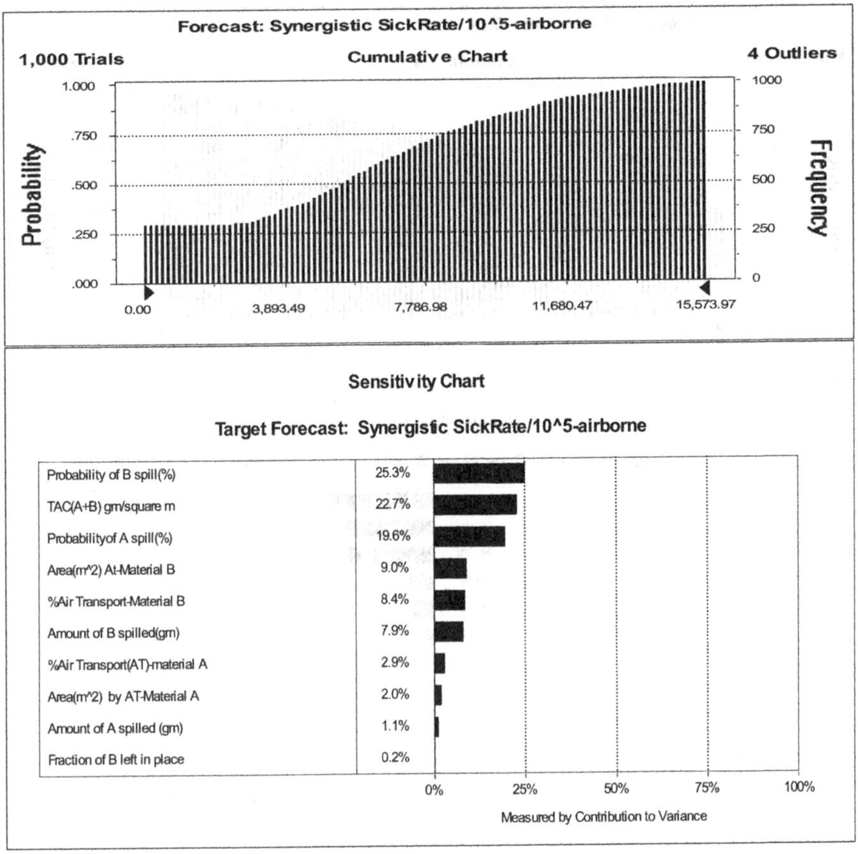

Fig. 9.14. Cumulative probability of synergistic death rate per 100,000 population for airborne contaminant fractions allowing for uncertainty in parameter values as mentioned in text; relative contribution of the uncertain parameters to the variance of the synergistic death rate. These results relate to uniform distribution choices for the uncertain parameters

The point is if having less uncertainty on the death rates for in-place, airborne, and waterborne synergistic effects are considered necessary, then the uncertainty on the TAC for synergism needs better quantification overall with the other factors of lesser but not insignificant importance. In this way determining not only the particular synergism mode causing the greatest death rate at a given probability level is possible, but also which parameters to focus on to reduce uncertainty. Concentrating on sharpening the parameter ranges that contribute little to total uncertainty is useless if the dominant uncertainty causes are not addressed first. This observation leads naturally to consideration of RI.

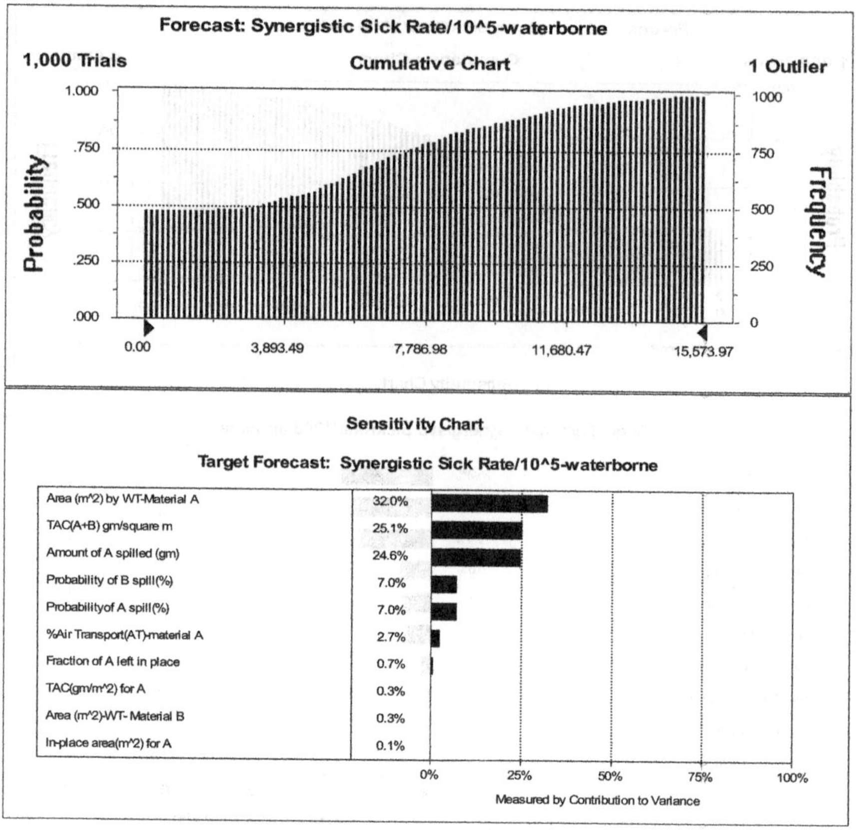

Fig. 9.15. Cumulative probability of synergistic death rate per 100,000 population for waterborne contaminant fractions allowing for uncertainty in parameter values as mentioned in text; relative contribution of the uncertain parameters to the variance of the synergistic death rate. These results relate to uniform distribution choices for the uncertain parameters

The RC discussion above demonstrates which variables are causing the greatest contributions to the uncertainty in the synergistic death rates but does not illustrate the magnitude of the contributions, which is the function of RI. As noted in Appendices A and B, some measure of volatility related to the standard deviation, σ, divided by the mean value, E, and measures the stability and accuracy of the mean value. Crystal Ball® not only produces the output probabilities but also a report detailing the statistical information describing each output distribution. The three death rates given above are σ /E (in-place) = 4.47/4.89; σ /E (airborne) = 3.85/2.27; and, σ /E (waterborne) = 4.38/3.68. In these cases the volatility measure is approximately 1 to 1.5, suggesting that little confidence should be placed in the mean value because the uncertainty is large. Thus, to attempt to narrow the range of the dominant variables is important because they contribute a significant RI (as measured by a volatility of order unity) in the lack of accuracy of the corresponding mean values. In addition, the mean values per 100,000 population are E (in-place) = 4890,E (airborne) = 2270, E (waterborne) = 3680. These values will be compared with corresponding values arising from the choices of triangular distributions for the basic uncertain variables.

Using similar arguments, the cumulative probabilities of sickness caused by the synergistic effects and dominant causes of the corresponding uncertainties can also be investigated. Figures 9.13 through 9.15 display these results for in-place, airborne, and waterborne transport, respectively. For the in-place sickness rate, an approximately 67% chance that no one will become sick and approximately 75% chance less than about 6,700 per 100,000 population will become sick is shown (Figure 13). Likewise, Figure 9.14 shows that an approximately 30% chance no one will become sick from airborne contamination and approximately 75% chance less than about 9,800 will sicken. For the waterborne contamination, Figure 9.15 shows that an approximately 40% chance no one will sicken and a 75% chance that less than 7,800 will sicken per 100,000 population. Correspondingly, the RC components most influencing the uncertainty of these cumulative probability results show that the two dominant contributors to the uncertainty are: 1) the TAC for synergism and the material A fraction left in place for the in-place sickness rate, with contributions of 37% and 19%, respectively; and, 2) the material B spill probability and the TAC for synergism for the airborne sickness rate, with contributions of 25% and 22%, respectively (the of material A spill probability is a close third at almost 20%). For waterborne sickness, the area contaminated by waterborne transport for material A, the TAC for synergism, and the material A spill amount cause the dominant contributions to the uncertainty (32%, 25%, and 24.6%, respectively). Despite their uncertainties, all other factors contribute relatively much less to the uncertainty of the sickness rates.

Again, Crystal Ball® provides the volatility measure σ /E for each sickness component and determine if the ratio is small or large compared to unity. A small ratio implies an accurate estimate is being provided while a large ratio compared to unity implies the relative importance to narrow the dominant contributor ranges to the three sickness rate uncertainties. For in-place sickness, σ /E = 4.13/2.91; airborne σ /E = 4.36/5.30; and, waterborne sickness rate σ /E = 4.43/3.29. Again, the volatility measures are comparable to unity for all three sickness rates, implying the relative importance to better assess the dominant variables ranges contributing to uncertainty in the sickness probability assessments in order to place greater reliance on the assessed values. The mean values for the sickness rates per 100,000 population are: E(in-place) = 2910; E(airborne) = 5300; and, E(waterborne) = 3290. These mean values will be compared against the corresponding values arising from the choices of triangular distributions for the basic uncertain variables below.

4.2 Triangular Distributions, Aggressive Synergism

The choice of uniform distributions for the underlying variables from which the Monte Carlo computations have drawn their various random samples is arbitrary. There is no reason that other choices of underlying distributions cannot be invoked without further information to the contrary. Part of the RS sense of argument is to note that if the choice made for each underlying distribution sensitively influences the outputs, both in magnitude and also in variance contributions (for both RC and RI), then a way to obtain a better description of the variables for ranges as given is needed. If, on the other hand, the choices of the distributions have only a small contribution to the total uncertainty, then the choice of distribution that is influencing the outputs and their uncertainties is less important than the dynamic ranges of dominant variables contributing to the variance. Thus it is important to determine RS so the need to concentrate on better describing the underlying distributions or whether to concentrate on narrowing the dynamic ranges is known. In this way, the focus can be on the dominant sensitive contributors to the uncertainty without spending an inordinate amount of time and effort on factors that are not as significant.

This component of the problem is illustrated using precisely the same suite of variables as in the previous subsection, but each variable is assumed to be drawn from a triangular distribution, symmetrically centered on the nominal values given in Appendix D and with the triangular probabilities set to zero at the ±50% points. Figures 9.16 through 9.21 display the cumulative probability and associated relative contributions to the synergistic death rates (Figures 9.16 through 9.18) and synergistic sickness rates (Figures 9.19 through 9.21), respectively.

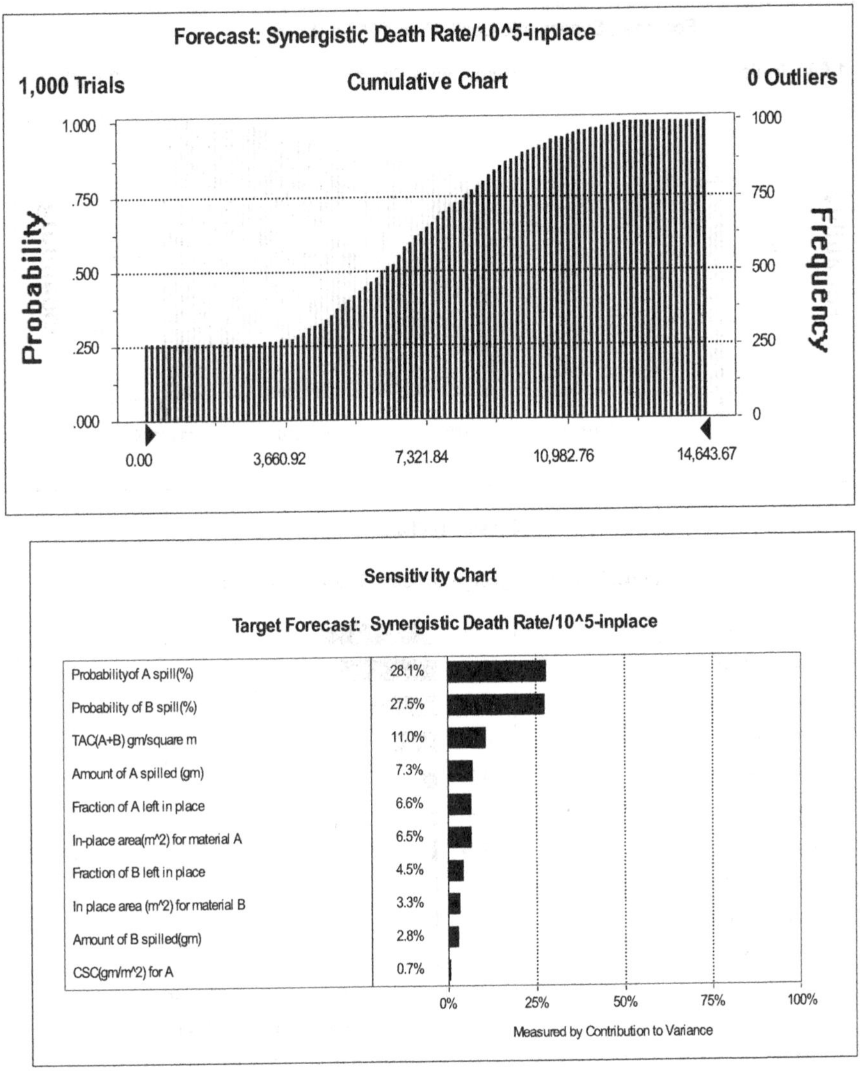

Fig. 9.16. Cumulative probability of synergistic death rate per 100,000 population for in-place contaminant fractions allowing for uncertainty in parameter values as mentioned in text; relative contribution of the uncertain parameters to the variance of the synergistic death rate. These results relate to triangular distribution choices for the uncertain parameters

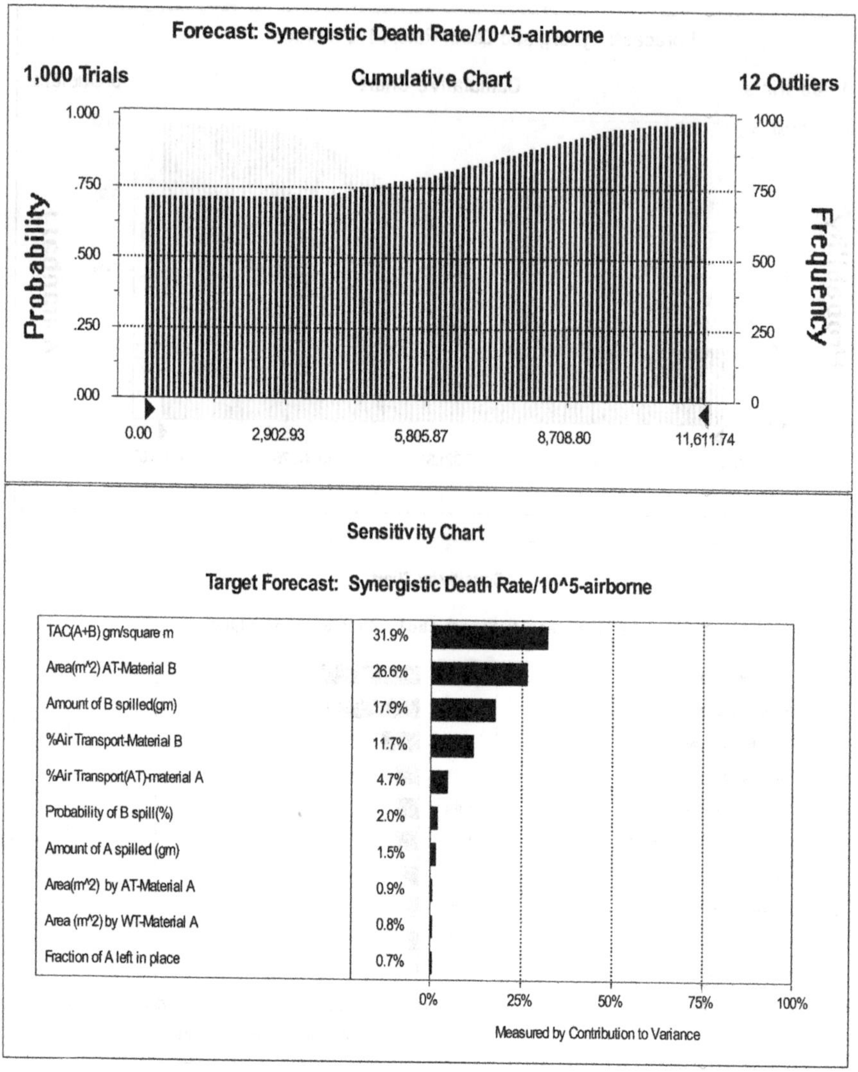

Fig. 9.17. Cumulative probability of synergistic death rate per 100,000 population for airborne contaminant fractions allowing for uncertainty in parameter values as mentioned in text; relative contribution of the uncertain parameters to the variance of the synergistic death rate. These results relate to triangular distribution choices for the uncertain parameters

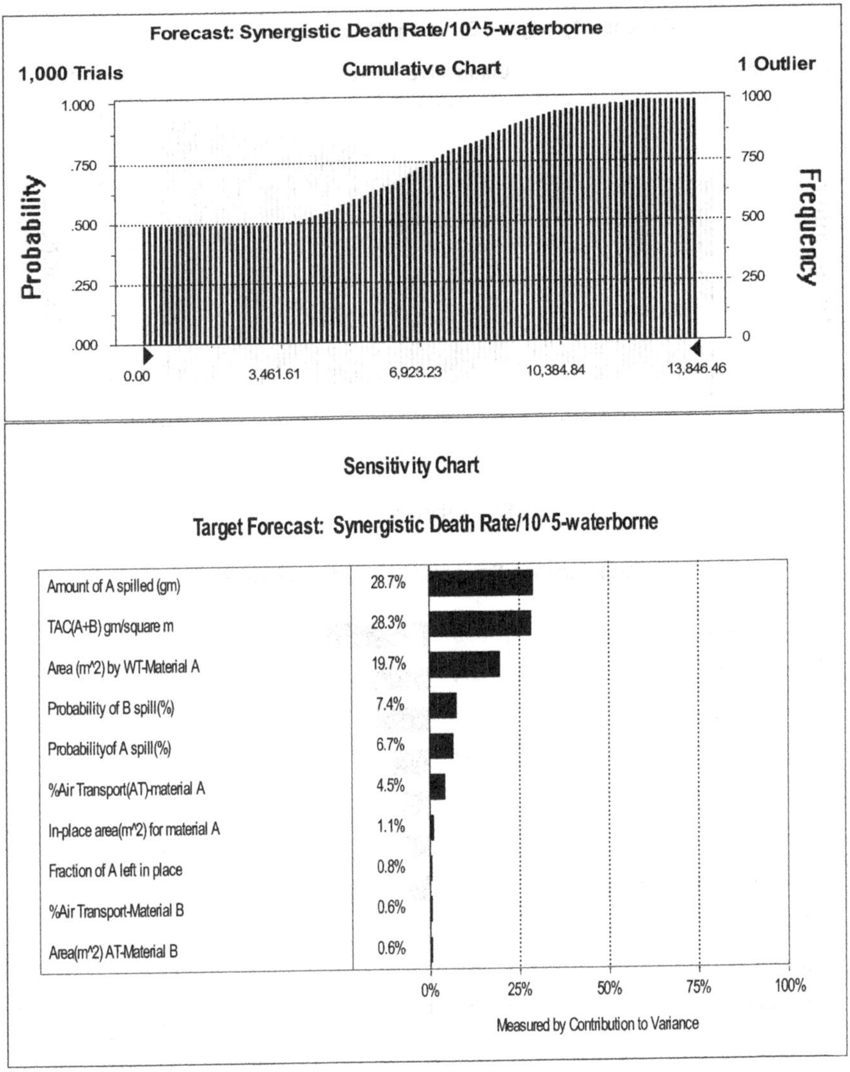

Fig. 9.18. Cumulative probability of synergistic death rate per 100,000 population for waterborne contaminant fractions allowing for uncertainty in parameter values as mentioned in text; relative contribution of the uncertain parameters to the variance of the synergistic death rate. These results relate to triangular distribution choices for the uncertain parameters

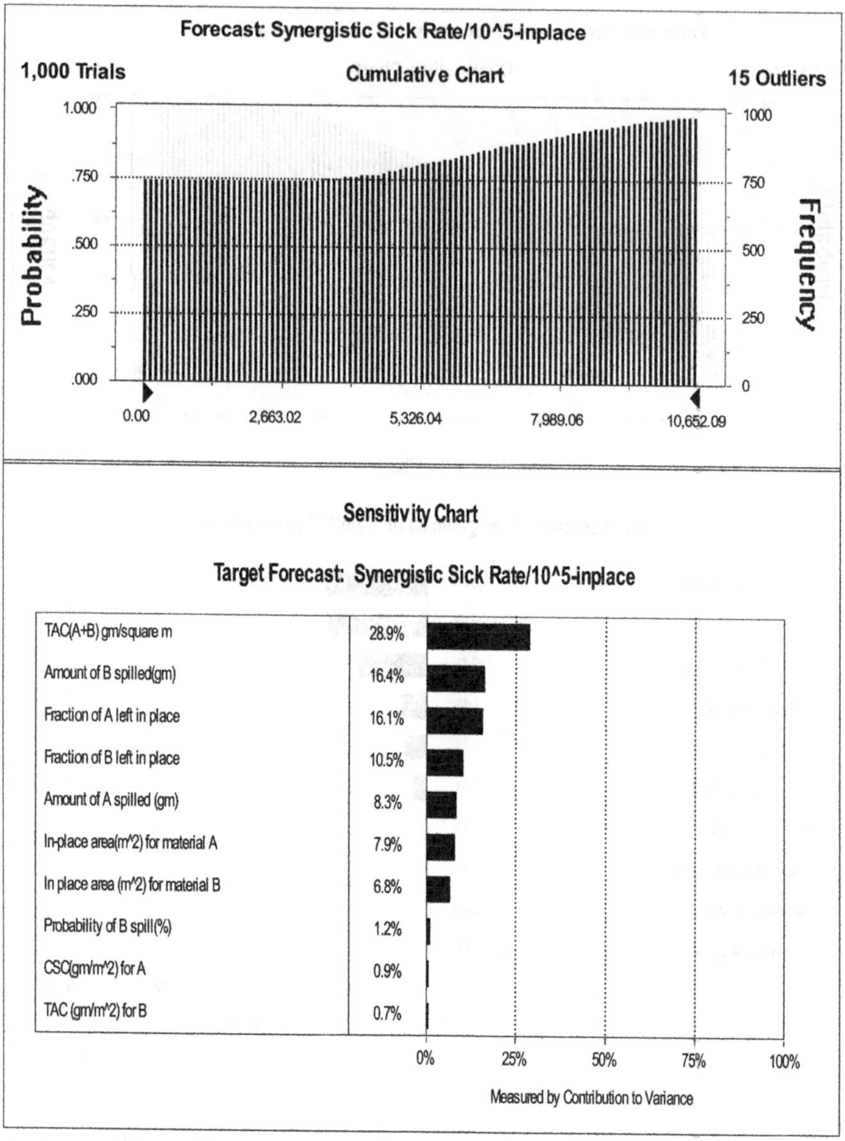

Fig. 9.19. Cumulative probability of synergistic sickness rate per 100,000 population for in-place contaminant fractions allowing for uncertainty in parameter values as mentioned in text; relative contribution of the uncertain parameters to the variance of the synergistic death rate. These results relate to triangular distribution choices for the uncertain parameters

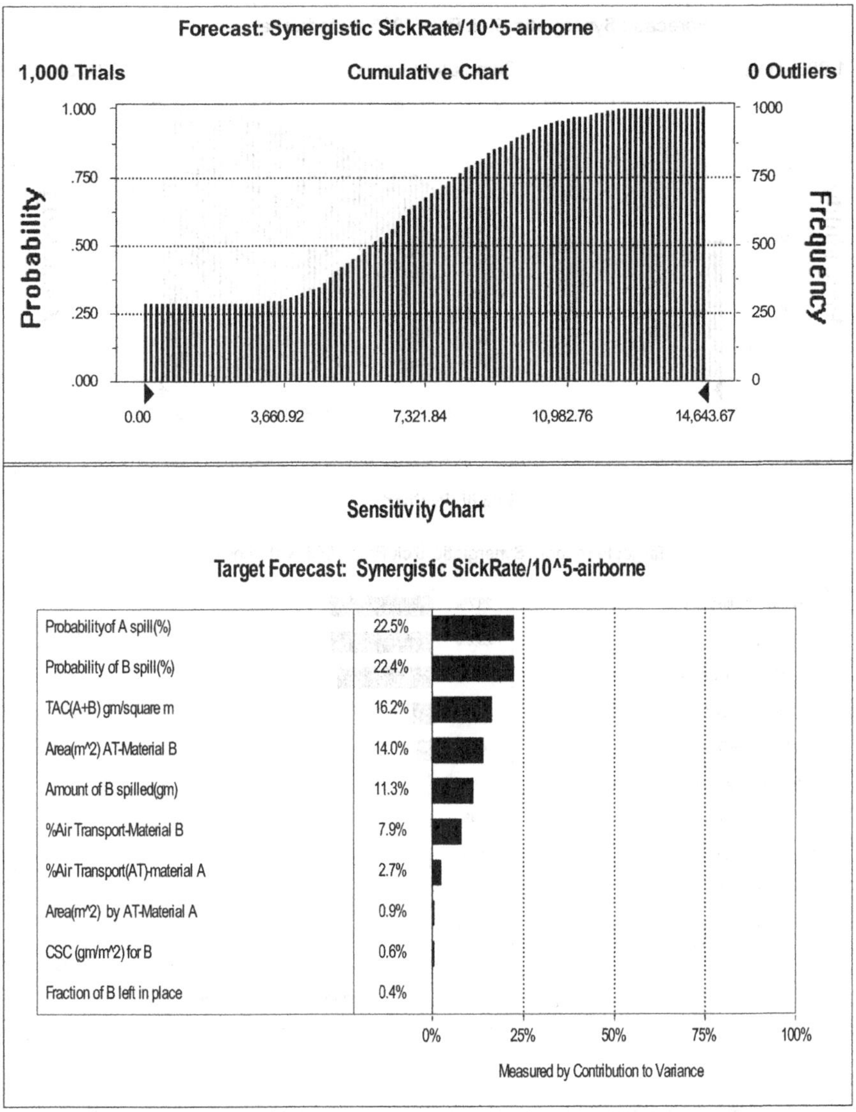

Fig. 9.20. Cumulative probability of synergistic sickness rate per 100,000 population for airborne contaminant fractions allowing for uncertainty in parameter values as mentioned in text; relative contribution of the uncertain parameters to the variance of the synergistic death rate. These results relate to triangular distribution choices for the uncertain parameters

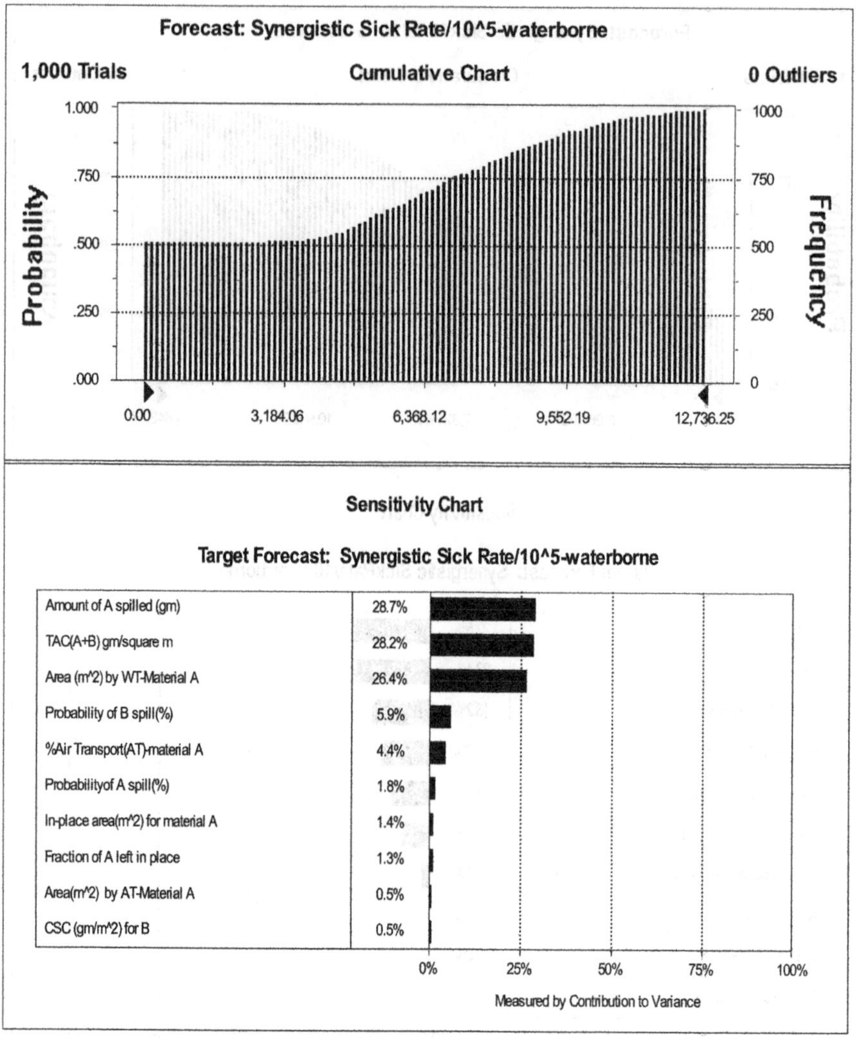

Fig. 9.21. Cumulative probability of synergistic sickness rate per 100,000 population for waterborne contaminant fractions allowing for uncertainty in parameter values as mentioned in text; relative contribution of the uncertain parameters to the variance of the synergistic death rate. These results relate to triangular distribution choices for the uncertain parameters

In comparison to Figures 9.10 through 9.12 for the uniform choices of underlying distributions, inspection of Figures 9.16 through 9.18 now shows a 26% chance no one will die because of in-place synergism, 68% chance no one will die because of airborne synergistic contamination, and 50% chance no one will die because of waterborne contaminant synergistic effects (versus 37%, 70%, and 52%, respectively, for the uniform case). In addition, the two most dominant components for the RC to the variance of each cumulative probability death rate plots are: for in-place considerations, the material A spill probability at 28% contribution to the variance, and the material B spill probability at 27.5% contribution (in the uniform case the dominance was TAC for synergism at 25% and material A spill probability at 18%); for airborne considerations, the TAC for synergism at 32% and the area contaminated by airborne transport of material B at 27% (in the uniform case the dominance was TAC for synergism at 33% and area contaminated by airborne transport of material B at 23%); for waterborne considerations, the variance is dominated almost equally by the amount of material A spilled at 28.7% and the synergistic TAC at 28.3% (in the uniform case the two dominant factors were the area contaminated by waterborne transport of material A, at 35% and the amount of material A spilled at 27%, with synergistic TAC third at 22%). In short, the type of statistical distribution used alters the RC values to the synergistic death rates. Indeed, using the Crystal Ball® report for the triangular situation yields volatility measures in the three death rate cases of: σ /E (in-place) = 3.8/5.6; σ /E (airborne) = 3.6/2.1; and σ /E (waterborne) = 4.1/3.8 together with the mean value assessments per 100,000 population of E (in-place) = 5,600; E(airborne) = 2,100; and, E(waterborne) = 3,800.

Note that corresponding mean values in the uniform distribution choice situation are E(in-place) = 4890, E (airborne) = 2270, E (waterborne) = 3680. Thus, the in-place synergistic death estimate shows approximately 17% difference with the remaining airborne and waterborne synergistic mean death rates very close in either the uniform or triangular situation. But the differences in all three situations are well within one standard error range around the mean values (due largely to the high volatilities in both the uniform and triangular situations). The implication is that trying to improve the RS of the results to the choice of distribution is useless until the RS to the large dynamic ranges of the dominant variables causing the high volatilities can be narrowed down further.

Figures 9.19 through 9.21 show that precisely similar arguments can be undertaken for the synergistic sickness rates leading again to conclude that trying to better determine the distribution of each uncertain variable is futile until the dynamic ranges of the dominant variables in the variance can be narrowed to the point that the volatility is reduced considerably below the region of unity where it currently is sited for the illustrations used. The RS then allows focus on the relevant components of the problem needing better resolution (e.g., the dominant variance contributions and not the distribution types).

4.3 Uniform Distributions, Benign Synergism

While the previous two sections have considered only the synergistic component of the likely death and sickness toxic spill problem under aggressive conditions (i.e., where synergism provides for much higher death rates than either material would on its own), by no means are these conditions the only factors of concern. First, synergism depends on having two materials that interact to produce TAC and CSC different values than their respective single TAC and CSC values. A spill with only a single material may occur, so that direct death rates and sickness rates from the single material are more appropriate considerations than synergistic effects. Second, while a synergistic effect may be highly toxic, the probability of spilling both materials that interact synergistically is remote so considering each individual material as though no toxic synergism occurred is recommended. Third, one toxic material may be transported by water and another either remains in place at the spill site or is transported by air so that the two materials never get a chance to be synergistically involved. Fourth, if synergism occurs it may convert two materials that are inherently highly toxic into a material with very benign characteristics (e.g., sodium and chlorine are individually extremely toxic but their product, sodium chloride or common salt, is not toxic).

Thus, the direct likely death and sickness rates for individual materials is important to consider as well as their synergistic mix. The pattern of such calculations follows the same outline as the two sections above but with one major difference- now no dependence of toxicity effects on parameters for both materials taken together exists, rather the dependence is only for single material uncertain parameters. This simplicity has the effect that each material can be treated separately and the possibility of a double spill is not necessary but only the probability of each single spill, so the chances are changed of direct likely death and sickness rates compared to those for the synergistic events described above.

To demonstrate these computations, the same parameters and their uncertainties have been retained as for the uniform distribution case described above with two exceptions. The TAC and CSC for synergism are both raised to extremely high values compared to the corresponding values for materials A and/or B so that basically there is no synergism; instead direct death and sickness rates for each material are used. Now only the case of material B is considered to illustrate the method. The reason for choosing material B rather than material A is the amount of material B likely to be spilled if a spill occurs is much in excess of material A (see Appendix D). This factor provides for a clearer depiction of the extremes of direct death and sickness rate probabilities. The results of running the Crystal Ball® Monte Carlo situations and concentrating on the likely direct death and sickness probabilistic results for material B are shown together with the RC to the variance of the six outputs (Figures 9.22 through 9.27).

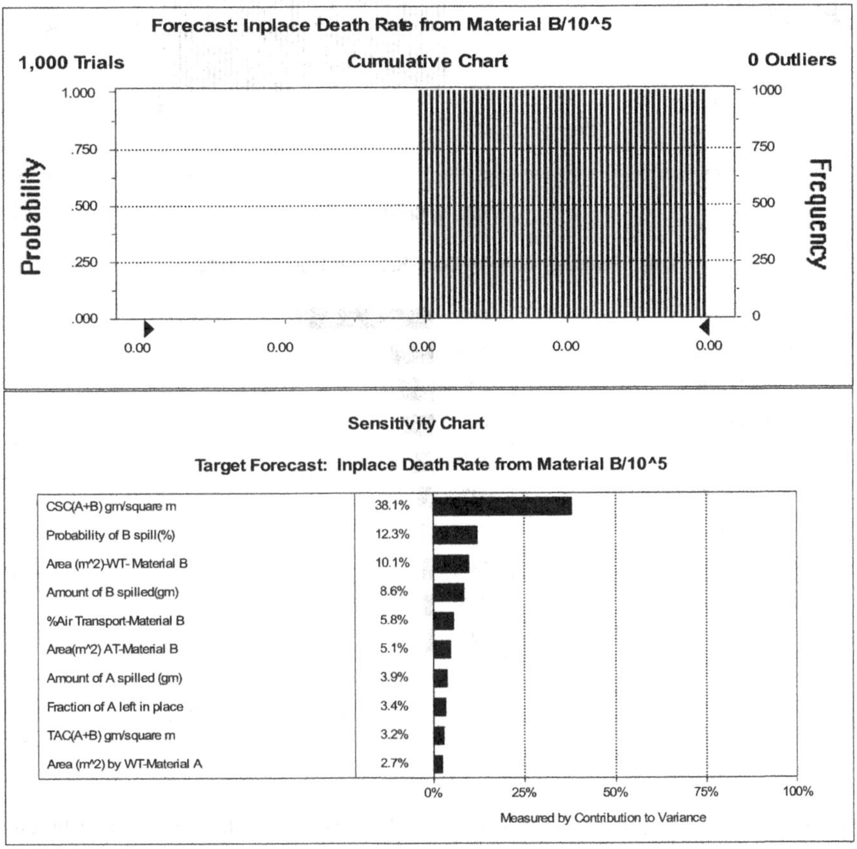

Fig. 9.22. Cumulative probability of material B death rate per 100,000 population for in-place contaminant fractions allowing for uncertainty in parameter values as mentioned in text and when there is no synergistic effect; relative contribution of the uncertain parameters to the variance of the synergistic death rate. These results relate to uniform distribution choices for the uncertain parameters

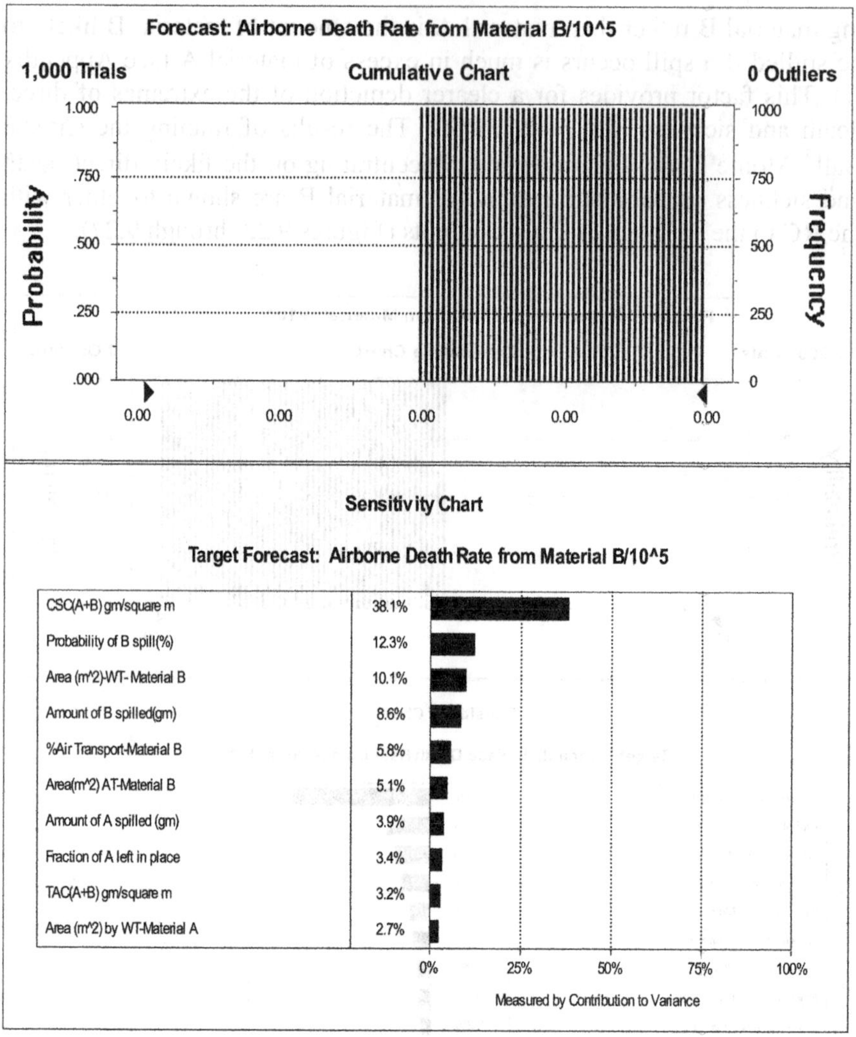

Fig. 9.23. Cumulative probability of material B death rate per 100,000 population for airborne contaminant fractions allowing for uncertainty in parameter values as mentioned in text and when there is no synergistic effect; relative contribution of the uncertain parameters to the variance of the synergistic death rate. These results relate to uniform distribution choices for the uncertain parameters

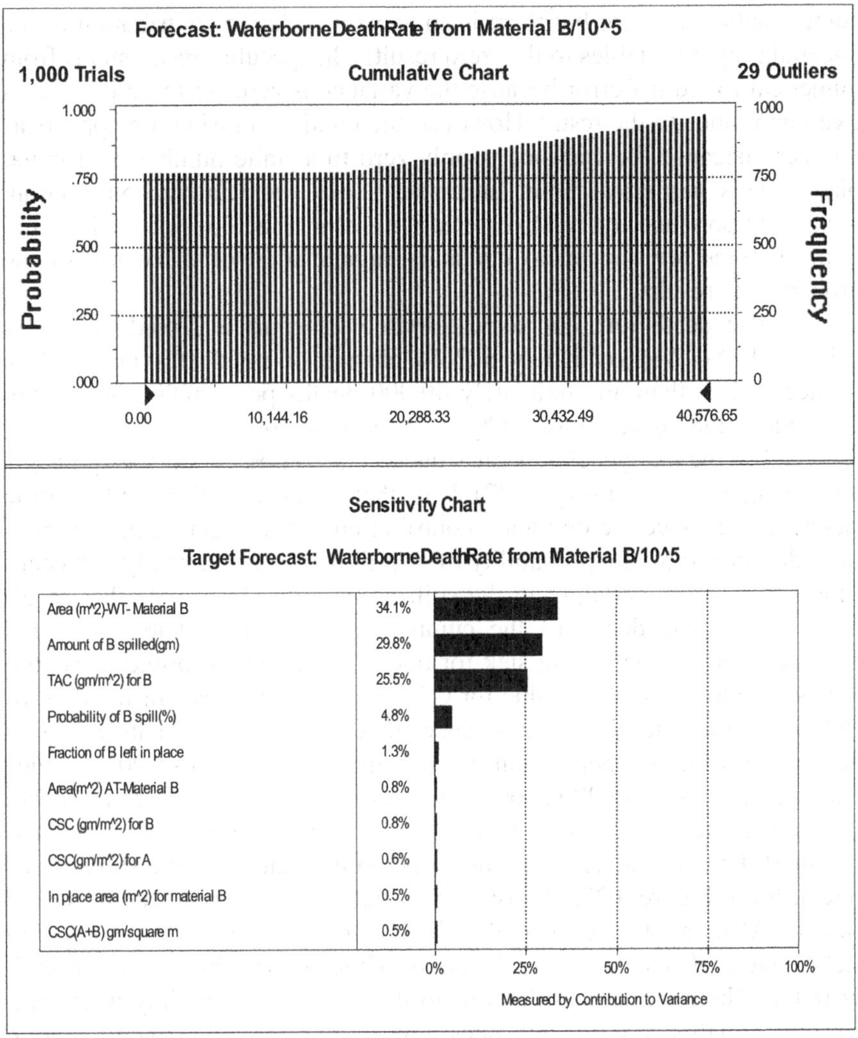

Fig. 9.24. Cumulative probability of material B death rate per 100,000 population for waterborne contaminant fractions allowing for uncertainty in parameter values as mentioned in text and when there is no synergistic effect; relative contribution of the uncertain parameters to the variance of the synergistic death rate. These results relate to uniform distribution choices for the uncertain parameters

Consider first the death rates for the three modes of transport of material B (Figures 9.22 through 9.24). The first important point to note from this extreme situation is the death rates for both in-place and airborne contamination are zero but there is an apparent relative contribution of the uncertain input variables to this zero result! This peculiar result stems from numerical round-off error because the variance is zero, so there is no relative importance to the result. However, the relative contribution appears as non-zero because the variance is only zero to a finite number of decimal places. This numerical point underscores that the relative contribution, relative importance, or relative sensitivity cannot be considered in isolation; all three are needed to obtain an appreciation of the worth of the probabilistic results.

The death rate due to waterborne transport is slightly different (Figure 9.24). Approximately a 76% chance of no deaths and a 90% chance of less than approximately 40,000 deaths per 100,000 population exist due to the higher transport by water of material B.

For the likely sickness rates due to material B the story is quite different (Figures 9.25 through 9.27). Note that both the in-place and airborne sickness rates have the dominant contribution to the uncertainties in those rates dominated by the probability of a spill, both approximately 98% contribution of the uncertainty in the spill probability. Also, note that an almost linear dependence of the cumulative sickness probability on the number of people becoming sick for both in-place and airborne rates continues because of the low value for CSC for direct sickness for material B. Both rates indicate almost no chance of less than approximately 15,000 people per 100, 000 population not becoming sick, but also 90% chance that less than about 45,000 people will become sick irrespective of whether they are located at the in-place spill site domain or subjected to airborne transport. For the waterborne contamination by material B the story is different. From Figure 9.27, almost a 25% chance no one will become sick is shown. Also, a 50% chance that less than approximately 25,000 will sicken and a 90% chance less than approximately 45,000 will sicken is illustrated. The relative contributions to the variance uncertainty in the waterborne situation for sickness rates indicate that the dominance is due to the material B spill probability (at approximately 27.5% contribution), the area contaminated by waterborne transport of material B (at approximately 26.7% contribution), the amount of material B spilled (at approximately 22.5% contribution), and the TAC for material B (at approximately 18.7% contribution); all other factors contribute negligibly to the variance. Surprisingly, the CSC for material B hardly plays a role because the effective concentrations of material B are sufficiently high compared to the CSC (even allowing for its uncertainty) that a large fraction of the population would sicken. This fact is also seen in the high sickness rate probabilities recorded in Figure 9.27.

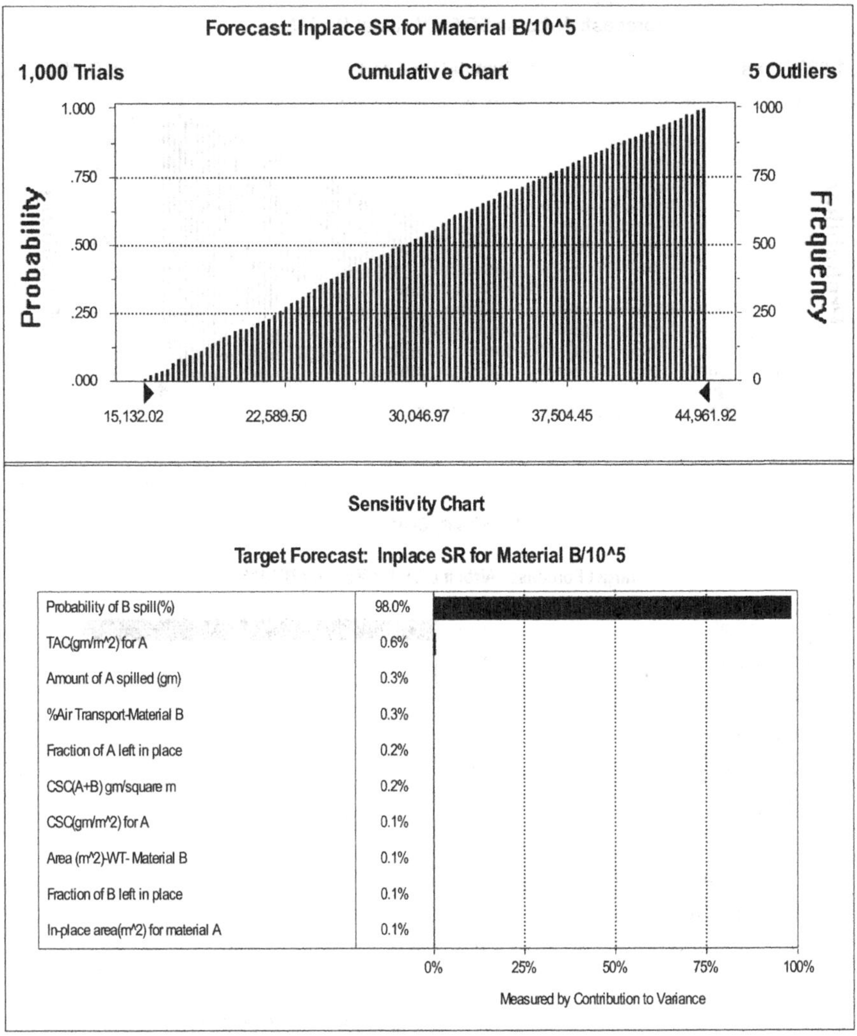

Fig. 9.25. Cumulative probability of material B sickness rate per 100,000 population for in-place contaminant fractions allowing for uncertainty in parameter values as mentioned in text and when there is no synergistic effect; relative contribution of the uncertain parameters to the variance of the synergistic death rate. These results relate to uniform distribution choices for the uncertain parameters

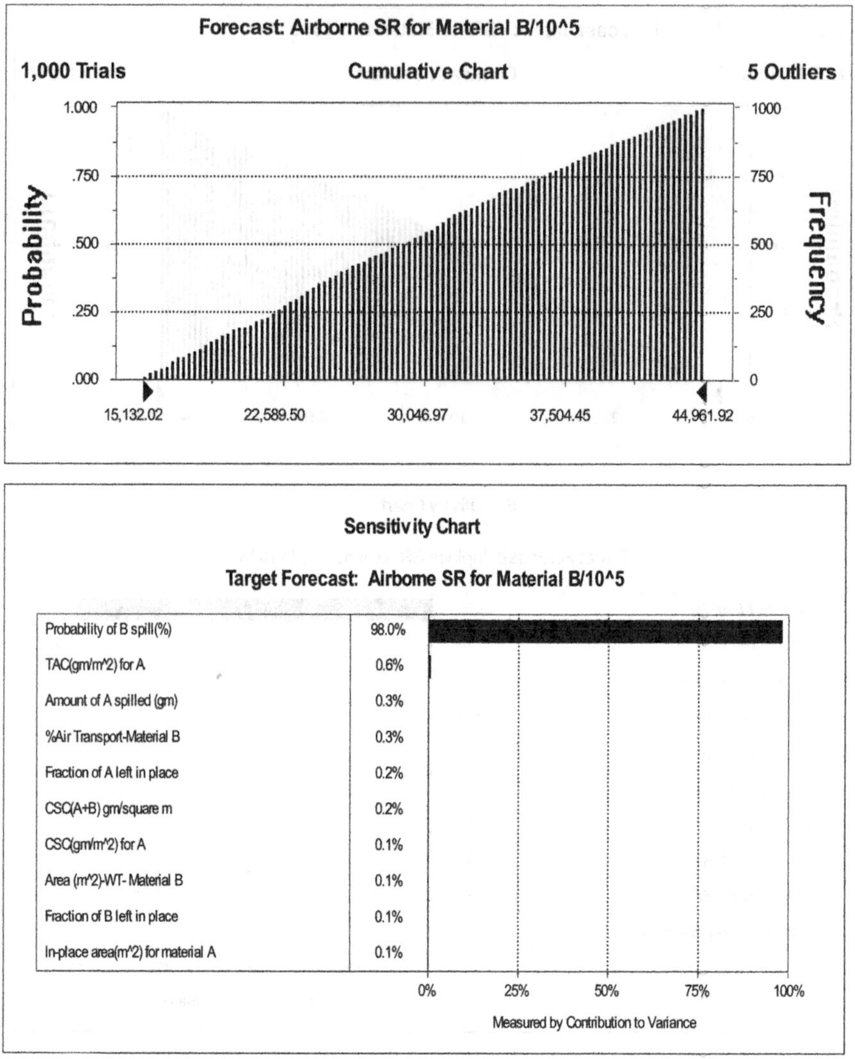

Fig. 9.26. Cumulative probability of material B sickness rate per 100,000 population for airborne contaminant fractions allowing for uncertainty in parameter values as mentioned in text and when there is no synergistic effect; relative contribution of the uncertain parameters to the variance of the synergistic death rate. These results relate to uniform distribution choices for the uncertain parameters

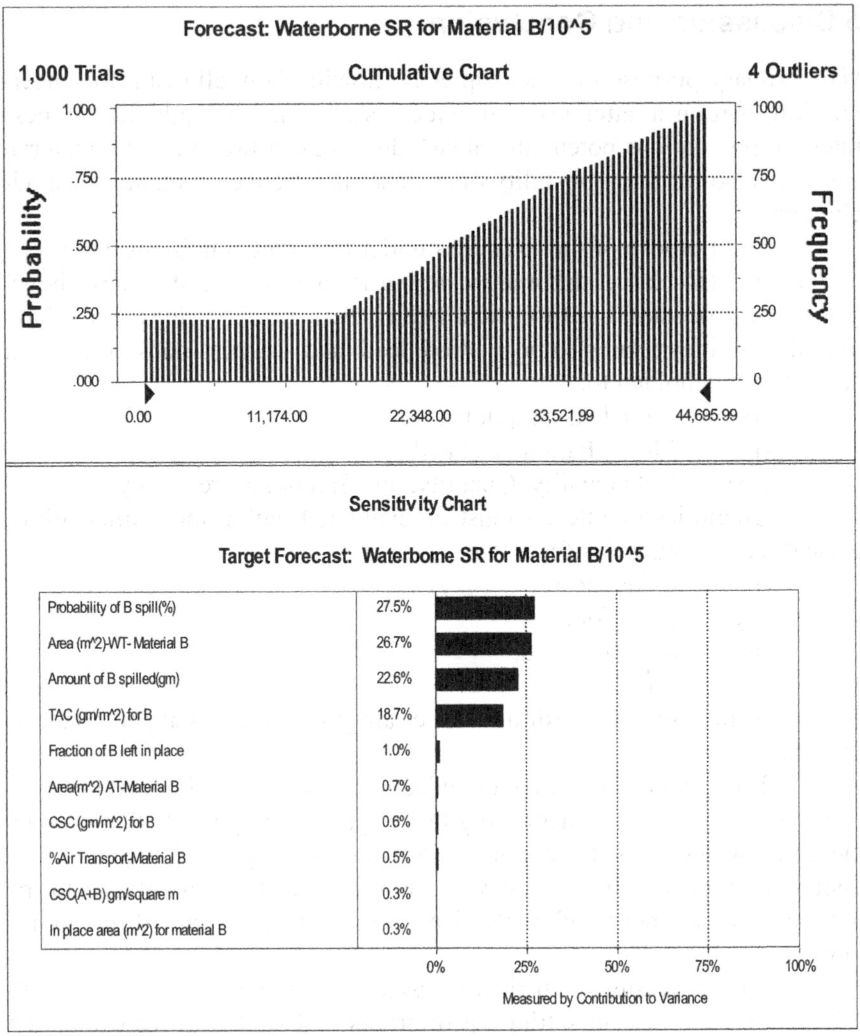

Fig. 9.27. Cumulative probability of material B sickness rate per 100,000 population for waterborne contaminant fractions allowing for uncertainty in parameter values as mentioned in text and when there is no synergistic effect; relative contribution of the uncertain parameters to the variance of the synergistic death rate. These results relate to uniform distribution choices for the uncertain parameters

5 Discussion and Conclusion

\The primary purpose of this chapter is showing how allowing for uncertain information in attempts to provide assessments of death and sickness rates for populations potentially at risk due to exposure to a toxic material spill and also for the probability of such spills enhances contaminant modeling code.

The purpose of the examples under different conditions and transport modes (in-place, airborne, waterborne, and synergistic) has shown that every procedure used to scientifically evaluate the values and behaviors of quantities related to population death and sickness rates assessment are variably beholden to:

(i) Model Assumptions;
(ii) Model Parameters; and,
(iii) Data Quality, Quantity, and Sampling Frequency.

Quantities of interest must be evaluated within the framework of these three dependencies for:

(i) uniqueness;
(ii) precision;
(iii) accuracy; and,
(iv) sensitivity.

so that an idea of the worth and uncertainty of the assessments made is known.

Then, with quantitative knowledge of the degree of uncertainty, efforts can be focused on identifying those system components that provide the greatest degree of uncertainty and using the degree of uncertainty to assign some quantitative measures of risk and relative importance to the individual assessments of likely death and sickness under different regimes.

These measures can then be used in economic and health worth calculations and also in setting up priorities and strategies for evaluating options within a set of scenarios to minimize likely fiscal loss as well as likely death and/or sickness.

The specific examples were tailored to show several major points. First, just because a parameter is uncertain does not necessitate its need for improvement. Indeed, Examples where groups of uncertain parameters did not have influence on outputs of interest have been shown above as well as examples where uncertain parameters had large influence on outputs.

Some of these output uncertainties were insensitive to the type of underlying distributions chosen for the input parameters while others showed a greater degree of relative sensitivity, so in general checking each situation to know when an enhanced understanding of the distribution type appropriate for each input parameter is required.

Some of the advantage to a Monte Carlo calculation type is the ability to determine the relative contribution of each uncertain parameter to the percentage uncertainty of an output. This enables a rapid way to decide which parameters need to be improved if desired. The volatility provides a measure of the relative importance of deciding whether to improve the absolute uncertainty and, once this decision is made, the relative contribution and relative sensitivity resolve which input parameters to improve.

A further advantage provided by the illustrative examples is using as many uncertain outputs and inputs as desired without increasing the basic Monte Carlo run time. This advantage has herein enabled the demonstration of the method for quantitatively handling uncertainty in the relative importance, sensitivity, and contribution of toxic contamination computations.

APPENDIX A

APPROXIMATE RISK MEASURES

Although an effective method for assessing error, a Monte Carlo procedure suffers from the drawback that a series of runs must be performed to assess error range on a physically required parameter. Such a numerical investigation can be extremely time consuming if the basic analysis code is slow and may be causal for limited error determination use. The very fast MS Excel basin modeling code described herein is intended to eliminate this problem. The Monte Carlo procedure is favourably completely general and can be applied to any response equation once the range and underlying probability distributions for each parameter and measurement uncertainty have been provided.

The question addressed is whether a probabilistic assessment of error and uncertainty on quantities of physical interest can be produced without performing the tedious, time-consuming, Monte Carlo calculations for each measurement value with assigned ranges for intrinsic parameters and measurement uncertainty. Clearly, if such a procedure can be developed then it becomes a relatively simple matter to assign error ranges to quantities of relevance.

In order to make full use of the powerful techniques of probability measures, knowledge would be required of the intrinsic probability of obtaining an event. By and large this knowledge is based on data relating to scientific conditions. But data available are often very limited at the assessment stage, usually imprecise, derived by analogy, or dependent on conditions about which little knowledge (as opposed to surmise) is available. For these reasons, approximations and assumptions are introduced in attempts to obtain relatively robust estimates of data-related quantities from which some form of assessment can be made.

For the most part, parameters or variables that are not too well known are usually treated as randomly varying in some manner around a mean value. The random component is customarily represented by a frequency distribution histogram that provides the relative number of times a parameter was observed in a given interval range compared to all interval ranges. Customarily, the area under the histogram is normalized to unity so that the frequency distribution provides an approximate empirical assessment of the probability of occurrence of a parameter in each interval range based on the data.

Often, interest centers not so much on the frequency distribution but on the cumulative frequency distribution. In one case (greater than), the cumulative frequency distribution histogram is usually the residual fractional area still lying beyond a parameter value. In the other case (less than), the measure is the fractional area contained up to the parameter value. This set is mutually exclusive (i.e., the probability of a value >V plus the probability of a value <V must sum to unity). Thus, if p(x)dx measures the frequency distribution (normalized) of occurrence of x in the range x to x + dx, then

$$P(y > x) = \int_{y}^{\infty} p(x)dx \qquad \text{(A1)}$$

measures the cumulative frequency distribution (i.e., the chance of exceeding a particular value y), while

$$P(y<x) = \int_{o}^{y} p(x)dx = \int_{o}^{\infty} p(x)dx - \int_{y}^{\infty} p(x)dx = 1 - P(y>x) \qquad \text{(A2)}$$

measures the chance of not exceeding a particular value y.

Providing the frequency distribution (even in rough form) is often difficult for a large number of circumstances in discussing assessments. For that reason moments of the underlying distribution are often used as approximations. The mean value, $E_1(x)$, of x for a frequency distribution p(x)dx is

$$E_1(x) = \int_{-\infty}^{\infty} xp(x)dx \qquad \text{(A3)}$$

while the mean square value $E_2(x)$ is

$$E_2(x) = \int_{-\infty}^{\infty} x^2 p(x) dx. \tag{A4}$$

The variance, σ^2, around the mean is given by

$$\sigma^2 = E_2(x) - E_1(x)^2 \geq 0 \tag{A5}$$

where σ is the standard deviation.

In many situations, only multiple powers of distributions will be needed, defined by

$$E_j(x) = \int_{-\infty}^{\infty} x^j p(x) dx. \tag{A6}$$

Also, the median value, $x_{1/2}$, of a frequency distribution is defined as that value of x such that

$$P(y < x) = P(y > x) = 1/2. \tag{A7}$$

and the mode x_m (for a unimodal distribution) as that value of x at which p(x) has its maximum value.

The lognormal distribution occurs physically in many situations ranging from the areal size distribution of sunspots to lease sale bid distributions. The normal distribution cannot be appropriate when there is a constraint on a variable (e.g., area cannot be negative- bid values must be positive). Under such conditions, empirical evidence suggests that an approximate measure of cumulative frequency distribution is provided by a lognormal behavior with

$$P(x|x_{1/2},\mu) = \frac{1}{2}\,[1 + \mathrm{erf}(\ln(x/x_{1/2})/2^{1/2}\mu)] \tag{A8}$$

with the mean value of x, $E_1(x)$, given through

$$E_1(x) = x_{1/2} \exp(\mu^2/2), \tag{A9a}$$

the mode value by

$$x_m = x_{1/2} \exp(-\mu), \tag{A9b}$$

and the variance in x, $E_2(x) - E_1(x)^2 = \sigma^2$, given by

$$\sigma^2 = E_1(x)^2[\exp(\mu^2) - 1], \tag{A9c}$$

where $x_{1/2}$ is the median value.

Empirically, at the assessment stage obtaining enough information is often difficult if not impossible to determine the precise shape of the frequency distribution of a particular parameter or variable. Often, a fairly good achievement is to be able to estimate a likely minimum, x_{min}, maximum, x_{max}, and most probable value, x_p, for a parameter.

Performing equations (A9) in reverse to obtain estimates of μ, x, $x_{1/2}$ and x_m is possible if the variable of interest is approximately log normally distributed. Thus from equation (A9c)

$$\mu = [\ \ln \{1 + \sigma^2/E_1(x)^2\}]^{1/2} \tag{A10a}$$

and

$$x_{1/2} = E_1(x)\exp(-\mu^2/2) = E_1(x)[1 + \sigma^2/E_1(x)^2]^{-1/2} < E_1(x) \tag{A10b}$$

and, from equation (A9b),

$$x_m = x_{1/2} \exp(-\mu). \tag{A10c}$$

while

$$x_\sigma = x_{1/2} \exp[(\ln(1 + \sigma^2/E_1(x)^2))^{1/2}] > x_{1/2} \tag{A10d}$$

Then, from equations (A10) one has the estimate

$$\mu^2 = \ln(E_2/E_1^2) \tag{A11}$$

Clearly, some assessment is called for the manner in which each estimate is made, the sensitivity of output results to changes in assumptions, parameter values, and the quality, quantity, and sampling frequency of control information. Confidence in any individual estimate can only occur with this analysis.

The important points about a cumulative log-normal cumulative distribution are: 1) that the expected value, E(x), of a quantity x occurs close to 68% (i.e., a 2/3 chance of the actual value being less than or equal to E(x)); 2) that the 10% and 90% cumulative probability values are stable measures of the range within which an exactly log-normal distribution satisfies as best it can a set of histogram values (Feller 1957). Accordingly, the 10% and 90% cumulative probability points are often used to provide a measure of "volatility" of an estimate of a quantity, x.

A volatility measure, v, is often written (Warren, 1978) as

$$v = [P(90) - P(10)]/P(68) \tag{A12}$$

which provides an estimate of uncertainty in the P(68) value. This volatility factor is extremely useful: if the volatility is small compared to unity then the estimate is fairly accurate because the P(90) and P(10) values (which bracket the P(68) value) are then close, so the fractional error in P(68) is small; if the volatility is large compared to unity then a large fractional error exists in the P(68) estimate, and that estimate is then not as reliable as one with a low volatility.

Two simple calculations can be performed on a set of categories. Suppose N categories are being investigated each with its expected value $E(x_i)$, corresponding to P(68), and each with its volatility, v_i.

Then a risk-weighted measure can be made from each category as

$$R_i = (E(x_i)/v_i) \Big/ \sum_{i=1}^{N} (1/v_i) \tag{A13}$$

with the total as

$$R = \sum_{i=1}^{N} R_i \tag{A14}$$

which weights the contribution of each category to the total according to its relative volatility.

The second simple calculation that can be made is the risk-weighted relative importance, RI(i) (in %) of the ith category in relation to the other categories, with

$$RI(i) = 100\,(E(x_i)/v_i) \Big/ \left(\sum_{i=1}^{N} E(x_i)/v_i \right) \tag{A15}$$

This relative importance factor measures the likelihood of a particular category contributing compared to all categories. As such, RI is a useful measure of where most effort should be placed to enhance the likelihood of realizing an improvement in death and sickness assessments with allowance made for the speculative nature of individual categories.

APPENDIX B

SOME PROPERTIES OF A LOG NORMAL DISTRIBUTION

a. Exact Statements

Consider the lognormal probability distribution

$$p(x|a, \mu)\, dx \;\alpha\; \exp\,[-(\ln x-a)^2/2\mu^2]\, dx/x \qquad \text{(B1)}$$

where a, μ are fixed parameters and the variable x ranges in $0 \leq x \leq \infty$; and, where $p(x|a,\mu)$ is the differential probability of finding the value x in the range x to x+dx.

With the normalization

$$\int_0^\infty p(x|a,\mu)\, dx = 1 \qquad \text{(B2)}$$

then

$$p(x|a,b)\, dx = \frac{1}{2}\, \pi^{-1/2}\, \mu^{-1}\, \exp\,[-(\ln x-a)^2/2\mu^2]\, dx/x \qquad \text{(B3)}$$

The mean value of x^n, $E_n(x)$, is given by

$$E_n(x) = \int_0^\infty x^n p(x|a,\mu)\, dx = \exp\,(an + n^2\mu^2/2) \qquad \text{(B4)}$$

Thus:

$$E_1(x) = \exp(a+\mu^2/2) \quad \text{(B5a)}$$

$$E_2(x) = \exp(2a+2\mu^2) \quad \text{(B5b)}$$

Hence

$$\mu^2 = \ln (E_2(x)/E_1(x)^2) \quad \text{(B6)}$$

Then also the value $x_{1/2}$ at which

$$\int_0^{x_{1/2}} p(x|a,\mu)\, dx = 0.5 \quad \text{(B7)}$$

is given by

$$a = \ln x_{1/2} \quad \text{(B8)}$$

so that

$$E_1(x) = x_{1/2} \exp (\mu^2/2) > x_{1/2} \quad \text{(B9)}$$

Consider the cumulative probability

$$P(x|a,\mu) = \int_0^x p(u|a,\mu)\, du \quad \text{(B10)}$$

that provides the probability of obtaining a value less than or equal to x. On $x_1 = x_{1/2} \exp(\mu)$, $P(x_1|a,\mu)$ is given through

$$P(x_1|a,\mu) = \frac{1}{2} + (2\pi)^{-1/2} \int_0^{2^{-1/2}} \exp(-s^2)\, ds \cong 0.84 = P(84) \quad \text{(B11a)}$$

that is independent of a and μ; while on $x_2 = x_{1/2} \exp(-\mu)$, $P(x_2|a,\mu)$ is given through

$$P(x_2|a,\mu) = 1-P(x_1|a,\mu) \cong 0.16 = P(16) \quad \text{(B11b)}$$

that is also independent of a and μ.
Thus,

for P(16)	$x = E_1(x)\exp(-\mu - \mu^2/2) < x_{1/2}$	(B12a)
for P(84)	$x = E_1(x)\exp(\mu - \mu^2/2) > x_{1/2}$	(B12b)
for P(50)	$x = x_{1/2} = E_1(x)\exp(-\mu^2/2) < E_1(x)$	(B12c)

Note that for $\mu < 2$ (i.e. $E_2(x) < e^4 E_1(x)^2$), $E_1(x)$ occurs between P(50) and P(84) while for $\mu > 2$, $E_1(x)$ occurs at greater than P(84). For $\mu \lesssim 2$, a good pragmatic approximation is for $x = E_1(x)$, $P(E_1(x)|a,\mu)$ occurs at approximately 68% (i.e., approximately midway between P(50) and P(84)). A slightly better approximation is $E_1(x)$ occurs at approximately $(50 + 17\mu)$ %.

b. Approximate Statements

Empirically, obtaining enough information to determine the precise frequency distribution shape of a particular parameter or variable is often difficult if not impossible. Often, a fairly good achievement is to be able to estimate a likely minimum, x_{min}, maximum, x_{max}, and most probable value, x_p, for a parameter. An approximate idea of relevant mean and variance can then be obtained from Simpson's triangular rule. Therefore,

$$E_1(x) \cong \frac{1}{3}(x_{min} + x_p + x_{max}) \quad \text{(B13a)}$$

$$E_2(x) = E_1(x)^2 + \sigma^2. \quad \text{(B13b)}$$

with

$$\sigma^2 \cong \frac{1}{2} E_1(x)^2 - \frac{1}{6}[x_{min} x_{max} + x_p (x_{min} + x_{max})]$$
$$= \frac{1}{18}\left\{[x_p - \frac{1}{2}(x_{min}+x_{max})]^2 + \frac{3}{4}(x_{max}-x_{min})^2\right\} \quad \text{(B13c)}$$

Further, assuming that the variable is approximately log normally distributed, then

$$\mu \cong \{\ln [1 + \sigma^2/E_1(x)^2]\}^{1/2} \tag{B14}$$

so that an equivalent log normal approximate distribution can be constructed using equations (B12) together with estimates of $E_1(x)$ and μ from equations (B13a) and (B14), respectively.

c. Multiple Parameter Distributions

When assessing economic and health objectives as well as basic scientific analysis outputs, many parameters occur alone or in combination with other parameters so that each parameter has its own uncertainty. Practical procedures are needed for estimating the combined effects of parameter uncertainty for a toxic death and sickness project.

Two types of fundamental parameter combinations seem to be prevalent: 1) sums of parameters; and, 2) products or rates of parameters.

(i) Sums of Parameters

Two or more independent random variables A and B both with n-ormal distributions combine to give a sum (A±B) that is also precisely normally distributed with mean value

$$E_1(A \pm B) = E_1(A) \pm E_1(B) \tag{B15a}$$

and variance

$$\sigma(A \pm B)^2 = \sigma(A)^2 + \sigma(B)^2 \tag{B15b}$$

Empirically, N independent random variables from any frequency distributions (not necessarily normally distributed) appear to add to yield a sum $S_N(+ x_1 \pm x_2 \pm x_3 \pm \ldots\ldots \pm x_N)$ that is approximately normally distributed as N becomes large, with mean value

$$E_1(S_N) \cong E_1(x_1) \pm E_1(x_2) \pm \ldots\ldots\ldots \pm E_1(x_N) \tag{B16a}$$

and variance

$$\sigma(S_N)^2 = \sum_{i=1}^{N} \sigma(x_i)^2 \tag{B16b}$$

(ii) Products of Parameters

Multiple independent random variables X, Y, Z, ... from log-normal distributions combine in generic product form $X^aY^bZ^c$ .... to provide a product distribution that is also precisely log-normally distributed with mean value

$$E_1(X^aY^bZ^c \,.......) = E_1(X^a)\, E_1(Y^b)\, E_1(Z^c)....... \qquad \text{(B17a)}$$

second moment

$$E_2(X^aY^bZ^c \,.......) = E_2(X^a)\, E_2(Y^b)\, E_2(Z^c)....... \qquad \text{(B17b)}$$

and scale factor μ given through

$$\mu^2 = \ln\,[E_2(X^aY^bZ^c \,.......) \,/\, E_1(X^aY^bZ^c \,.......)^2] \qquad \text{(B17c)}$$

Empirically, N independent random variables from any frequency distributions (not necessarily log-normally distributed) tend to combine to produce a product $P_N\,(+X_1^a\, X_2^b\, X_3^c \,......X_N^d\,)$ that is approximately lognormally distributed as N becomes large, with mean value

$$E_1(P_N) \cong E_1(X_1^a\,)\, E_1(X_2^b\,) \,........., \qquad \text{(B18a)}$$

and scale parameter

$$\mu^2 \cong \ln\,[E_2(P_N)/E_1(P_N)^2] = \sum_{i=1}^{N} \mu_i^2 \qquad \text{(B18b)}$$

where

$$\mu_i^2 = \ln\,[E_2(X_i^p\,)/E_1(X_i^p\,)^2] \qquad \text{(B18c)}$$

APPENDIX C

RELATIVE IMPORTANCE, RELATIVE CONTRIBUTION, AND RELATIVE SENSITIVITY DEFINITIONS

Because of parameter uncertainty, a useful approach to analyze the system's uncertainty behavior is to treat the parameters as random variables that follow some probability distribution (not necessarily the same distribution for each variable). The distributions commonly used are lognormal, exponential, uniform, and triangular with the mean, variance, and range of each parameter defined. The first two distributions are often chosen because they often fit data at several sites. The uniform distribution is often selected because commonly only the range of a parameter is known. The triangular is often selected because in addition to the minimum and maximum values, information about the most commonly occurring value is known. By selecting a value for each location from a specific distribution –i- for the parameter set, the properties can be defined throughout and a solution obtained for the output variable of interest, labeled as ψ.

Repeating the process N times (a series of N-Monte Carlo computations), N profiles of ψ can be created. The N values of ψ can be averaged to obtain the mean and variance σ^2 of ψ that apply to this point for a specific distribution i:

$$< \psi >_i = \frac{1}{N}\sum_{j=1}^{N} \psi_j$$

$$\sigma_i^2 = \frac{1}{N-1}\sum_{j=1}^{N}(\psi_j - < \psi >_i)^2 . \qquad \text{(C1)}$$

At each point, the global mean $<\psi>_G$, the arithmetic mean of the expected values from the M probability distributions chosen, the global variance $< \sigma^2 >_G$, and the arithmetic mean of the variances obtained from each distribution can be calculated:

$$< \psi >_G = \frac{1}{M}\sum_{i=1}^{M} < \psi >_i$$

$$< \sigma^2 >_G = \frac{1}{M}\sum_{i=1}^{M} \sigma_i^2 . \qquad \text{(C2)}$$

By calculating the quantity $< \rho^2 >_T$,

$$<\rho^2>_T = \frac{1}{M}\sum_{i=1}^{M}\left(<\psi>_i - <\psi>_G\right)^2, \tag{C3}$$

which is the variance of the means of the distributions from the global mean, the total uncertainty at each point can be obtained:

$$\sigma_T^2 = <\rho^2>_T + <\sigma^2>_G. \tag{C4}$$

Here $<\rho^2>_T$ is a measure of the uncertainty in the mean ψ-behavior because of the uncertainty in the type of distribution, and $<\sigma^2>_G$ is the average fluctuation around the mean ψ-behavior irrespective of distribution.

For every point examined, the relative contribution of each distribution i toward the global mean can be expressed as:

$$RC_p(i) = \frac{\left(<\psi>_i - <\psi>_G\right)^2}{\sum_{m=1}^{M}\left(<\psi>_m - <\psi>_G\right)^2} \tag{C5}$$

the relative importance of each distribution i toward the average variance can also be calculated:

$$RC_{\sigma^2}(i) = \frac{\sigma_i^2}{\sum_{m=1}^{M}\sigma_m^2}. \tag{C6}$$

Finally by calculating the ratios

$$\frac{<\rho^2>_T}{\sigma_T^2} \quad \text{and} \quad \frac{<\sigma^2>_G}{\sigma_T^2} \tag{C7}$$

the degree that total uncertainty is dominated by the lack of knowledge in the type of distribution or by fluctuations around the mean values can be evaluated. A large first ratio value indicates the choice of probability distribution model is critical in total uncertainty and, hence, more data are needed for a clear determination of the distribution shape. In contrast, a large second ratio value indicates that fluctuations around the mean behavior of ψ are dominating the total system uncertainty; therefore, the parameter ranges need to be defined more sharply.

Now hold all but one of the variables at their mean values and vary the last according to a distribution i with mean, variance, range given for that parameter. By performing Monte Carlo simulations one obtains $<\psi>_{i,k}$, the mean, and $\sigma_{i,k}^2$, the variance, due to fluctuations in the k-th random parameter according to an i-th distribution. By repeating the procedure for all R of the parameters one evaluates the relative importance towards the mean of each parameter k for every distribution i:

$$\mathrm{RI}_{i,k}^{<\psi>} = \frac{<\psi>_{i,k}}{\sum_{m=1}^{M} <\psi>_{i,m}} \tag{C8}$$

as well as the relative importance toward the variance of each parameter k and distribution i:

$$\mathrm{RI}_{i,k}^{\sigma^2} = \frac{\sigma_{i,k}^2}{\sum_{m=1}^{M} \sigma_{i,m}^2}. \tag{C9}$$

Clearly, this process can be repeated for all distributions (i= 1…N) and then, for each parameter k, one can calculate the relative importance toward the mean:

$$\mathrm{RIC}_{k}^{<\psi>} = \frac{\sum_{i=1}^{N} <\psi>_{i,k}}{\sum_{i=1}^{N} \sum_{m=1}^{M} <\psi>_{i,m}}, \tag{C10}$$

and the variance:

$$\mathrm{RI}_{i,k}^{\sigma^2} = \frac{\sum_{i=1}^{N} \sigma_{i,k}^2}{\sum_{i=1}^{N} \sum_{m=1}^{M} \sigma_{i,m}^2}, \tag{C11}$$

irrespective of distribution. Thus, the above analysis can provide a ranking of the importance of each parameter in the evaluation of the mean and variance of an output for a specific distribution and also irrespective of the choice of distribution.

The RS can be defined equivalently. Each parameter has a range as well as a central value. Now if either, or both, of the upper or lower ends of the range is changed then so, too, are the contributions of that parameter to both the average value as well contributing to the uncertainty of each particular output variable for a given choice of distribution of that parameter. The simplest way to handle this problem is to allow each end-point of the range of a parameter to also be a random variable centered on its own mean value. By performing Monte Carlo simulations with these "new" variables, then determines how sensitive the mean value of an output is to the uncertainty in the end-points of a range choice for each parameter. In a sense one repeats computation of the measures of RI and RC just given, but now including the extra variables of end-point variations. In this way one determines how sensitive the output and its uncertainty are to the sensitivity of the range end-point uncertainty choices in comparison to the uncertainties brought about by the incomplete knowledge of all other parameters in the calculation.

APPENDIX D
EXCEL PROGRAM FOR TOXIC DEATH AND SICKNESS CALCULATIONS

Properties of Contaminants

		TAC		CSC	
Name	Material #	Terminal Activity Concentration(gm/m^2)		Critical Sickness Concentration (gm/m^2)	
A	1	100	Comments columns	0.1	NOTE:CSC entries are usually
B	2	200		0.2	less than those in the TAC
C	3	300		0.3	columns
D	4	400		0.4	
E	5	50		0.5	
F	6	10		0.0006	
G	7	900		0.9	
H	8	1300		13	
J	9	25		0.025	
K	10	71		1	

Event Conditions of Primary Contaminant Spill

Name	Material #	Amount spilled (gm)	Probabilityofspill(%)	Fraction left in place	In-place area(m^2)	Fraction Transported
A	1	10000	25	0.1	100	0.9
B	2	50000	30	0.2	1000	0.8
C	3	3000	19	0.3	250	0.7
D	4	5000	12	0.01	700	0.99
E	5	7300	0.1	0.9	25	0.1
F	6	12	17	0.02	30000	0.98
G	7	19	1	0.3	450	0.7
H	8	83000	0.001	0.7	2200	0.3
J	9	96000	0.0001	0.3	195	0.7
K	10	1470	1.2	0.99	27	0.01

Direct Death Rates/100,000 Population in Contaminated Areas (excludes spill probability)

Name	Material#	In-place Death Rate	Airborne Death Rate	WaterborneDeathRate
A	1	0	0	0
B	2	0	0	0
C	3	0	0	0
D	4	0	0	0
E	5	100000	0	0
F	6	0	0	0
G	7	0	0	0
H	8	0	0	0
J	9	100000	100000	100000
K	10	0	0	0

Direct Sickness Rate(SR)/100,000 Population in Contaminated Areas (excludes spill probablity)

Name	Material#	In-place SR	Airborne SR	Waterborne SR
A	1	100000	100000	100000
B	2	100000	100000	100000
C	3	100000	0	100000
D	4	0	100000	100000
E	5	0	0	0
F	6	0	100000	100000
G	7	0	100000	0
H	8	100000	0	0
J	9	0	0	0
K	10	100000	0	0

Likely Direct Sickness Rate per 100,000 population (includes spill probabilities)

Name	Material#	In-place SR	Airborne SR	Waterborne SR
A	1	25000	25000	25000
B	2	30000	30000	30000
C	3	19000	0	0
D	4	0	12000	12000
E	5	0	0	0
F	6	0	17000	17000
G	7	0	1000	1000
H	8	1	0	0
J	9	0	0	0
K	10	1200	0	0

	Event Conditions of Transport				
Name	Material #	%Air Transport(AT)	Area(m^2) by AT	%Water Transp(WT)	Area (m^2) by WT
A	1	35	400	65	930
B	2	12	900	88	270
C	3	18	2500	82	4500
D	4	24	1700	76	2700
E	5	23	2100	77	1300
F	6	46	200	54	9500
G	7	91	10	9	2100
H	8	100	30000	0	350
J	9	73	1600	27	27.39
K	10	94	70000	6	120500

Chapter 10

Quantitative Risks of Death and Sickness from Toxic Contamination: Age-Dependent Toxic Sickness/Death Exposure Limits

Summary

This chapter explores the likely death and sickness rates caused by release of a toxic contaminant with a resistance to sickness that is age dependent. In addition, up to ten different sicknesses can be allowed for with each released contaminant. The transport of portions of the released contaminant can be by airborne or waterborne methods, leading to different contaminant levels and so different death and sickness rates. Gender specific sicknesses, such as prostate cancer or ovarian cancer, can also be included, making the quantitative procedure developed here of very broad use. The influence of uncertainty in any and all of the parameters having to do with either the age dependence model for sickness and/or death rates, or having to do with the age distribution model of the population, or having to do with the spill and transport of toxic contaminant, can easily be handled using a risking program, such as Crystal Ball ™, in conjunction with the basic death and sickness rate program. Several numerical illustrations are given to show how one can assess likely death and sickness rates under all such effects.

1 Introduction

The influence of toxic materials (chemical, biological, radioactive, and physical) on the health and survivability of mankind is one of the most pressing problems that modern society has to face. Many environmental consequences follow from the release (either accidental or with purpose) of toxic contaminants, but here we restrict the discussion to the effects on humankind and do not consider the toxic effects on the land, sea and air quality for plants, mammals, fish etc.

The main theme of any contaminant release is the deleterious effects such release can produce in humans in terms of long-term sickness and/or death. Each release of material has four basic possibilities for causing illness and/or death. First, the released material may remain in place at the release site and so cause death and/or sickness from the concentration at such a site; second a fraction of the released material may be transported away from the release site by water transport (ground water, rivers, rain run off) and so cause death and/or sickness over a wider spread area than the local effects at the contaminant release site; third a fraction of the released material may be transported by air transport and so be strewn over an area determined by wind currents and eventual deposition of the material. Fourth, there can be a synergistic effect at all three locations (in-place, water transport, air transport) in which released material can interact with either previously released material or with naturally occurring material to produce a more virulent toxic behavior or a less toxic behavior, thereby changing the death and sickness rates that would occur with just a single sort of released material. Considerations related to death and/or sickness involving the above points have been presented in the previous chapter in terms of Terminal Activity Concentrations (TAC) and Critical Sickness Concentrations (CSC) for each material released and or synergistically active.

However, not considered in Chapter 9 were the types of death and/or sickness that can result from release of toxic materials. For instance, release of radioactive material (as in the case of Chernobyl or Windscale) can cause leukaemia, bone marrow cancers, brain tumors, direct radiation burns, etc. Or the release of chemical pollutants can lead to lung cancer, to ovarian cancers, to kidney and liver problems, etc. Thus the types of sicknesses and death modes are very much dependent on precisely what type of material is released and the multiple potential types of death and sickness each such single material can cause. In addition, the TAC and CSC values for each type of illness (whether it is terminal or not) are often dependent on the ages and sexes of the population fraction affected by such toxic materials. For instance, the release of ^{90}Sr from the Windscale reactor had a disproportionate influence on babies and small children because of the fact that the ^{90}Sr was spread by air transport to grass where it was eaten by cows and concentrated in the produced milk that, in turn, was drunk more by babies than adults. Thus one must allow for the impact of different TAC and CSC values on different population fractions. The purpose of this chapter is to show how to take these considerations into account when investigating the death and/or sickness rates of a population caused by toxic contaminants.

2 Specific TAC and CSC Effects for One Toxic Material/Multiple Illnesses.

Consider the release of just one toxic material but with the capability of producing a multiple set of sickness types in individuals that may also lead to death by various paths. The reason for considering just one toxic material is that many toxic components can then be handled as the superposition of several individual toxic materials. If synergistic effects were also to occur, then one can regard the synergistically produced toxic material as a new "release" and so consider it as a single toxic contaminant in its own rite that could also be transported further by air or water mechanisms. Thus the basic methodology to be developed in this section of the chapter also applies to synergistically produced toxic materials.

For the single toxic material released we assume that it is capable of causing up to ten possible sicknesses and/or death modes. For each of these modes one has to specify a TAC and a CSC as shown in figure 10.1. If there is no effect then one sets the TAC and/or the CSC to high values so that the release of the material will not then trigger an effect of the particular sickness. Figures 10.2a and 10.2b show charts of the TAC and CSC values for each sickness so that one can immediately compare how toxic each is in relation to the remaining sicknesses.

	Properties of Contaminants				
		TAC		CSC	
Name	Sickness #	Terminal Activity Concentration(gm/m^2)		Critical Sickness Concentration (gm/m^2)	
A	1	10	Comments columns	9	NOTE:CSC entries are usually
B	2	9		8	less than those in the TAC
C	3	8		7	columns
D	4	7		6	
E	5	6		5	
F	6	5		0.6	
G	7	5.5		0.9	
H	8	6.5		1.3	
J	9	7.5		0.25	
K	10	8.5		1	

Fig. 10.1. Terminal Activity Concentrations (TAC) and Critical Sickness Concentrations (CSC) for ten sicknesses that can be caused by release of a single type of toxic material

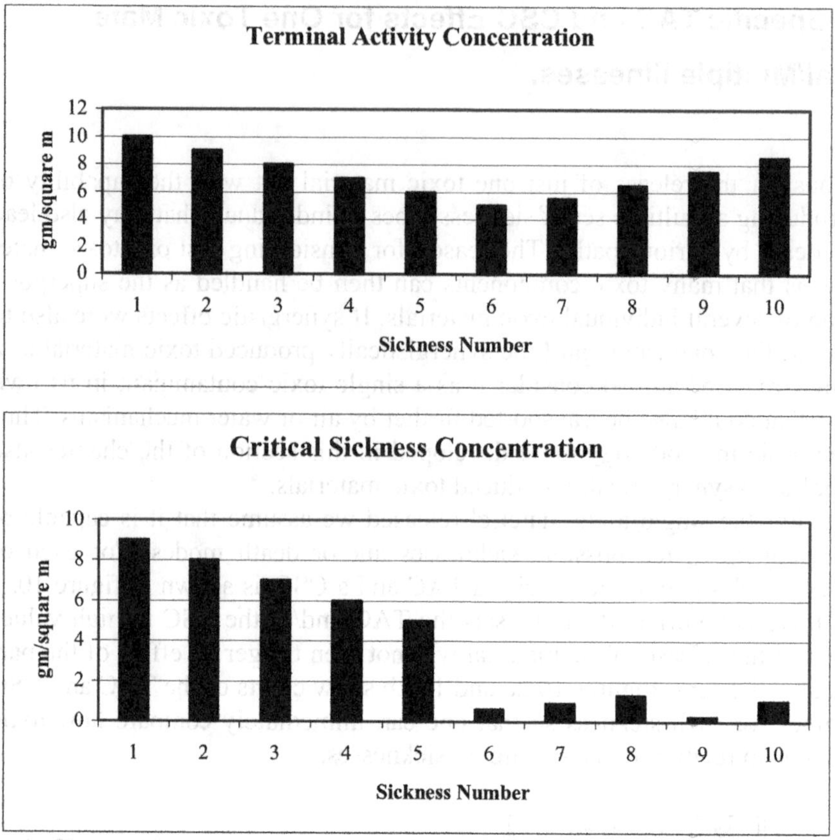

Fig. 10.2. Plots of the TAC and CSC for each of the ten sicknesses caused by a single toxic material

As was done in chapter 9, one now gives the probability of a toxic material release, the amount of material released, the fraction transported by air, and the fraction transported by water, together with the areas covered by the residual in-place material, by the water transported component, and by the air transported component. In this way one has the physical concentrations of each of the three parts: the in-place, the air transported fraction, and the water transported fraction. And these three concentrations can be given including or excluding the probability of contaminant release. Such a grouping of results is shown in Figures 10.3a-c based on the fractions transported by air and water and the areas covered by the air and water transport as shown in figure 10.4a-d that are, in turn, based on the total released amount, the probability of a release and the in-place area covered by the contaminant remaining in-place, as shown in figure 10.5a-c.

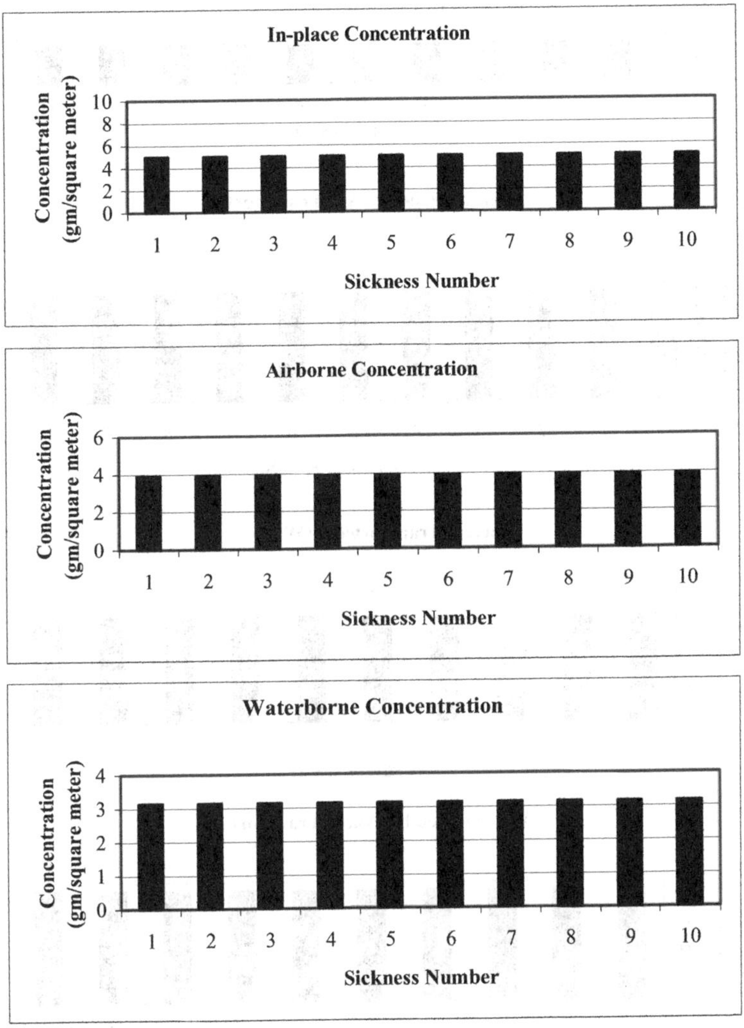

Fig. 10.3 Plots of in-place, airborne, and waterborne concentrations of a toxic material spilled at the in-place site

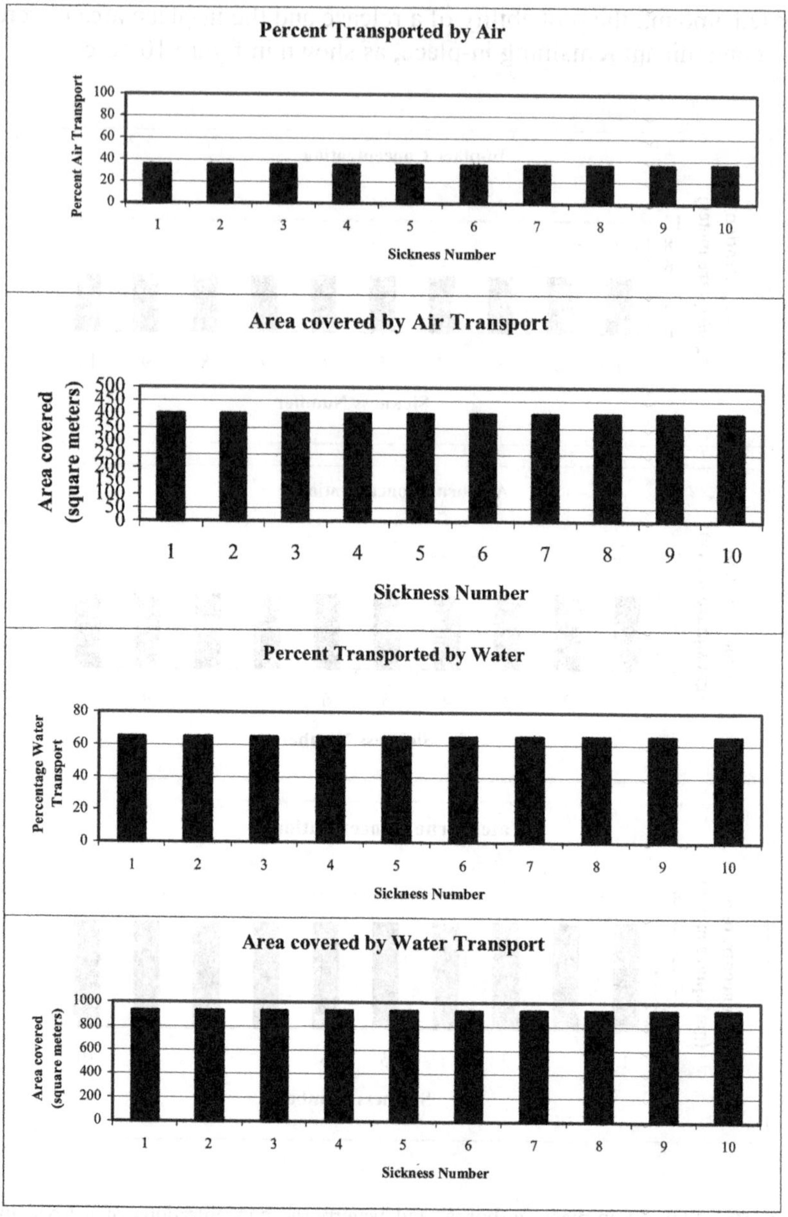

Fig. 10.4. a. Percentage of toxic material transported by air; b. Area covered by the air transported toxic material; c. Percentage of the toxic material transported by water; d. Area covered by the water transported material

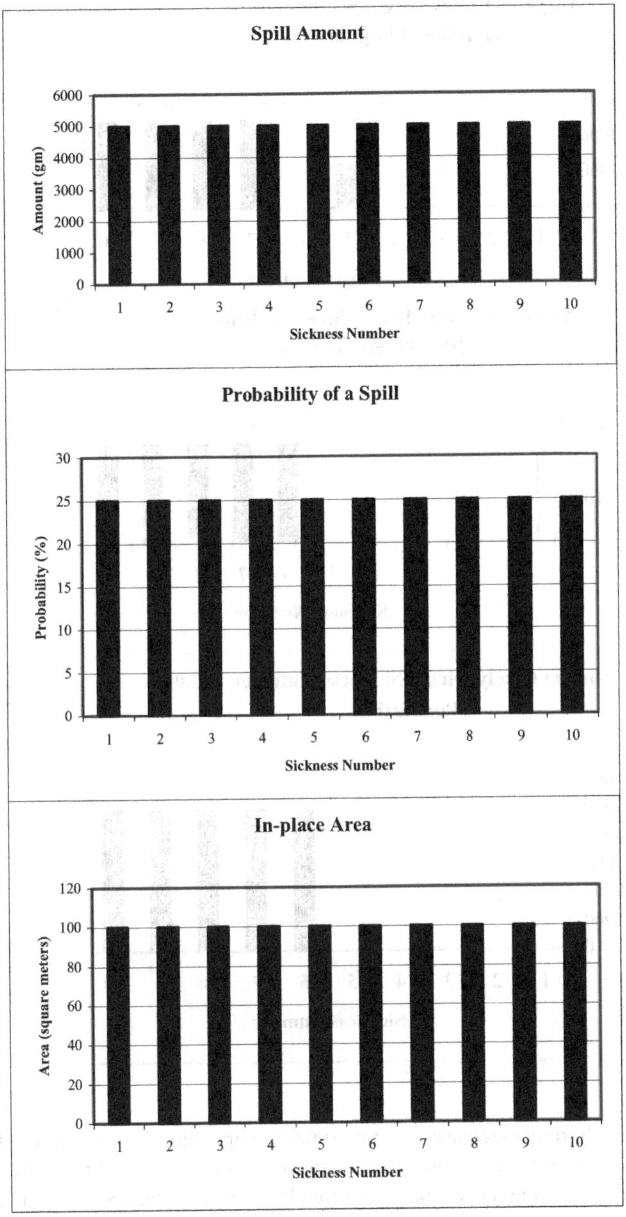

Fig. 10.5. a. Spill amount of toxic material; b. Probability of a spill; c. In-place area covered by the spilled material that is not transported by air or water

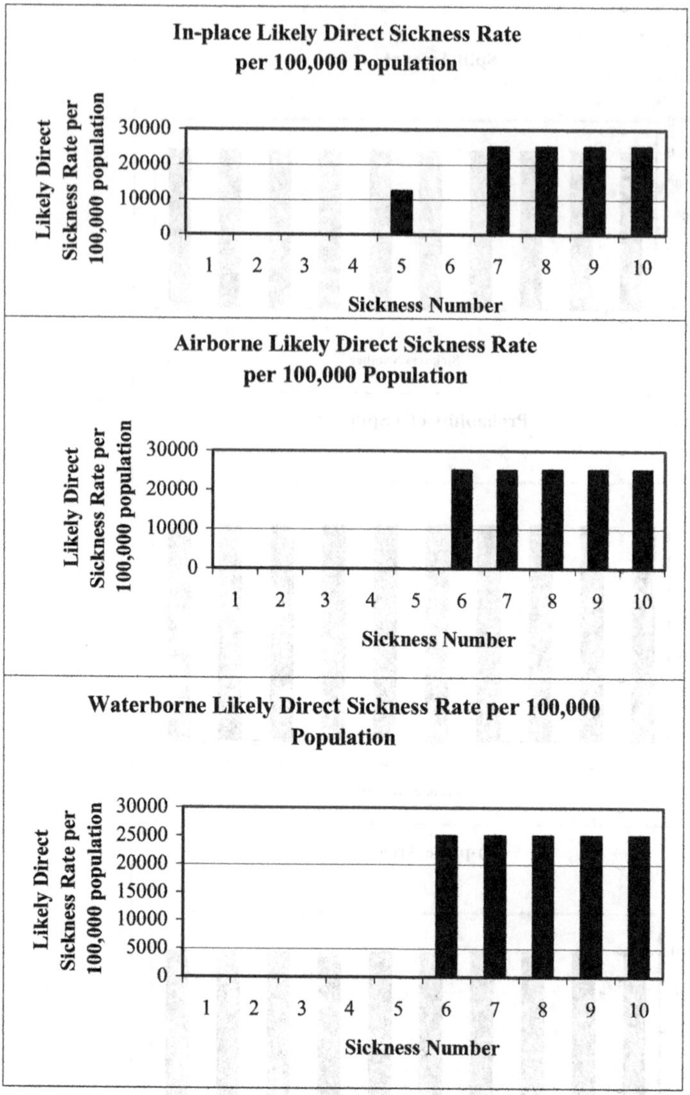

Fig. 10.6. a. In-place likely direct sickness rate per 100,000 population for each of the ten sicknesses caused by release of the toxic material; b. Airborne likely direct sickness rate per 100,000 population for each of the ten sicknesses caused by release of the toxic material; c. Waterborne likely direct sickness rate per 100,000 population for each of the ten sicknesses caused by release of the toxic material

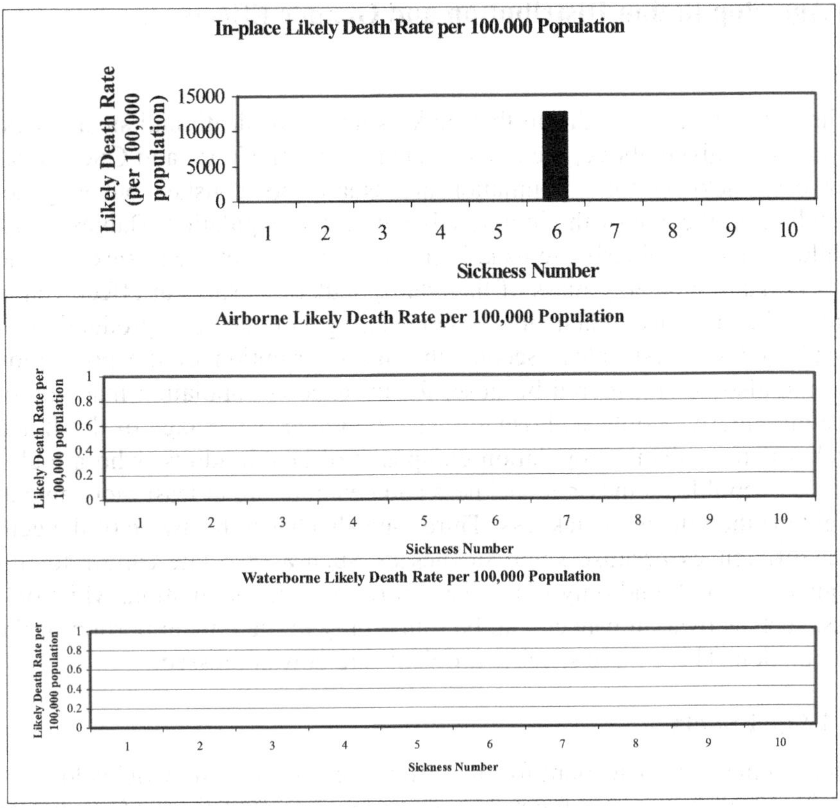

Fig. 10.7. a. In-place likely death rate per 100,000 population for each of the ten sicknesses caused by release of the toxic material; b. Airborne likely death rate per 100,000 population for each of the ten sicknesses caused by release of the toxic material; c. Waterborne likely death rate per 100,000 population for each of the ten sicknesses caused by release of the toxic material

Relative to the TAC and CSC for each of the sickness and death modes given in Figure 10.1, one can then compute the likely sickness and death rates per 100,000 population. These results are shown in figure 10.6a-c for likely sickness rates, and in figure 10.7a-c for likely death rates for each of ten sicknesses, labelled 1-10. To be noted in this simple example is that sickness number 6 is quite deadly with a likely death rate of 12,500 per 100,000 population. In addition, the sickness rates for several of the other illnesses are well above the critical limits set by the corresponding CSC values.

3 Age, Population Distribution and Gender Effects

Three major factors indicate that bulk estimates of death and sickness estimates, as given above, are not sufficient. First, the TAC and CSC values for each sickness for a population are usually not constant but vary depending on the age of the individuals within the population. Babies, small children and the elderly are usually much more susceptible to sickness and death than the healthy mass of the adult population. Such an effect should surely be included when one is considering contaminant production of death and sickness rates. Second the age distribution of the population should also be considered because, for instance, a population made up of mainly elderly people will usually have a greater percentage of death and sickness rates than a population composed of health adults, who are also less susceptible en masse to sickness and death, or are at least more able to fight off the effects of sickness. Third, one also has to be aware of the gender differences because some sicknesses, such as prostate cancer for instance, can be found only in the male portion of the population, while others, such as ovarian cancer, can be found only in the female portion of the population. These three specific problems are now addressed.

3.1 Age Effects

One simple way to account for the differences in TAC and CSC values for each sort of sickness as functions of age, A, is to split the TAC values as follows:

$$TAC(young) = TAC1+(TAC2-TAC1)*(A/A1) \text{ for } 0<A<A1;$$
$$TAC(adult) = TAC2+(TAC3-TAC2)(A-A1)/(A2-A1) \text{ for } A1<A<A2;$$
$$TAC(elderly) = TAC3exp(-scale*(A/A2-1)) \text{ for } A>A2.$$

The ages A1, A2, and A3 represent the young group, the healthy adult group, and the elderly group, respectively, while TAC1, TAC2 and TAC2 represent fixed scaling TAC values for each age group. The functional form of this TAC dependence on age is exhibited in figure 10.8. The TAC at birth (TAC1) is taken to increase with age (linearly) representing the increased resistance of growing young people to diseases and sicknesses; after reaching adulthood (at a TAC value of TAC2) it is taken that the resistance to sickness and/or death varies (also linearly with age) until one reaches the elderly age group (at a TAC value of TAC3); thereafter the TAC is taken to decrease exponentially with age, reflecting the lessened resistance of elderly people to stave off sickness and/or death. The exponential scaling at which the decay sets in can be adjusted through the value "scale" in the equations above so that one can treat with hardy groups of elderly people as well as with more susceptible groups. One could further refine the TAC values into narrower divisions if it is considered appropriate, but for pedagogical purposes it is sufficient to choose just three groups.

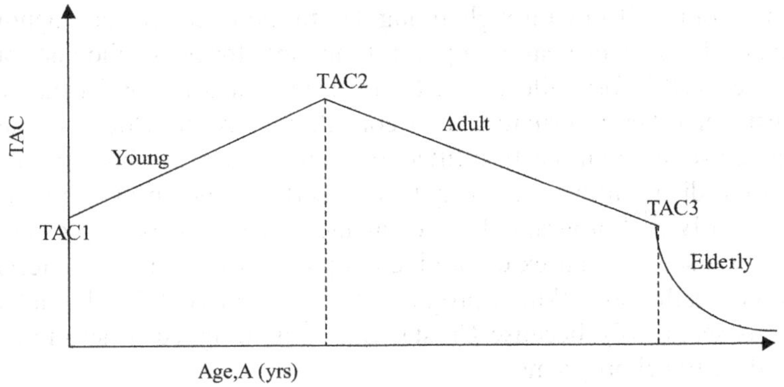

Fig. 10.8. Sketch of the population distribution dependence on TAC values as a function of age

Because the CSC values in each group must be less than the TAC values, a simple way to accomplish this demand is to set CSC1= f_1 TAC(young); CSC2=f_2TAC(adult) and CSC3 = f_3TAC(elderly), where f_1, f_2 and f_3 are all fractions less than unity and which may differ for each sickness being considered. In this case one also has to specify the age group (young, adult or elderly) of the population being considered in order to determine the likely death and sickness rates per 100,000 population of that age group.

We now combine these age dependent TAC and CSC effects with the population distribution before we provide numerical illustrations.

3.2 Population Age Distribution Effects

The distribution with age of a population at risk from a contaminant spill is a critical ingredient in assessing the potential impact in terms of death and sickness rates for the possible sicknesses that could occur. This criticality arises because of the different TAC and CSC values for different segments of the population for each sickness. Thus a population biased towards the young or elderly is likely to be more at risk in terms of percentages of the population that will die or become critically sick as a consequence of the release of toxic material.

This particular aspect of the general problem is most easily addressed using Crystal Ball™ in conjunction with the main toxic program because Crystal Ball™ is designed to handle precisely the problem of different parameter distributions (in this case the age distribution of a population). There are twelve choices possible in Crystal Ball™ for the distribution, ranging from uniform through triangular to exponential and lognormal. Eleven of the choices deal with preset functional forms for the distribution while the twelfth form allows one to construct a distribution based on data or whatever other information one considers relevant. One also has the choice to use distributions best fitted to a suite of data. In short, for a given population distribution one can quickly insert the known distribution into the Crystal Ball™ program. The advantage so obtained is now clear. One can rapidly see to what extent a given population is at risk by interfacing the basic death and sickness program with Crystal Ball™. This interface proceeds seamlessly because Crystal Ball™ is designed to so mesh with any and all Excel programs.

One can also very quickly determine the fraction of the population at highest risk (young, adult, or elderly) to a particular contaminant release and disease because all one has to do is to set the TAC and CSC values extremely high for two fractions of the population, while retaining the population age distribution, and then only the third fraction of the population will be at risk. By repeating this procedure for each fraction in turn one can, with just three runs of Crystal Ball ™, determine the relative "at risk" components of each population.

Another advantage of using Crystal Ball™ is that one can also allow any and all of the TAC and CSC values to be uncertain, in addition to the contaminant release and transport parameters, and so determine which parameters need to be better determined in order to lower the risk of death and/or critical sickness for each and every sickness caused by the contaminant.

This particular aspect was methodically investigated in chapter 9 for one sort of sickness but with multiple contaminant sources, and the methodology remains the same here. There seems to be little pedagogical point in repeating such an analysis although one would, in practical situations, evaluate each such cause to determine the dominant parameters requiring better definition.

To illustrate one such example we regard the population age distribution as being roughly triangular as shown in figure 10.9, with a peak at 20 yrs and an average age of 36 years, commencing at zero years and ending at 90 years. Then one runs Crystal Ball™ in conjunction with the basic death and sickness program and reports the corresponding death and sickness rates. Here we do so for just the case of sickness 1 of Appendix B because all other sicknesses and death rates follow a similar methodology.

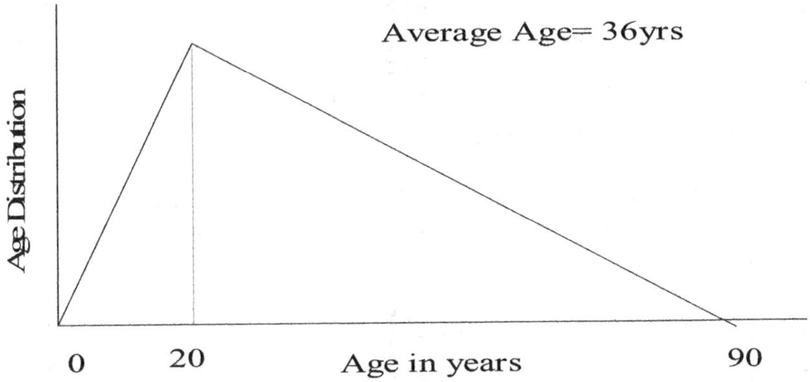

Fig. 10.9. Population distribution with age used in the numerical illustrations given in text

Figure 10.10, panels a-c, shows representations of the likely death rates per 100, 000 population for in-place, waterborne and airborne contaminants respectively, while figure 10.10, panels d-f, shows corresponding sickness rates for in-place, waterborne and air borne contaminant products. We show the different death and sickness rates in a variety of guises to emphasize how one can view the statistical patterns.

For the respective death rates, the in-place likely death rate shown in figure 10.10a is the frequency distribution of likely deaths per 100,000 and indicates there is about an 88% chance of no-one dying of the population and about 22% chance that 25,000 will die from sickness 1.For the air borne component of the toxic contaminant, figure 10.10b shows the reverse cumulative probability. This different distribution to that for figure 10.10a is chosen because there is an almost uniform distribution of very small chances that people will die (with a cumulative sum of around 12%) ranging from zero to about 22,000 per 100,000 population, and also a most likely probability of about 88% that no-one will die from the airborne transport of toxic contaminant. This reverse cumulative plot has been drawn to show the contrast in results in relation to the likely deaths from waterborne transport, as shown in figure 10c using the direct frequency distribution as also used in figure 10.10a. From figure 10c, visually it appears that there is 98.2 % chance of no one dying by waterborne transport but there seems (visually) not to be any representation of the "missing" 1.8%. In fact the missing 1.8% is uniformly distributed between zero

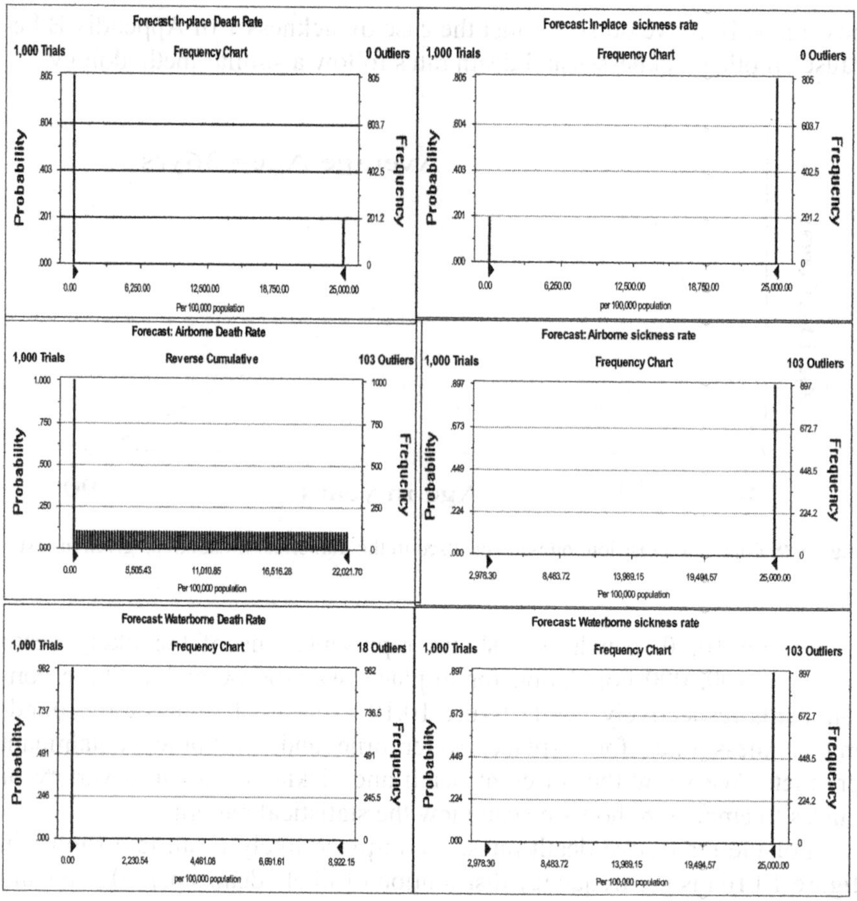

Fig. 10.10. a-f. Plots of likely death and sickness rates per 100,000 population for the population distribution of Figure 10.9 and with the TAC and CSC values at the age boundaries given in Appendix B

deaths and 8,922 deaths (the axis of figure 10.10c) but each value is so small it does not show visually on the figure- which is why we showed a reverse cumulative chart for figure 10.10b, which has the same visual problem. In other words, emphasis of particular aspects of the various death and sickness rates can be enhanced visually so that one can see immediately what is happening.

In the cases where no death rate occurs it can still happen that a fraction of the population can sicken once the toxic concentration climbs above the CSC value appropriate for different age groups of the population.

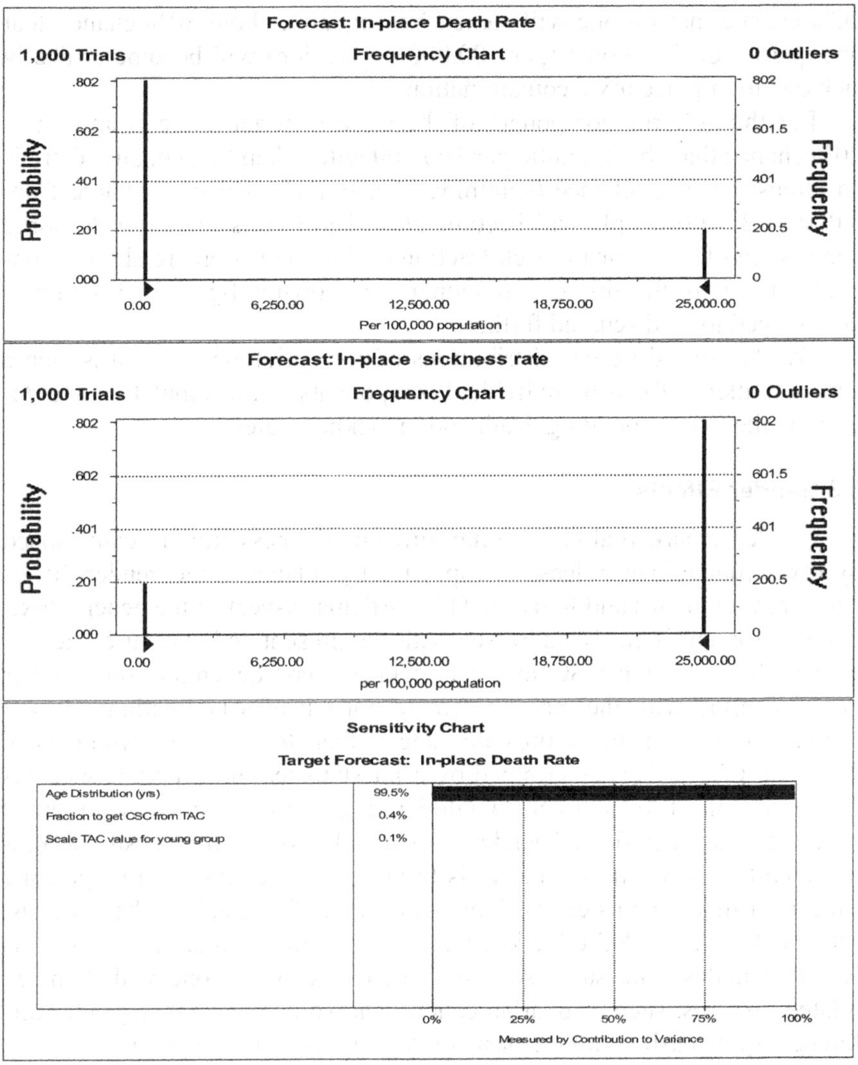

Fig. 10.11. Probability of likely in-place death (figure 10.11a) and sickness rates (figure 10.11b) when not only an age distribution of the population is considered, but also when the marker values for TAC and CSC at the age boundary between young and adult populations are allowed to be uncertain as given in text. Figure 10.11c provides the dominant contributions to the mean value of the in-place death rate, showing that the wide range of the population distribution is the most important factor contributing to the uncertainty

Shown in figure 10.10d, therefore, is the distribution of sickness rates per 100,000 population for in-place contaminant. Note that there is about a

20% chance that no one will sicken but, equally, about 80% chance that one quarter of the people (per 100,000 population) will become critically sick due to in-place toxic contamination.

For the airborne component of the toxic material, there is an almost 90% chance that about 25,000 per 100,000 will sicken but, equally, there is an almost uniform chance (summing to about 10%) that anywhere from 3,000 to 25,000 people will become sick. The results for the waterborne component of the critically sick fraction of the population are almost identical to those for the airborne component, as shown in figure 10.10f, and so do not need to be discussed further.

By choosing different distributions of population with age, it is then a simple matter to focus on individual components of the population and determine their corresponding death and/or sickness rates.

3.3 Gender Effects

As remarked already, some forms of sickness and or death caused by toxic contaminant release are specifically related to the gender difference between males and females. This particular aspect of the general toxicology problem is easily addressed with the present code because one just adjusts the TAC and CSC for each sickness type depending on whether one is dealing with the male or female portion of a population. Thus if sickness 1 is taken to be prostate cancer, then for men one would have TAC and CSC values determined by prior statistical information, whereas for females the TAC and CSC would be set extremely high so that there is no effect. The specifics of gender-related sicknesses can be easily accommodated this way. An alternative is to take, say, sickness 1 to be prostate cancer for males, and sickness 2 prostate cancer for females. Then one just sets the TAC and CSC of sickness 2 at high values and so ignores the effect in females. The same sort of maneuver can be done with female–related diseases, such as ovarian cancer, and so one can cover gender differences quite easily with the basic toxic death and sickness code.

4 Relative Importance of Hazards and Risks

There is a large number of parameters that one needs to quantify in carrying out an estimate of death and sickness rates for a population due to toxic contaminant release. The general method of handling such uncertainties has been spelled out in Lerche and Foth (2003). Accordingly, here we can be somewhat briefer and refer the interested reader to that paper and also to chapter 9 for a more complete description.

The parameters split into several groups almost automatically. The first group consists of parameters describing the TAC and CSC values for each sickness at the limiting points of the different age groups, together with the age points at which one shifts from one age group to the next, and the scaling rate determining the inability of the elderly group of the population to withstand the sicknesses. In addition, one could also include a different dependence of TAC and CSC on age (we chose a linear dependence for the young and adult groups and an exponential dependence for the elderly group) as an extra set of model and functional behavior parameters. In addition, one can change these parameters for gender dependence and also one can change the CSC behavior (currently modelled as a fixed fraction of the TAC for each age group and each sickness).

The second group of parameters relate to the age distribution of the population considered at risk due to contaminant release. We have presented results for a triangular distribution of a population, centered at a peak age of 20 years and tailing linearly to zero at age zero and also at age 90 years. But there is no reason that one cannot choose either alternative functional forms of such population ages or parameter values describing the population for a given distribution (i.e. change the peak age of 20 years, the youngest age of zero years, and also the oldest age of 90 years for a triangular distribution). This second group of parameters then is also uncertain depending on how well one really thinks one knows the distribution of the population at risk.

The third group of parameters have to do with the toxic release itself. The amount of released material, the probability of a spill, the fraction transported by water and also by air, the in-place area contaminated, the airborne area contaminated, and the waterborne area contaminated are all parameters that are somewhat uncertain.

One of the main concerns in attempting to obtain a trustworthy assessment of the hazard to a population should there be a spill of contaminant is the relative importance of knowing which of the many parameters one should determine better in order to be more precise about the likely death and sickness rates for each sickness that could be caused by toxic release. In the case of release of multiple toxic materials, or synergistic combinations, that could cause death, chapter 9 shows how to determine which are the parameters causing the dominant uncertainties .The same procedure can be used for the present study that includes age distribution, gender, and multiple sickness effects. To illustrate this point we consider just two uncertain variables and discuss their relative impact on sickness1 death and sickness rates.

We let the TAC value separating the young group of the population from the adult group be uncertain with ±50% variation around its nominal value of 3, so that it ranges between 1.5 and 4.5, and we also let it have a uniform distribution between the end of range points for illustrative purposes. We also take the fraction in Appendix B of 0.1 for the young age group that determines the CSC from the TAC value to be uncertain with again ±50% variation around its nominal value, so that it ranges on 0.05 to 0.15 with, once more, a uniform distribution choice. We already have a distribution for the population age so that there are, in fact, three uncertain variables contributing to the uncertainty. With these three uncertain parameters as given, we again ran Crystal Ball and considered the death and sickness rates for in-place, airborne and waterborne transport. In the interests of saving space, we show here only the results for the in-place death and sickness rates. Figure 10.11 presents the frequency distributions of the likely death and sickness rates for sickness 1 together with the relative contributions each of the three uncertain variables makes to the average value of the death rate. Perhaps the most salient point to note from figure 10.11 is that 99.5% of the uncertainty on the death rate is caused by the population distribution and not by either of the other two parameters. Thus there is little point in trying to narrow the range of uncertainty of either of the two parameters until one can either better determine the population distribution (or deal with a narrower age "slice" of the total population) or until one can decide that the death and sickness rates form an acceptable risk for the population. This determination could be based on the fact that, as shown in figure 10.11, the likely death rate distribution indicates that there is about 80% chance of no-one dying and only 20% chance of a likely death rate of 25,000. Correspondingly, there is about a 20% of no-one getting sick and about 80% chance around 25,000 of the population could become sick.

5 Discussion and Conclusion

The major point of this chapter has been to develop a procedure to allow for multiple sickness types and their corresponding death and sickness rates for release of toxic contaminant material. Population distributions, resistance to the toxic contaminant as a function of age group of a population, and also gender dependent sicknesses caused by the toxic contaminant, all play roles in allowing assessments of the likely death and sickness rates, either at the spill site (the in-place death and sickness rates) or as transported by water or air mechanisms. The chapter shows how such assessments can be carried through for different choices of relevant parameter values and also allowing uncertainty in any and all parameter values used to quantify the system.

The major aim of both the previous chapter and the current chapter has been to present a stable quantitative methodology so that one can see what factors are of the greatest significance when one is attempting to assess the damage that release of toxic material can bring to a population in terms of likely death and sickness rates.

Appendix A. TAC and CSC Fractions Used

TAC Values for each Age group

Age group:	Young	Adult	Adult	Elderly
Age limit	<14	14	65	>65 years
	TAC Values			
A	7	14		6
B	3	6		4
C	4	8		5
D	5	10		6
E	6	12		7
F	8	16		5
G	9	18		3
H	2	5		1
J	4	7		4
K	3	10		6

Fractions (<1) to get CSC values from TAC values

Age group:	Young	Adult	Adult	Elderly
Age limit	<14	14	65	>65 years
A	0.1	0.2		0.15
B	0.2	0.4		0.25
C	0.3	0.6		0.35
D	0.4	0.8		0.45
E	0.5	0.99		0.55
F	0.6	0.89		0.65
G	0.7	0.79		0.75
H	0.8	0.69		0.85
J	0.9	0.59		0.87
K	0.95	0.49		0.17

Appendix B. Minimum TAC Values marking the shift between Age Groups

Minimum TAC Values for each Age group					
Age group	Young	Adult	Adult	Elderly	
Age limit	14	14	65	65	years
	TAC1	TAC2		TAC3	
A	3	10		4	
B	3	6		4	
C	4	8		5	
D	5	10		6	
E	6	12		7	
F	8	16		5	
G	9	18		3	
H	2	5		1	
J	4	7		4	
K	3	10		6	

Appendix B. Minimum TAC Values marking the shift between Age Groups

	Minimum TAC Values for each Age group				
Age group	Child	Adult		Elderly	
Age limit	14	14	65	65	years
	TAC1	TAC2		TAC3	
A	5	10		4	
B	3	6		4	
C	4	8		5	
D	5	10		6	
E	6	12		7	
F	8	16		5	
G	9	18		5	
H	2	5		1	
J	4	7		4	
K	[illegible]	10		6	

Chapter 11

Methods for Estimating Associated Risks of Sinkhole Occurrences with data from the Ruhr Valley Region of Germany

Summary

The generation of sinkholes in the Ruhr valley region in Germany is caused by many centuries of coal mining at various depths and, at the least, by residual underground open spaces that do not stay stable with time. Due to the unknown distribution of underground spaces and their undefined, decreasing stability with time, probability calculations provide a somewhat successful approach in order to obtain information about risks of sinkhole generation and about possible relations to structural, geological, mining or other influences.

We demonstrate here procedures and strategies for probabilistic calculations of sinkhole generation. In order to explain the basic steps of data organization and calculation we provide an application of some of the quantitative method using an example from a local data set (sinkholes from a map of the Muddental region) combined with the corresponding structural geological map. This example illustrates the basic methods and clarifies the application of probabilistic methods. Due to the small data base available, it was not possible to develop further the results and so to provide a final unequivocal classification of the dominant causes to understand the risks of sinkholes occurrences. Nevertheless, should further information be made accessible by the relevant authorities, one will be able to improve and statistically sharpen the arguments given here using the probabilistic methods and so more clearly define the best possible correlations for estimation and prediction of sinkhole occurrences. The main point made, even with the limited data available to date, is that the methods provide powerful tools to help describe systematically the causes and risks of sinkhole occurrences. Current information from the Ruhr Valley region would indicate that preventative methods are not being used on a routine basis, rather post-sinkhole occurrence filling with cement and renovation of badly damaged structures is the norm. It is to be hoped that the methods given will help control the sinkhole problem and dangers.

1 Introduction

In the extended area surrounding the Ruhr Valley region of North-Rhine Westphalia, Germany there has been centuries of mining for coal. As a consequence the subsurface is riddled with shafts and tunnels, some of which are known from historical records, while others were either illegal or records lost, so that today the pattern of such tunnels and shafts is, at best, incomplete and only partly known. And, even when the locations of shafts and tunnels are known, what are still unknown are the depths, heights, widths and tortuosities of such underground openings. In addition, there do not seem to be available records detailing the rock types and compositions of rock formations overlying and underlying underground openings, nor of the strength of the surrounding rock masses. In the shafts and tunnels themselves, there is even less of a record available of the variations of residual rock types along the tunnels after economic mineral removal.

In addition, since their creation at times ranging to centuries prior to the present-day, the shafts and tunnels have been subjected to a variety of changing natural and anthropogenic influences, such as ground water infiltration, surface building construction, autobahn construction and fluctuating traffic loads, and rail track construction with fluctuating train loads with time, to name but a few. As a consequence, from time to time, sinkholes appear at the surface as do sink depressions, the latter caused by underground collapses not breaching the surficial layers.

The sinkholes have various sizes, ranging from fractions of a meter across to several tens of meters. Unfortunately, as far as we are aware there appears not to be available a systematic record of lateral shape, depth, and time of occurrence of such sinkholes, and even less of a record of the characteristics of the sink depressions, although the spatial locations of recorded sink events are catalogued. The connection with the anthropogenic mining activities is clearly present, but to what extent there are naturally occurring sinkholes due to intrinsic weakness of rock formations, and to what extent such weaknesses have been influenced by mining, and so exacerbating the sinkhole occurrence rate, is also an unknown.

In terms of discovery of sinkholes, and their reporting, two factors stand out. First it is unlikely that any large sinkholes are overlooked because of their large size in relation to population density. However for small sinkholes, say of the order of a meter or so across to fix ideas, it is indeed possible that there is serious under-reporting. This may occur because a small sinkhole can easily fill with dirt during rains and storms, and so is again filled and no longer visible, or it may become so overgrown with vegetation that it is just not visible unless stumbled across by accident. In addition, small sinkholes act as conduits for rain water to actively and efficiently infiltrate the shallow subsurface, and so provide waters for further erosion of deeper-lying rock and tunnel structures, leading to even more sinkholes in the course of time.

The potential for catastrophic damage to property and people of such sinkholes and sink depression events is enormous. For instance, the newspaper "Stadtspiegel Bochum" (29.July 2000) reported that the regional train line Bochum - Essen had to be closed until the 1st August due to a sinkhole. Costs of more than $100, 000 were involved for repairs. On 28th June 2001 a regional train line had to be closed at least until 5th of July (for the second time since October 2000) because of track deformation caused by such a sinkhole (newspaper WAZ no. 147 from 28th July 2001). In the summer of 2000, the main high-speed train line through Bochum had to be shut because two hours before an InterCity Express (ICE) train would have crossed the track it was, fortunately, noticed that the track had distorted due to a sink depression. This latter case cost over $120,000 to fill the hollowed out section under the track bed with cement, on an emergency basis. The price of the regional train track repair is still unclear but will cost at least several hundred thousand dollars.

The point about these examples is that, even though no serious damage was done to equipment nor were lives of people lost, it was purely by serendipity that the events were noticed in time. And the ancillary costs in lost time of commuters and train re-routing are also not included in such estimates of costs. Eventually such serendipity will fail with a concomitant disaster.

The long range problem is not just to perform remediation after an event (which could be catastrophic) occurs, but rather to develop methods to assess the likely areas with the highest risk of potential sinkhole and sink depression occurrences, so that more intensive investigation and preventative procedures can be undertaken to minimize either the occurrence chances or the putative catastrophic damage chances. Such methods would allow one to focus efforts on the highest risk of dangerous occurrences and also to suggest appropriate long term solutions rather than responding with emergency cement fill after each event. And in this regard, while cement fill cures the immediate location problem, the stress state of the tunnel is then changed, as is also the ground water flow direction and magnitude, so that it is not at all obvious that one has done anything except exacerbate the long range problem even more, with future occurrences likely to be larger and more devastating than the remediation of the immediate problem.

What is clearly needed is a systematic set of quantitatively reproducible methods that (i) can also be updated as more sinkholes and sink depressions become available, (ii) allow one to provide some probability estimate of areas of highest risk potential, and that (iii) also allow one to identify the major influences causing sinkholes and sink depressions, so that one can concentrate on those types of high risk areas as more important to remediate prior to catastrophic events. To date, even with the limited quality and quantity of sinkhole data available, there seems not to be available any such attempts to grade for risk, nor any understanding of the relative importance of different geological factors and anthropogenic influences in causing or enhancing the appearance of such potential economic and human disaster scenarios. However, a start on this problem has been made in the volume edited by Maund and Eddleston (1998), and, in particular, in the paper in that volume by Scott and Statham (1998), and also by Cole (1988), although, In our opinion, there is still a considerable way to go before we can understand and mitigate potential risks and hazards from such situations. In that regard that the absence of critical data to aid in such investigations is a major drawback and one that led to the efforts here to provide probabilistic procedures to address to some extent the basic problems.

The purpose here is to provide some procedures to show how one goes about setting up such risking methods and to show how they use data available to quantify the probability of occurrence, and also how new data can be used to update the patterns of behavior. In addition, the relative importance, relative sensitivity, and relative contribution of different factors in assessing risk areas and potential for disaster are also shown to be similarly quantifiable. In this way one has, at the least, some of the tools in place that one needs in attempts to understand the causative agencies and their dominance in creating sinkholes and sink depressions, and, hopefully, with enough data an estimate of recurrence times for each area. In this way one could not only provide high-risk area maps but also high-risk timing of areas.

Because of the variation in size of the known sinkholes it is distinctly possible that more than one type of subsidence is active in the region. As a reviewer of an earlier version of this chapter has accurately noted "Coal-mine related subsidence in cases where partial extraction mining has been conducted can generally be classified as either sag (also referred to as trough) or pit (also referred to as pothole) subsidence. Sag subsidence is generally characterized by the settling of a relatively broad area of the surface (referred to as a sag or trough), while pit subsidence is generally characterized by the formation of a relatively small and localized collapse feature (referred to as a chimney, crownhole, pit, or pothole). Sag subsidence is caused by the failure of coal pillars, either when pillars are crushed under the weight of overburden, or when the overburden weight causes pillars to punch into an underlying weak material. In the case of pit subsidence, coal pillars are relatively stable, and subsidence at the surface results from the upward migration of collapse features that develop due to the mine roof rock falling between the coal pillars. Pit subsidence features often correlate with the location of mine entry or working intersection locations (where roof rock spans a large extracted area), with the adjacent pillars and overburden remaining unaffected"

We are in full agreement with this assessment and further information on different types of coal-mine subsidence is to be found in Speck and Bruhn (1995), Waltham (1989), Whittaker and Reddish(1989), and Scott and Statham (1998), to whom we refer the reader. However, as we shall see in the next section, there is just not enough information available from the public authorities to determine if one or more active mechanisms dominate. Indeed, even so basic a quantity as the overburden thickness at each sinkhole occurrence is not recorded, making it extremely difficult to form any strong position on occurrence mechanisms and their dominance.

2 Geological Knowledge, Statistics and Properties

There are three basic factors that go into attempts to quantify the risk of subsidence in sinkhole problems. First, one needs some idea of the underlying geology, lateral and vertical lithological variations, faults and their variations, topography, water accumulation areas on the surface, and also how this background has been disturbed over the years by anthropogenic influences. Second, one would like to have available intrinsic properties of each and every sinkhole: such as width, length, depth, volume, time of occurrence, correlation with geological formations and also with man-made tunnels, and structural weaknesses in the geological formations. Third, one would like to construct a procedure using such data to provide predictions of where, when and how large any future sinkholes could be. This latter component can be tested by using part of the data base and seeing to what extent the remaining components can be predicted with such constructions and therefore one determines the accuracy, reproducibility and so the predictability of such quantitative methods. One also so determines the quality, quantity and resolution of data needed to be more precise with the predictions. Consider each component in turn with particular emphasis on conditions for the Ruhr Valley region of Germany.

2.1 Geological Background

The Ruhr Valley region is part of the Variscan continental crust, the so-called Rhenish block. The northern molasse basin of the Variscan orogenic belt contains successions of greywacke sediments with sandstone or silt-stone character. This sequence also contains repeated cycles of coal seams that represent the changing conditions with more or less marine or terrestrial predominance in the former coastal region. The whole Variscan molasse succession that forms the Ruhr valley region was moderately folded after deposition. Consequently, the Ruhr valley region is characterized by regular synclines and anticlines with a trending strike direction ENE-WSW. Additionally, the folded sequence was also faulted, mostly by thrust faults parallel to the fold axes but also by faults mainly perpendicular to the fold axes. Due to later erosion processes, parts of the anticline structures are not recorded today and we find incomplete sequences of folds often with "air-anticlines".

Additionally, the thrust faults caused reductions of the successions in some profile cuts. The coal seams within the folded molasse sequence were excavated, more or less systematically, by mining industries. Due to mining, a widespread system of shafts, drift mines and tunnels had developed over the centuries. The coal was mined by different kinds of methods and procedures with time, with the consequence of leaving vacant spaces underground. Especially during the early days of coal mining, as well as during war times, mining activities took place without the supervising control of a mining administration. The consequence is that, today, we have incomplete records and partly unknown positions of shafts or drift mines. Such uncertainties exist predominantly in the former central Ruhr valley region with relatively near surface coal seams. Therefore, in this region the majority of shafts and tunnels cause sinkhole depressions at the surface. The underground openings lose their strength with time and overburden fills out these spaces partly and irregularly. That is why we find sinkholes in the Ruhr valley region today and will also continue to do so in the future with unknown consequences.

2.2 Statistics

In order to provide a basis to start the quantitative procedures of the next sub-section, we have constructed a method for classifying the different structural properties of sinkholes.

We have taken data as available, for example on a sinkhole map of the region around Muddental (Muddental, M 1/10 000, 2000), and classified them, so far as is possible, by correlation of this map with the corresponding geological map, i.e. with structural properties. Because the quality of a scale-reduced print is extremely difficult to read, we do not show the composite map here (but a copy can be had by contacting C. Lempp at the University of Halle-Wittenberg). However, the large maps to their original scale 1/10 000 and 1/25 000, respectively, were the basic tool for evaluation of the structural, geological and spatial correlations as shown under the characteristic features scheme of figure 11.1.

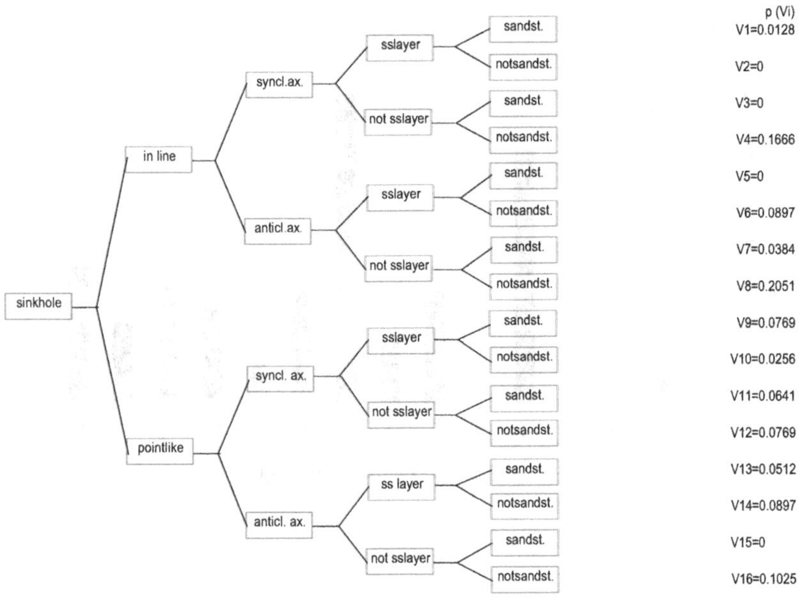

Fig.11.1. Category distribution of sinkhole characteristics in the Muddental region based on the known 78 sinkholes

This composite map shows clearly some interconnection of the structural geological characteristics and the spatial distribution of sinkholes. This framework has been used to create a classification that is as objective as possible and, furthermore, is extremely useful to demonstrate the quantitative probabilistic procedures. Accordingly, we have characterized the sinkholes within the framework of such groupings. This depiction is given in Figure 11.1. While not the only way one could categories the sinkhole properties, the scheme given in Figure 11.1 has the advantage that it points out where one has lacunae in the data so that either one cannot make a serious effort to involve such groupings of sinkholes (because the data are too sparse) in a quantitative risk and probability analysis, or one makes a more determined effort to improve the data base by a more systematic investigation of all properties of sinkholes than appears to be available in the current data base.

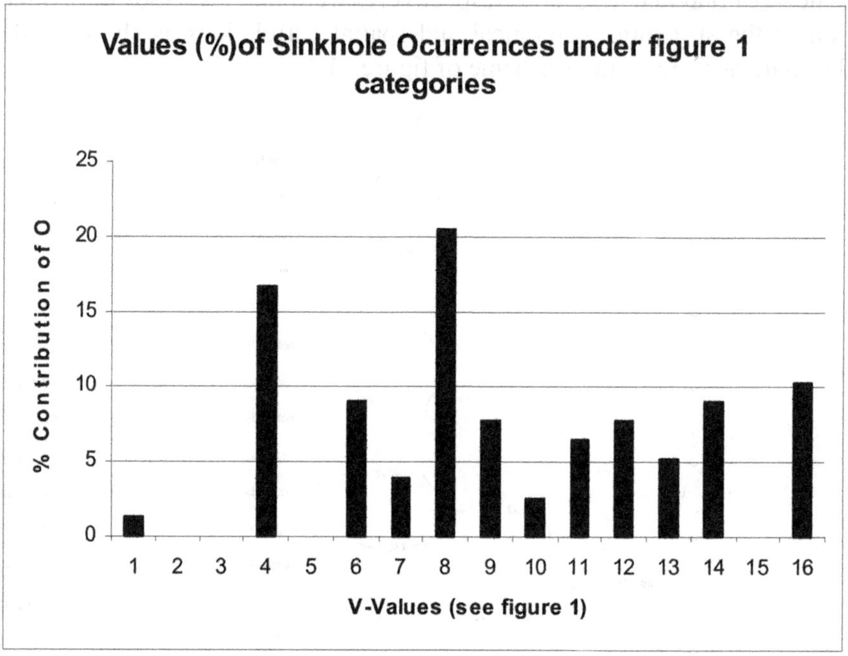

Fig. 11.2. Scaled distribution of sinkhole characteristics in block form showing the relative percentages in each of the categories

This distribution of the available data is shown in figure 11.2, where it can be seen that while two categories dominate (V4 at 17% and V8 at 21%), in total they amount to only about 38% of all V-categories. The remaining categories contain lower percentages of contributions (ranging from about 3% to 11%) , except for V2,V3 and V5 that have no measured sinkhole occurrences for the Muddental region. Whether the same is true of other sinkhole areas is unknown due to lack of data. The total number of sinkholes on the Muddental map is 78. What is clear from the distribution of the available data is that particular components are either poorly represented, because of less than complete collection of sinkhole properties despite the existence of such sinkholes, or that the available sinkholes indeed are pauperate in sinkholes of the poorly represented classes. But, in either event, what one has available is an estimate of the most dominant components of the sinkhole properties, as represented through the distributions of figures 11.1and 11.2. For example one can evaluate that the probability of occurrence of sinkholes in groups along lines (V1+V2+V3+V4+V5+V6+V7+V8) is 0.51, while at individual points (V9+...+V16) the corresponding probability is 0.49, nearly equal. Similarly, the probability of occurrence of sinkholes near syncline axes (V1+V2+V3+V4+ V9+V10+V11+V12) is 0.42, while near anticline axes (V5+V6+V7+V8+V13+V14+V15+V16) the corresponding value is 0.58, only about 33% larger based on the limited statistics available for the Muddental region.

In contrast, the distribution of sinkholes differs more clearly if one compares their probability of occurrence either on bedding layers (i.e. ss-layers: V1+V2+V5+V6+V9+V10+V13+V14), which has a value 0.35, or not on any dominant bedding layer (V3+V4+V7+V8+V11+V12+V15+V16), with a worth of 0.65- representing an almost 2:1 contrast. Consequently, bedding layers appear not as dominant discontinuities that take part in sinkhole generation on a major basis, although they do interact about a third or so of the time in such sinkhole occurrences. The last structural or geological attribute (i.e. the lithology in which the sinkholes occur) is also unequally distributed between sandstone layers and not sandstone layers as they are recorded in the geological map. Within sandstone layers only about a quarter of the sinkholes occur (V1+V3+V5+V7+V9+V11+V13+V15) with a value 0.24, whereas about three quarters occur in siltstones, claystones or coal seams (V2+V4+V6+V8+V10+V12+V14+V16), with a value 0.76. Such lithological relationships may arise from simple differentiations on the geological map.

The next subsection shows how one can use such information to build predictive risking factors for sinkhole occurrences.

2.3 Properties

Apart from the direct distribution of known sinkholes, as represented by the above distributions, there are also indirect influences on the sinkholes that, while they may not be so apparent, can play a significant role in the appearance of sinkholes, even when the basic distributions given above are available. The point here is that the appearance of sinkholes is also tied to other than structural-geological conditions of a region in respect of mining influence and weakening of rock formations due to private or public buildings (i.e. streets, bridges, houses), in respect of sinkholes in the local neighborhood with their increased influence on weakening of rock mass in their surroundings, or in respect of intrinsic weaknesses of rock formations, as exhibited through faults and fracture domains. A host of other geological variables can also play a role, having to do with the strength of materials in juxtaposition, geomorphologic variations, thickness of overburden, and disturbance of natural conditions in the subsurface by the anthropogenic activities of tunnel construction. Later surficial activities, such as reservoir impoundment, road construction and associated traffic dynamic forces, train-track construction and the associated dynamic pounding by regularly scheduled trains, all have long-term impacts on the forces influencing the stability of the subsurface, and so on the creation of sinkholes. A list of all such factors that could influence sinkhole production, and so hazard and risk in a given area, would make for a very long chapter indeed and one that would defeat the purpose of showing how to identify the dominant attributes and conditions from limited data and then how to use such determinations to set up probabilistic assessments of likely occurrence risks and conditions. We are, nevertheless, painfully aware that conditions that are appropriate for the Ruhr Valley may not be so relevant for, say, West Virginia or the Black Warrior Basin of Alabama. What is urgently needed is a compendium of many case histories so that general rues of governance can be established, of relevance to all such probabilistic methods of assessment of risk. This chapter provides at least one avenue of approach to setting up such a comparison and, hopefully, will enable others, more gifted than we are, to rapidly facilitate such a compendium and establish governance rules.

Figure 11.3 provides a means to assess, from the known sinkholes, which sort of factors have a dominant influence on the otherwise pristine conditions surrounding sinkhole creation as exhibited in figures 11.1 and 11.2. Again, just as for figure 11.2, components that have but poor representation in figure 11.3 from the known data base are due either to the true absence of such categories of dynamically influencing conditions on sinkhole production, or due to the poor data base in terms of providing enough examples to allow a particular component to be considered as distinctly dominant. In either event one learns where the data need either better control or where one needs more data in order to improve the statistical reliability of each component branch of figure 11.3. But, that being said, one has at least the beginnings of a procedure for classifying the sinkhole data in terms of their direct geological properties and also influencing conditions such as mining activity, settlement of buildings or being in the neighborhood of other sinkholes.

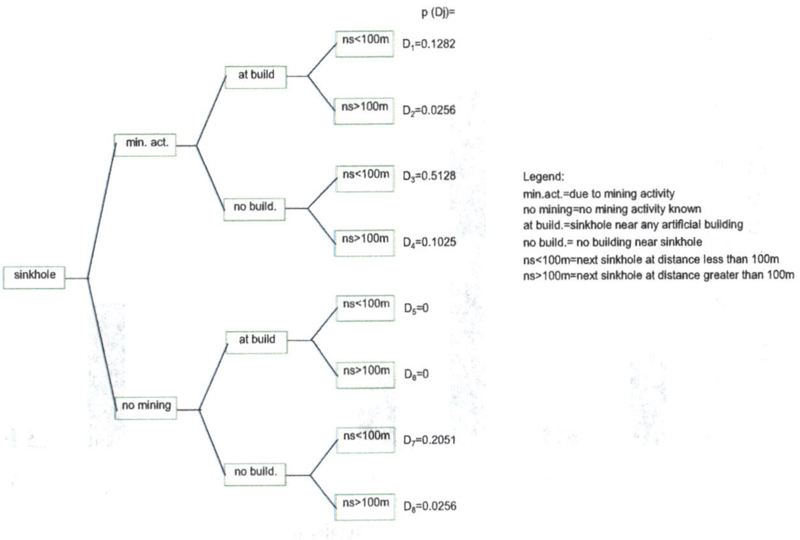

Fig. 11.3. Property distribution of geological influences on sinkhole occurrences in the Muddental region for the 78 known sinkholes

One can take the frequency of occurrence of each component branch and reduce it to a relative probability by dividing the number of occurrences for each branch by the sum total of all occurrences- as depicted in figure 11.4. What is apparent from figure 11.4 is that one D-category (D3) alone contains over 50% of all sinkhole occurrences, with the next highest D-class being D7 at a much lower 21%, and the sum of the remaining D-classes providing the residual 19%, with some D-classes having no representation of observed events (D5 and D6). Clearly, one strong message being conveyed by the Muddental data is that concern should be focused predominantly on areas satisfying conditions of class D3 because, based on the data to hand, there is over 50% probability a sinkhole will appear under such conditions as opposed to elsewhere under different external conditions. Secondary consideration should focus on areas with class D7 characteristics where the next highest probability of occurrence is likely (21%).

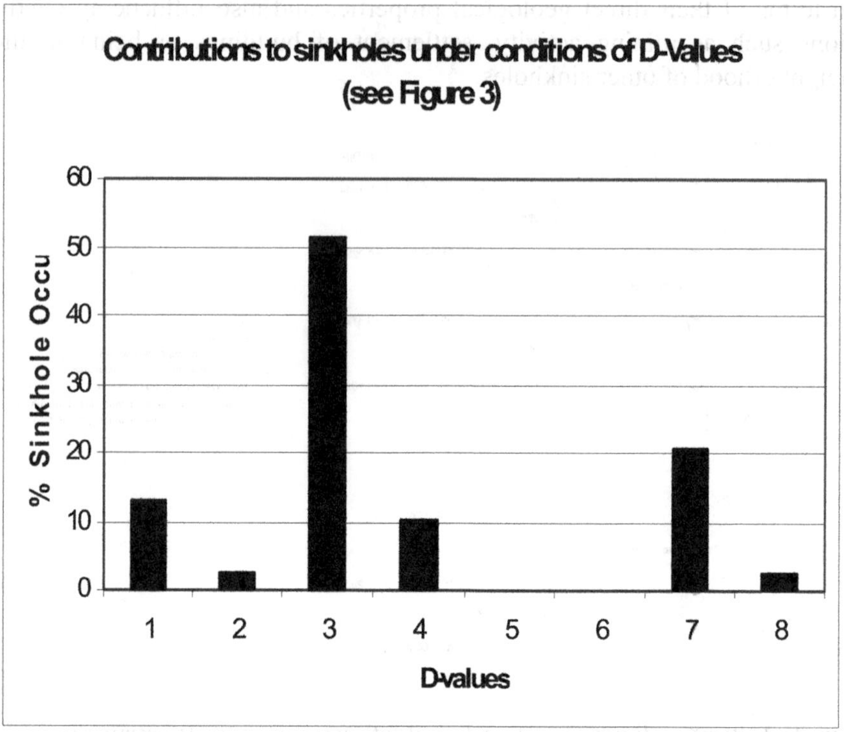

Fig. 11.4. Scaled distribution of sinkhole influencing properties in block form showing the dominance of conditions D3 and D7

Again, one can deduce some relations in our example from the Muddental sinkhole map: Mining activity is supposed to be a decisive reason of sinkhole occurrence. Comparing entries in categories influenced by mining from Figure 11.3 (D1+D2+D3+D4), which has a total probability of occurrence of 0.77, and entries with no nearby known mining activity (D5+D6+D7+D8), which provide a total probability of 0.23, it would indeed seem that mining activity was an important factor for sinkhole generation. However, the fact that nearly a quarter of the sinkholes have no known association with any mining activity is also of interest.

To check the influence of buildings in the neighborhood of sinkholes, one can compare entries in the group (D1+D2+D5+D6), yielding a probability of occurrence of just 0.15, with the group not under the influence of any neighboring buildings (D3+D4+D7+D8), which yields a probability of occurrence of 0.85. Thus, the influence of buildings appears to be of only minor importance for the Muddental region. In contrast, the influence of the direct neighborhood of another sinkhole, (i.e. at distance less than about 100m to the next sinkhole) appears to be dominant in most of the cases in the example (D1+D3+D5+D7) with a probability of occurrence of 0.85, whereas at a distance of more than 100m to the next sinkhole (D2+D4+D6+D8) the corresponding probability of occurrence is only 0.15, implying that there is a dominant correlation of cause and effect for neighboring sinkhole occurrence.

Structural properties (see fig.1) →		V1	V4	V6	V7	V8	V9	V10	V11	V12	V13	V14	V16
External influences (see fig. 3) ↓							1.28						
	Sums: Row / Column →	1.28	16.66	8.98	3.85	20.48	7.68	2.56	6.4	7.68	5.12	8.97	10.25
D1	12.82	0	2.56	0	0	1.28	1.28	1.28	2.56	2.56	0	0	1.28
D2	2.56	0	0	0	0	0	1.28	0	0	1.28	0	0	0
D3	51.28	0	5.13	5.13	3.85	19.2	0	1.28	2.56	0	2.56	5.13	6.41
D4	10.25	0	0	0	0	0	2.56	0	1.28	0	2.56	1.28	2.56
D5	0	0	0	0	0	0	0	0	0	0	0	0	0
D6	0	0	0	0	0	0	0	0	0	0	0	0	0
D7	20.51	1.28	8.97	3.85	0	0	1.28	0	0	2.56	0	2.56	0
D8	2.56	0	0	0	0	0	1.28	0	0	1.28	0	0	0

Note: Entries in each cell are in %; there are zero entries for V2, V3,V5 and V15

Fig. 11.5. Combined distributions of sinkhole properties and influencing conditions for the Muddental sinkhole data. Note the absence of entries in several categories indicating that no known such sinkholes exist

With the data so organized one can now begin to put together the procedure for evaluating the probabilities of occurrence of new sinkholes under particular conditions.

3 Probability Estimates of Sinkhole Occurrences

The data now sit in two functional forms, one having to do with the basic properties of the geological position of each sinkhole, as depicted in Figures 11.1 and 11.2, the second having to do with the influence of various geological, anthropogenic and external conditions on each sinkhole. What is now needed is to combine these two groups to obtain the probability of occurrence of particular sinkholes under specific geological, external and/or anthropogenic conditions. Such a combination then allows one to develop probabilistic methods of risk and occurrence. The simplest way to achieve the desired information is to construct a table as depicted in figure 11.5. The values in rows indicate the external conditions of each type of sinkhole while the columns indicate the influence of each type of structural geological condition on the particular sinkhole type. Thus for each column-row pair one inserts the number of occurrences measured and then one reduces the two dimensional array to probability statements by dividing each entry by the sum of all entries. In this way one constructs a relative probability table of occurrence chances for a sinkhole type under the influence of particular conditions. By summing rows one then recovers figure 11.2 while by summing columns one recovers figure 11.4.

This particular way of representing all the data now allows one to provide estimates of probability of occurrence of any type of sinkhole under known or unknown geological conditions and, indeed, one can also describe the probability of occurrence of sinkholes with particular attributes. If the data on occurrence times were to be available, one could also provide the expected waiting time until the next sinkhole occurrence of a particular type, size, etc. We demonstrate how to construct these various procedures using several illustrations.

3.1 Illustration1

The procedure for making such estimates operates as follows. Denote by $p(V_i \mid D_j)$ the probability of obtaining sinkholes of property type V_i under external conditions D_j, as available from Figure 11.5, based on the data to hand. Suppose, further, that in a given area where no sinkhole has hitherto fore occurred one is not too sure of the external influencing conditions. For instance, it may be that one group of specialists consider mining influences as dominant, but another group is more of the opinion that it is not the mining activity per se that has a crucial effect on the sinkhole generation but the influence of prior sinkholes in the immediate neighborhood (distance < 100 m) as the dominant factor to take into account. Consider only these two factors in the example to follow, so that they form a mutually exclusive set. Denote by p(m) the probability that mining influences dominate, and by p(s) the probability that sinkhole neighborhood is paramount. Because we treat with only these two factors, then $p(m) + p(s) = 1$. From the illustration above we have relative probabilities of 0.77 and 0.85 for mining and sinkhole neighborhood dominance respectively, *all other factors being considered equal.* Thus if one wishes to consider just these two classes of factors then one would initially assign a value $p(m)=0.77/(0.77+0.85) = 0.48$, and so a value $p(s) = 0.52$.

Interest then centers on obtaining an estimate of the probability of a sinkhole occurring subject to the uncertainty of the external conditions. Referring to figure 11.3 we see that for the condition where no mining activity dominates but there is influence of nearby sinkholes we require the two values $p(D_5)$ and $p(D_7)$. Set $p(D_5) + p(D_7) = p(nbh) = 0.21$, so that the sum probability of obtaining a sinkhole of any sort without mining dominance is $p(s)= p(D_5) + p(D_6) + p(D_7) + p(D_8)$. If mining influences dominate but there is no effect of sinkhole neighborhood then we require the four values $p(D_1)$, $p(D_2)$, $p(D_3)$ and $p(D_4)$, which would give a total probability for the case where mining activity influences predominate of $p(m) = p(D_1) + p(D_2) + p(D_3) + p(D_4)$. Furthermore, let the probability of sinkhole occurrence with mining activity and of nearby sinkhole effect be $p(min; nbh) = p(D_1) + p(D_3) = 0.64$. But there is uncertainty on which of the two conditions (mining or not mining) dominates, so that the average probability, <P(av)>, that a sinkhole will occur, independent of which condition prevails, is then

$$<P(av)> = p(m)*p(nbh) +(1-p(m))*p(min;nbh).$$

In our example we obtain:

$$<P(av)> = 0.48 * 0.21 + (1- 0.48)*0.64 = 0.10 + 0.33 = 0.43$$

while the uncertainty on this value, as measured by the standard deviation, s, around the mean value is

$$s^2 = p(m)p(s)[p(nbh)-p(min;nbh)]^2$$

which for the example gives $s^2 = 0.48*0.52*(0.64-0.21)^2= 0.046$; i.e. $s =0.215$.

In terms of the trustworthiness of the mean value, <P(av)>, one can use the volatility measure $v = s/<P(av)>$ as a yardstick. A small ($v<<1$) value for the volatility indicates that there is but little uncertainty on the mean, while a value $v>>1$ indicates that little trust should be placed in the mean as an accurate descriptor of the chances of a sinkhole because of its associated large uncertainty. In the example using the Muddental data one has v=0.215/0.43=0.5, so that at one standard error the mean value could range as high as 0.645 and as low as 0.215, a dynamic range of 50% centered on the nominal mean. While not providing a sharp determination of the accuracy of the mean value, the point is the volatility provides an indication that the mean value has at least some meaning in terms of its uncertainty. In terms of the cumulative chance of obtaining a sinkhole, irrespective of whether it is the mining or neighborhood conditions that dominate, a useful measure is the cumulative probability, CP, of obtaining a value for the probability of occurrence in excess of a prescribed value P. Using only the mean value <P(av)> and the standard error of the mean, s, one can construct the CP as

$$CP = 50\{1-\tanh[(2/\pi)^{1/2}(P-<P(av)>)/s]\} \quad \%$$

Values for CP can then be risked, depending on the level of chance, P, one is prepared to accept that a sinkhole will or will not occur. For instance at a value $P = P_* = <P(av)> + s(\pi/2)^{1/2} = 0.43+0.215*1.26=0.70$, the CP is approximately 84%, indicating that there is only about 16% (100-84) chance that the probability of a sinkhole will be in excess of 70%. If the value of P_* is small compared to unity, one can then argue the risk of a sinkhole under the given conditions is small, so that one would downgrade such areas relative to others with higher values of P_* calculated under the same set of rules and quantitative procedures at a CP of 84%.

From figure 11.5, the highest probability is p (V8/D3) = 0.19 exceeding by factors of two and almost three, respectively, the lower probabilities of p(V4/D7) = 0.09 and p(V16/D3) = 0.07; all other probabilities are even lower than these three values . Note that these three high values occur under external conditions D3 and D7 , which are indeed the dominant occurrence conditions for the Muddental region based on the available data. Physically, the implication is that about each fifth sinkhole occurs away from sandstone layers but in a position such that nearby (< 100 m) one will find another sinkhole.. Without mining activity, the lower probability p (V4/D7) = 0.09 just underscores the extremely strong effect of neighboring sinkholes.

3.2 Illustration 2

In the above illustration one dealt with the chances of obtaining a sinkhole under uncertain external conditions, as exemplified by the probabilities p(m) and p(s) that either mining activity influences were dominant or that the influence of other sinkholes in the neighborhood prevailed, and one must prescribe the value of the estimated p(m) (or p(s)). Without further information there is no way to improve the estimate that either mining activity or neighborhood conditions is of dominant concern. However, if a sinkhole does occur in the uncertain domain, then one can use Bayesian updating methods to improve the likelihood that either mining activity or effects of the neighborhood of other sinkholes are the dominant concern.

We present the procedure for such an updating here so that one has collected in one place the major methods for analyzing sinkhole dominance characteristics. This updating procedure operates as follows. Again use the basic nomenclature developed in the first illustration. What can be shown from Bayes' Principle is that if an event A occurs under conditions B then the probability of conditions B being in force given that event A occurs is obtainable through $p(B|A)p(A) = p(A|B)p(B)$ where p(A) and p(B) are the probabilities of A and B occurring. Thus for the illustration given above one has, with B = mining, that

$$p(\text{mining}|\text{sinkhole}) = p(\text{sinkhole}|\text{mining})p(\text{mining})/p(\text{sinkhole}).$$

Hence one can insert from illustration 1 to obtain

$$p(\text{mining}|\text{sinkhole}) = p(\text{mining})p(m)/(p(m)p(\text{mining})+p(s)p(\text{neighboring sinkhole})).$$

Thus the chances that a sinkhole, should one occur, are caused more by mining conditions than by neighboring sinkhole dominance has been updated from p(m) to p(mining|sinkhole). The argument can be repeated each time a new sinkhole appears in order to determine a better estimate that the sinkhole is caused dominantly by either mining effects or neighboring sinkhole conditions, each time updating the intrinsic probabilities p(m) and p(s), as just described. In this way one uses the additional data available to narrow the uncertain conditions and causes in particular areas.

While the illustration given here is particularly simple in that it involves only two unknown factors (mining influence and neighboring sinkhole influence, each with an associated probability of being correct), the argument can easily be generalized to involve as many factors as are deemed necessary. Such a more complex illustration would just add technical details without really bringing any further insights into the basic Bayesian updating process and so has not been given here.

3.3 Illustration 3

Apart from the quantitative procedures given above to improve knowledge, or at least minimize risk, one is also interested in the likely time one must wait before another sinkhole will appear in a given region, and also on the likely size of the sinkhole. This information can be extracted from the basic database provided the attributes of each sinkhole are available. For instance, if just the time of occurrence, t_i, of each sinkhole is available (for $i = 1,2,\ldots N$) then one can construct the probability p(t) of obtaining a sinkhole in a given time interval from the frequency of occurrence of sinkholes- basically just by plotting the sinkhole times relative to the earliest as a zero time reference and then normalizing the histogram to unit total area. The average time between occurrences $<t>$, can then be obtained from

$$<t> = \sum t_i p(t_i)/\sum p(t_i)$$

where the sums are taken over all of the available data. Thus one can estimate the average time $<t>$ between occurrences of sinkholes of all types, irrespective of the geological or anthropogenic influences. By computing $<t^2>$ in the same way one can then compute the standard error, m, on the average waiting time from $m^2 = <t^2> - <t>^2$.

Then, just as for the cumulative probability argument given in illustration 1, one can compute the cumulative chance of obtaining a recurrence time greater than a particular time value, T, from $CP(T) = 50[1-\tanh\{(2/\pi)^{1/2}(T-<t>)/m\}]$ %. As T tends to a value very large in respect of $<t>$, there is less and less chance that one will have to wait for a further sinkhole and, by a time of $T = T^* = <t> + m(\pi/2)^{1/2}$, there is only 16% chance that the waiting time for another sinkhole will be longer than T^*.

If one wishes to concentrate on just sinkholes of greater than a given surficial area, then one sorts the data according to increasing area of sinkholes and then re-performs the estimate of average recurrence time, as above, for those sinkholes of a given size or greater. In this way one can construct a plot of the average recurrence time versus size, or indeed for any attribute. Should one also wish to involve the dynamical influence of particular geological conditions then, again, one sorts the data according to the conditions of interest, and once more computes the average waiting time for sinkholes satisfying such conditions.

3.4 Illustration 4

In addition to the basic patterns that can be extracted from the database of properties and geological influences as described above, it is also possible to estimate the lateral size of a sinkhole, provided enough data are available. Thus one can obtain not only an idea of when a sinkhole is likely to occur but also an estimate of how large it is likely to be. The general sense of the argument operates as follows. First, from the available data construct a histogram of lateral size distribution and normalize the histogram area to unity in order to obtain the probability of obtaining an area of a sinkhole of area A. Then, just as for the waiting time calculation above, one computes $<A>$ and $<A^2>$. and so the standard error on the mean area of a sinkhole from $q^2 = <A^2> - <A>^2$. The cumulative probability that the next sinkhole will have an area in excess of a fixed value A is then just

$$CP(A) = 50\{1- \tanh[(2/\pi)^{1/2}(A-<A>)/q]\}$$

so that for an area of $A^* = <A>+q(\pi/2)^{1/2}$ there is only 16% chance the area of a new sinkhole will be greater than A^*. Thus one can categorize not only the anticipated time before another sinkhole appears but also the chances of its size exceeding a specific value. Depending on the level of size of sinkholes in relation to the damage they are likely to cause to both infrastructure and human life, one can then plan accordingly for the chances of occurrence and timing of appearances of such damaging sinkholes.

By repeating the general argument for constrained data under particular conditions, one can calculate the classes of damaging sinkholes in relation to their basic properties and geological influences. In this way one can categorize those general areas where one is most likely to have the most damaging sinkholes of a particular size or greater, and also the probability of occurrence of such with time. The requirements in each and every case are a database containing the attributes and also a comparison of the data with regional and local geological maps so one can tie the occurrence of sinkholes to external geological conditions. The data from the Muddental region, although somewhat sparse, allow one to show how to proceed with such detailed considerations. And that has been the main point of presenting these illustrations.

4 Discussion and Conclusion

The time dependent, as well as multi-causal, occurrence of sinkholes means a considerable risk for infrastructure installations and for the people of the densely populated region of Ruhr valley. The occurrence of sinkholes in the former mining region of the Ruhr region was investigated within a small part of this region using an available map of recorded sinkholes in combination with the geological map of this part of the region. As a result, we developed a suite of structural geological characteristics related to the sinkholes recorded to have occurred, as well as asking for the influence of various anthropogenic or mining conditions that may have caused the sinkholes. Now, the main goal of this investigation was to illustrate how suitable data organization and data preparation allows one to generate probabilistic calculations as well as specific results pertinent to the available data set of the example area.

Even if the results reported end up being restricted to the domain of the available date base (which may happen because of the small data base made available for this study), nevertheless the probabilistic interpretation of sinkhole occurrences with this small database clearly demonstrates the methodology and provides reasons for such likely sinkhole occurrences. Presumably, as more data become available, this set of conditions or correlations form a basis to examine other parts of the Ruhr region in order to determine both large-scale regional as well as local patterns of conditions and influences. In general, the procedures advanced here yield a tool to obtain a quantitative comparison of different types of factors or influences that are responsible for sinkhole occurrences in a former mining area with abandoned underground openings. The degree of importance and relative contributions of these distinct influences can be quantified readily by means of such probabilistic methods.

Chapter 12

Environmental Concerns: Catastrophic Events and Insurance.

Summary

With ongoing reduction of engagement of public authorities in infrastructure development, private companies have to consider their responsibility in the case of catastrophic events, i.e. they must be aware of, and be able to calculate and finance, the kinds of risks that formerly were handled by the public authorities. This chapter discusses the environmental costs and insurance needed in the cases of catastrophic failure of a project due either to natural causes as the project is being developed or after it is underway in an operating mode. Anthropogenic catastrophes also can occur and these, too, are considered from the viewpoint of insurance cover needed. A third sort of catastrophe can occur if regulations are changed during development of a project or after the project is in operation. In all cases a corporation involved in such projects needs to figure out how much insurance to take out against the possibility of a catastrophe and also the premium it should pay. Using examples from the tunnel construction theater of operations, from oil exploration and field development, and from regulatory changes that can influence waste disposal procedures, this chapter shows how general procedures, developed to accommodate for such risk in operating systems in the hydrocarbon area, can be used almost as is in other fields and they also provide procedures for correctly allowing for catastrophic events to be included in assessing probable profitability and associated insurance needs at the same time. Numerical illustrations are given to illuminate the points being made as clearly as possible. The basic point made is that procedures developed for hydrocarbon risk assessment can be equally used, almost without change, in many other areas where risk is a concern.

1 Introduction

Two conditions can occur in environmental problems that require a corporation to consider some form of insurance. First is the possibility of a catastrophic event occurring during transport of waste or of chemical solutions (with environment pollutant consequences), for instance in a tunnel section leading through a karstified rock mass. Such leakage from a truck or railway freight wagon causes pollution, thereby involving massive and expensive cleanup costs. One can imagine two scenarios with different kinds of reasons for a catastrophic event: either the transportation corporation used insufficient (or poorly maintained) transportation carriers or a construction defect of the riding way becomes effective with the lapse of time. The first scenario would likely be a concern for the transporting corporation, while the second scenario would be of concern to the traffic net provider or its construction management. Any corporation would surely like to take out insurance against the catastrophic possibility.

Second is the possibility that the regulatory agencies may consider a unilateral change in the environmental stringency conditions for burial of toxic material after the project for waste deposition is underway. One can imagine, for example, a change of legislative conditions that prefers controlled burning of waste instead of waste deposition under technically controlled conditions. In this case the corporation could be involved in substantial further costs, thereby lowering potential profitability of a contract. The corporation would surely like to option against the possibility of such an event occurring prior to the chance that the contract conditions could be changed.

These two forms of insurance are not equivalent. In the catastrophic loss event situation one would like to pay an insurance premium to cover the unknown catastrophic costs should they occur. In the regulatory stringency situation, one usually knows ahead of time precisely how such more stringent conditions, if enacted, would influence the corporate profitability and one would like to have a contingency option operating that would be activated if and only if the regulatory agency does indeed enact the new more stringent or legislatively different conditions.

Lempp et al. (2003) considered the quantitative procedures specifically for hydrocarbon exploration and production. This chapter shows how the same procedures can be used in many arenas where risk is an important consideration. To provide a background of familiarity, we have taken many of the hydrocarbon examples of Lempp et al. (2003) and repeat them here with slight alterations so that one can see directly how the same basic procedures are relevant to a wide variety of disciplines. This "lifting", almost wholesale, of the various methods and procedures from Lempp et al.(2003) is deliberate- it permits one to see clearly that the basic patterns are not so much beholden to the specific discipline but rather are of a very generic nature with a profound capability for attacking a very large class of risk type problems in many arenas.

The purpose of this chapter is to provide some idea of the way such insurance costs can be figured so that a corporation is not so fiscally vulnerable to the catastrophic losses nor to the extra costs of a regulatory change to more stringent environmental conditions. There is, to be sure, a wealth of papers and books dealing with risk and benefit analysis under uncertain conditions. Perhaps the semi-quantitative book by Wilson and Crouch (2001), or the classic treatise by Raiffa (1968), or even some of our own more quantitative works (Lerche and Mackay 1999; Lerche and Bagirov 1999; Lerche and Paleologos 2001) provide a backdrop to the quantitative developments presented here; and they all certainly provide a plethora of references to risk types of problems in a variety of scientific, political, economic, medical, toxicological and environmental areas that the interested reader is encouraged to consult for a greater depth of understanding of the general principles and applications. Here we limit ourselves to the applications of such methods and procedures to catastrophic types of problems that seem to have not enjoyed so thorough a development over the years as have other aspects of risking types of problems.

We consider first the catastrophic loss problem. Two case histories will be outlined: "catastrophic failure" due to (a) a changing load on the tunnel construction; and (b) traffic accident with consequences for the construction.

(a) Consider a traffic tunnel (railway or motorway) through a karstified rock mass of soluble limestone that contains different kinds of natural open spaces generated by leaching over geological time. Despite careful rock mass investigations during construction of the tunnel tube there remains some uncertainty about unknown open spaces in the surrounding rock mass in the neighborhood of the tunnel. Even if such spaces should have been detected during the construction phase (either by systematic drilling operations and structural investigations from the newly generated underground opening or by suited geophysical monitoring from the surface or from underground), nevertheless, there is a distinct risk that one has overlooked relevant open spaces in the rock mass around the tunnel tube. In the case of active karstification or leaching -also maybe in combination with dynamic loading that is effective with the lapse of time due to transportation- it is more than possible that one or more undetected open spaces still exist that never could be technically treated or recognized in the construction of the tunnel lining. Such an unsupported space in the surrounding rock mass leads to an unbearable loading state of the tunnel construction that was not precalculated. Due to stress accumulation and changing stress distribution around the tunnel wall, the construction reacts with distinct dislocations and with sudden displacements of the riding pathway. Such kinds of irregularities can cause catastrophic failure of the transportation system. Therefore, one has to estimate the risk of overlooking a failure-generating rock structure in the influencing region around the tunnel.

(b) Again consider a traffic tunnel (railway or motorway) through a karstified rock mass of soluble limestone that contains different kinds of natural open spaces generated by leaching over geological times. Consequently, the rock mass shows high permeability as well as high pore space, i.e. high volume capacity to contain or restore fluids. Under standard operating conditions related to the construction of the tunnel tube, the lining is sealed against the surrounding rock mass and a separate inside drainage system is able to take all fluids out of the tunnel. Therefore, the vulnerable karstic groundwater system will not be affected by the normal use of the tunnel as a transportation way. The internal drainage system is able to accommodate and remove any input of fluids and also protects the groundwater system in case of leakage from containers filled with waste or chemical solutions. However, in the case of a failure of the liner construction and consequently a destroyed sealing system, the natural environment of the karstic rock mass is no longer protected against pollution. Such kind of failure with a destroyed tube sealing can be imagined as a consequence of burning or explosion of transported goods. In addition, extreme heat will locally destroy the seal provided by the concrete shell of the tunnel. This kind of failure can take place because of transportation accidents involving the tunnel construction and thus leading to environmental pollution. Therefore, one has to estimate the risks caused by transportation accidents with direct consequences for the stability of the tunnel construction.

Comparison of both catastrophic failure scenarios (a) and (b) with a tunnel in karstic rock shows a basic difference concerning the reasons why they take place. In both cases the construction of the tunnel will fail locally; consequently environmental pollution will take place and expensive remediation operations are necessary. But the causes for the generation of the catastrophic events differ. Therefore, the risks differ too. Case (a) involves an expected but undetected "weak zone" within the rock mass that becomes effective due to natural solution processes and due to dynamically induced fatigue generation of the porous rock mass. One can label this sort of cause as an "intrinsic risk". Case (b) involves a high-energy input to the construction that is abnormal and due to an accidental burn during transportation. One can label this sort of cause as an "external risk".

A prevalent concern that is becoming manifest in decision tree analyses for such projects is with environmental clean-up if something goes wrong. For instance, if a tunnel project is saddled with the small chance of a catastrophe and the associated high repair and clean-up costs, then it is all too easy to take what could have been a very worthwhile project and downgrade it to less than worthwhile. Equally, even if a catastrophe scenario is not included while a tunnel is under construction, a similar catastrophic loss condition is sometimes envisioned by management: economic tunnel development proceeds and, once into operation, there is the chance of a transportation spill with, again, enormous environmental clean-up costs. Accordingly, management has been known to use the catastrophic loss excuse to get out of (or never get into) good projects. Thus management bias can lead to a re-evaluation that takes, on average, a very good prospect and makes it seem very poor.

The point about all of these management biases is that if only the expected value (average value) is used to quantify the risk, then one is completely overlooking the fact that there is also an uncertainty on the expected value, which ameliorates the "catastrophic loss" effect. The purpose of this chapter is also to show how this amelioration operates in practice.

2 Catastrophic Loss Assessments

A catastrophic failure due to a natural cause while a tunnel is under construction or after it is in operation will lead to loss of equipment and massive environmental clean-up costs.

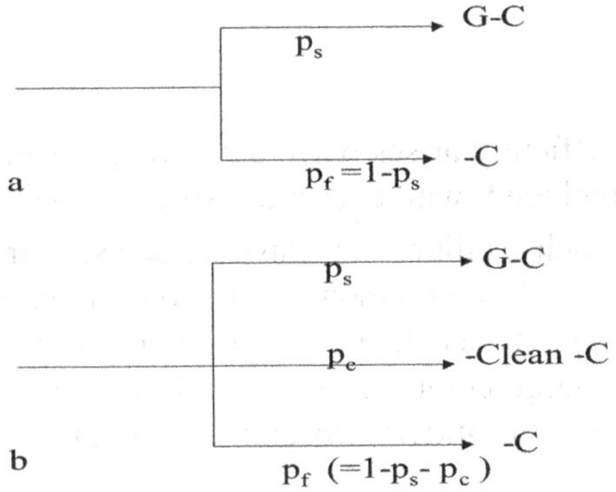

Fig. 12.1. Decision tree diagram illustrating: (a) a conventional evaluation scheme for an exploration opportunity; (b) changes brought about by inclusion of a "catastrophic failure" possibility

The classical way to assess the chance of catastrophic failure operates is as follows. First an evaluation of an opportunity is made, yielding an estimate of costs, C, potential gains, G, and chance, p_s, of successfully bringing the tunnel into operation so one realize the potential gains. If that were all that were to be done, then one can use the decision-tree diagram of Figure 12.1a to estimate an expected value of

$$E = p_s G - C \tag{12.1}$$

and a variance, σ^2, on the expected value of

$$\sigma^2 = p_s p_f G^2 \tag{12.2}$$

where p_f is the failure probability, $p_f = 1 - p_s$. Conventionally one would then estimate the probability, P_+, of obtaining an investment return greater than or equal to zero from

$$P_+ = \sigma^{-1}(2\pi)^{-1/2} \int_0^{\infty} \exp\left[-(x-E)^2/2\sigma^2\right] dx$$

$$\equiv \pi^{-1/2} \int_{-d}^{\infty} \exp(-u^2)\, du \tag{12.3a}$$

where

$$d = -E/(2^{1/2}\sigma) \tag{12.3b}$$

However, the difficulty arises when a mandated "catastrophic failure" is required to be included, with extremely small probability, p_c.A depiction of inclusion of such a failure in a classical decision-tree is given in figure 12.1b, where "Clean" is the amount one would have to pay for repair, clean-up and equipment loss if the catastrophic failure occurs. And "Clean" is usually very large compared to normal costs, C, of the project and, often, even compared to anticipated gains, G, if the project were to succeed.

In this case, the expected value, E, and the variance, σ^2, in the <u>absence</u> of the catastrophic loss are as given by equations (12.1) and (12.2) respectively, with the probability, P_+, of a positive return again given by equation (12.3b).

Inclusion of the catastrophic loss changes both the expected value E_C and variance, σ_C^2, respectively, to

$$E_C = p_S G - C - p_C Clean \equiv E - p_C\, Clean \quad (12.4a)$$

and

$$\sigma_C^2 = p_S G^2 + p_C\,(Clean)^2 - (p_S G - p_C Clean)^2. \quad (12.4b)$$

The probability, P_C, of making a return greater than zero on investment is now given by

$$P_C = \sigma_C^{-1}\,(2\pi)^{-1/2} \int_0^{\infty} \exp[-(x-E_C)^2/\sigma_c^2]\,dx$$

$$= \pi^{-1/2} \int_{-D_C}^{\infty} \exp(-u^2)\,du \quad (12.5a)$$

where

$$D_C = -E_C/(2^{1/2}\sigma_C) \quad (12.5b)$$

The point here is that both the expected value and its uncertainty (as measured through the variance) are used to assess the worth of the project. And the relevant comparison to make is between the probability of obtaining a positive return from the project both in the absence and presence of the "catastrophic failure" zone. Consider then, for illustrative purposes, the numerical example sketched in figures 12.2a and 12.2b, both in the absence (figure 12.2a) and presence (figure 12.2b) of the catastrophic loss scenario.

For the assessment in the absence of the catastrophic failure scenario one has, from figure 12.2a, that

$$E = 60, \quad \sigma = 0.55 \times 10^3, \quad E/(2^{1/2}\sigma) = 7.7 \times 10^{-2} \quad (12.6a)$$

while for the assessment including the catastrophic loss scenario one has, from figure 12.2b, that

$$E_C = -40, \qquad \sigma_C = 3.19 \times 10^3,\ E_C/(2^{1/2}\sigma_C) = -0.89 \times 10^{-2} \quad (12.6b)$$

Thus, in this case, while the expected value is shifted from positive to negative by the inclusion of the catastrophic option, the standard error in the mean is increased by a factor of 6, suggesting less accuracy in the mean.

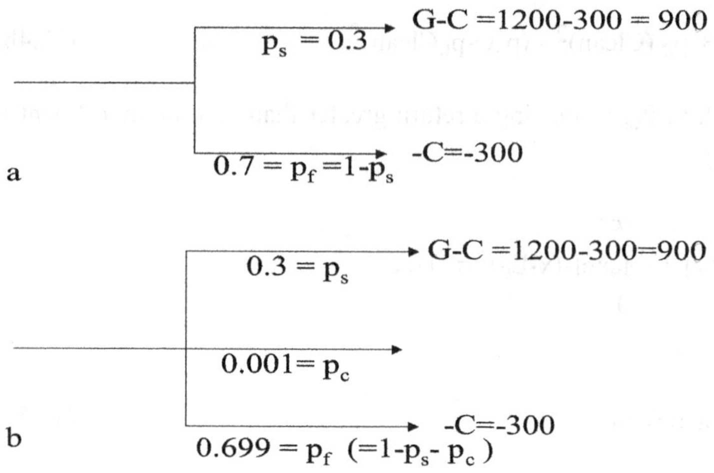

Fig. 12.2. Decision tree diagram illustrating: (a) numerical values for figure 12.1a; (b) numerical values for figure 12.1b

Indeed, a calculation of the probability of obtaining a positive return in the absence of the catastrophic option loss yields

$$P_{+} = 54.3\% \tag{12.7a}$$

while including the catastrophic loss yields the positive return probability of

$$P_{C} = 49.6\% \tag{12.7b}$$

Thus the catastrophic loss scenario is ameliorated correctly when inclusion of the variance on the mean is given. In the specific example given, just over a 4% drop occurs in the chance of being profitable, so that the small probability of a catastrophic loss is being included correctly - something which is not possible to have, or even visualize, if one operates with only the expected value as a measure of worth of the opportunity.

3 Catastrophic Loss After a Tunnel is in Operation

A different type of management concern is to argue that even if a tunnel is to be drilled and if it were to be successful, then there is the probability of an anthropogenic catastrophic event with attendant massive clean-up costs. Such arguments are again sometimes used to downgrade good projects. In this case the decision tree diagram is different than for the prior case. As sketched in figure 12.3a, the concern here is with catastrophic failure only after the tunnel is in operation.

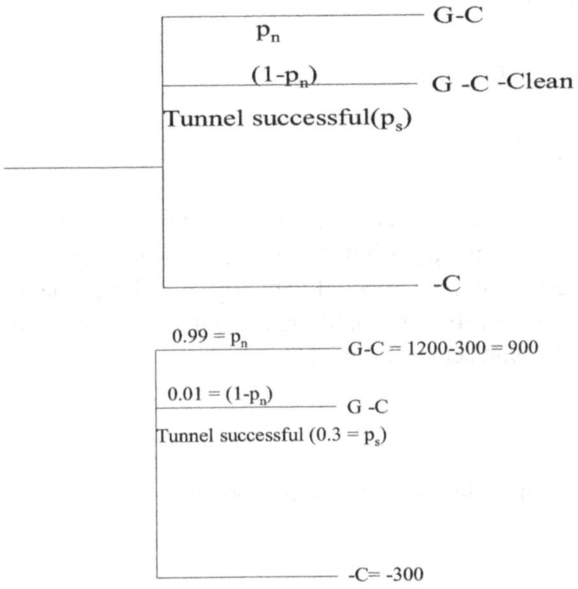

Fig. 12.3. Decision tree diagram illustrating: (a) how the probability of a catastrophe is included after a tunnel is in operation; (b) numerical values for the depiction of figure 12.3a

In this case, with p_n as the probability of no catastrophe and with "Clean" the cost of clean-up if an anthropogenic catastrophe does occur, from figure 12.3a one can write the expected value, E_c, and its variance in the presence of a catastrophic event as

$$E_c = p_s G\text{-}C - p_s(1 - p_n)\,\text{Clean} \qquad (12.8a)$$

and

$$\sigma_c^2 = p_s\{(1\text{-}p_s)G^2 - 2(1\text{-}p_s)\,(1\text{-}p_n)\,G \times \text{Clean}$$

$$+ (1-p_n)(1-p_s(1-p_n))Clean^2\} \qquad (12.8b)$$

Consider, for illustration, the numerical situation depicted in figure 12.3b. Then

$$E_c = -240, \text{ and } \sigma_c = 3.12.10^3 \qquad (12.9a)$$

so that

$$E_c /(2^{1/2}\sigma_c) = -5.44 \times 10^{-2} \qquad (12.9b)$$

For comparison, note that if the catastrophic scenario is omitted, then the corresponding values are

$$E = +60, \qquad \text{and } \sigma = 0.55.10^3 \qquad (12.10a)$$

so that

$$E/(2^{1/2}\sigma) = 7.64 \times 10^{-2}. \qquad (12.10b)$$

In this case, there is a reversal of the expected value (from profitable to non-profitable on average) but there is also an increase in the uncertainty (standard error) on the expected value by an order of magnitude. The corresponding probabilities of obtaining a positive return are:(a) in the absence of the catastrophic scenario one has

$$P \approx 54.3\% \qquad (12.11a)$$

(b) in the presence of the catastrophic scenario one has

$$P_c \approx 48.3\%. \qquad (12.11b)$$

Thus, inclusion of the variance (in the calculation of the influence of catastrophic event) lowers the chance of making a profit by only 6%. Amelioration is again properly taken into account.

4 Insurance Against Catastrophic Events

In the above examples with catastrophic tunnel events, the insurance costs must be paid either by the transportation net provider (in the case of a basic tunnel risk due to natural causes) or by the transportation company (in the case of transportation risk). Nevertheless, both companies have to optimize their potential profit and minimize their potential liabilities

Insurance is one of the costs that is becoming of increasing concern to both sorts of companies as they strive to limit corporate exposure to loss under any conditions, ranging from transportation hazards, tunnel maintenance, unanticipated karst space “break-outs” and allied corporate activities. Insurance is also one of the major costs that any endeavor handling hazardous material has to contend with. Here we illustrate how insurance costs can be estimated using problems from the hydrocarbon industry as a guide (the later tunnel construction examples then can be handled in a very similar manner as we will show) but the methods are of general applicability to a very wide set of disciplines as will become apparent from the specific illustrations.

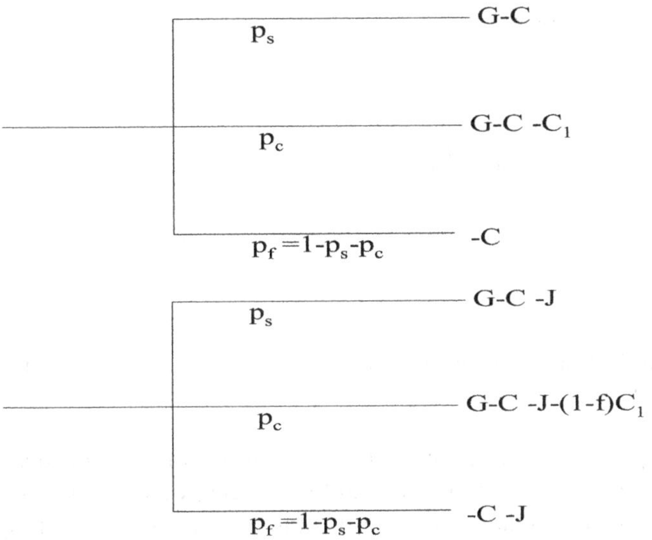

Fig. 12.4. Decision-tree diagrams for an exploration project showing the effects of (a) excluding insurance premium and coverage, (b) including insurance premium and coverage

On the petroleum exploration side, one of the hazards that insurance is designed to mitigate is the loss of an exploration rig due to adverse weather conditions (a hurricane), fire, a blowout, or other factors of low probability but high cost to the corporation if they occur. Thus an exploration project has associated with it not only the usual estimates of gains, G, normal exploration costs, C, and success probability, p_s, of encountering commercial hydrocarbons, but also the chance, p_c, of a catastrophic failure with costs C_1 that can be many times the normal exploration costs and also gains. This scenario is depicted in figure 12.4a. To mitigate against having to absorb the high costs, C_1, should the low event probability ever be realized, a company will insure a fraction, f, of such potential catastrophic losses at a premium J. Then should such a catastrophic event ever occur, corporate liability is limited to $(1\text{-}f)C_1$ plus the insurance cost J. This situation is sketched in figure 12.4b. The question of interest to the corporation is: What is the maximum insurance cost the company should pay?

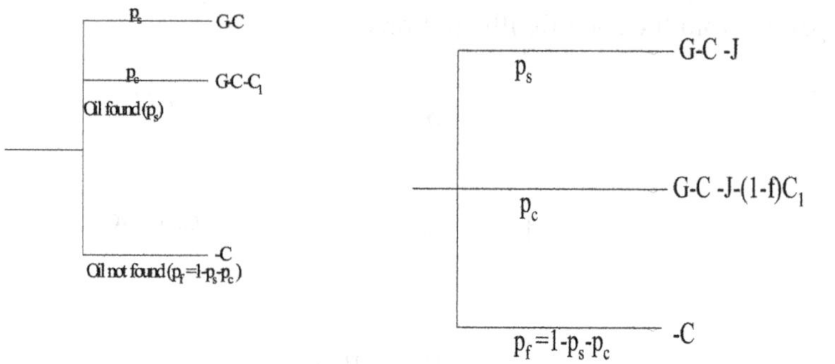

Fig. 12.5. As for figure 12.4 but only after oil is found so that an oil spill catastrophe is a real possibility

Equally, for a hydrocarbon development project there is the chance, p_s, that the wells involved in the project could yield a net gain, G, but there is also the chance that there could be a net loss, C, perhaps due to too many wells being involved or too low an oil sales price. But, in any event, part of the risk occurs after oil is found and development undertaken. There is always the chance, p_c, of an oil spill or a blowout with some attendant clean-up costs, C_1, which are usually sufficiently high that they can convert a likely profitable development project into a losing proposition. This situation is depicted in figure 12.5a. To mitigate against this potential for disaster a corporation will usually take out insurance, at a premium cost J, after oil has been found. The insurance company will then pick up the fraction f of the catastrophic costs C1 so that the corporate liability is limited to $(1\text{-}f)C_1$ if the catastrophe occurs. This insured situation is sketched in figure 12.5b. Once again the question is: How much should the corporation pay for insurance?

The two situations described above and in the decision-tree diagrams of figures 12.4 and 12.5 are not identical; neither, therefore, are their insurance costs. Indeed, a very large number of different types of decision-trees can be constructed to incorporate facets of insurance costs for different components of the various branches of the tree. The two examples described above are designed to show the differences that can arise in formulating insurance needs and costs.

We consider each of the examples in turn to develop the logic lines for handling such problems, and also give numerical illustrations to show practically the influence of insurance costs on the worth of a project, and how the amelioration of massive potential costs of catastrophic events can be integrated into project gains and normal costs.

4.1 General Catastrophic Loss Conditions

4.1.1 Mathematical Considerations

The first illustration to be considered is the catastrophic loss situation depicted in figure 12.4a (without insurance) and figure 12.4b (with insurance). The main effect of paying the insurance cost J is to cut the catastrophic loss the corporation would have to bear (if the catastrophic loss scenario occurred) from C_1 to $(1\text{-}f)\ C_1$.Thus the insurance company would pick up the costs fC_1 of the total. The question to be addressed is the price one should pay, J, for the insurance versus fractional cost fC_1 of the total catastrophic costs.

In the absence of insurance cover, as depicted in figure 12.4a, the expected value, E_0, of the project is

$$E_0 = p_sG\text{-}C(1\text{-}p_{fc})\text{-}p_{fc}C_1 \tag{12.12}$$

and the variance, σ_0^2, on this expected value is

$$\sigma_0^2 = p_s(1\text{-}p_s)G^2 + 2p_sp_{fc}G(C_1\text{-}C) + p_{fc}(1\text{-}p_{fc})(C_1\text{-}C)^2 \tag{12.13}$$

While the probability, p_{fc}, of catastrophic failure is usually estimated to be extremely small, the problem is that the catastrophic costs, C_1, if such a failure does occur, are so large that the product $p_{fc}C_1$ will usually dominate the expected value E_0. Thus, an otherwise highly profitable venture ($E_0>>0$) in the absence ($p_{fc} = 0$) of a catastrophic chance can be converted to an expected major cash drain ($E_0<<0$) on the corporation if $p_{fc}>0$- even for very small values of p_{fc}.

If an insurance premium, J, is paid so that the fraction, f, of the catastrophic costs would be paid by the insurance company, then the corporate risk is reduced to $(1-f)C_1$. Accordingly the corresponding expected value, E_1, is now

$$E_1 = E_0 + p_{fc} f C_1 - J \tag{12.14}$$

with variance, σ_1^2, around this expected value of

$$\sigma_1^2 = p_s(1-p_s)G^2 + 2p_s p_{fc} G[C_1(1-f)-C] + p_{fc}(1-p_{fc})[C_1(1-f)-C]^2 \tag{12.15}$$

which is smaller than σ_0^2 because of the lowering of the catastrophic costs charged to the corporation. In addition, the new expected value, E_1, is larger than the old expected value, E_0, by the amount $p_{fc}fC_1-J$, representing the pick-up of expected catastrophic costs by the insurance company rather than by the corporation.

As a first requirement, then, it would seem that one should pay an insurance premium, J, only when the fractional coverage, f, of catastrophic losses satisfies the inequality

$$p_{fc} f C_1 > J \tag{12.16}$$

because otherwise E_1 will be smaller than E_0, and so the expected value to the corporation would be reduced. In addition, because $\sigma_1^2 < \sigma_0^2$, the uncertainty on the E_1 value is less than that on E_0, i.e. E_1 is statistically sharper than E_0.

A further measure of worth of insurance is guided by the probability of making a profit, including the variance on the expected value. Thus, let P denote the probability of making a profit greater than or equal to zero. Then

$$P = \sigma^{-1}(2\pi)^{-1/2} \int_0^{\infty} \exp\left[-(x-E)^2/2\sigma^2\right] dx \tag{12.17}$$

which can be written as

$$P = \pi^{-1/2} \int_{-d}^{\infty} \exp(-u^2)\, du \tag{12.18}$$

where $d = E/(2^{1/2}\sigma)$.

Note that even if the expected value is negative, there is still a chance of turning a profit (P>0), but this chance declines as E becomes more negative at fixed variance. In the absence of taking out insurance the probability, P_0, of making a profit is

$$P_0 = \pi^{-1/2} \int_{-d_0}^{\infty} \exp(-u^2)\, du \qquad (12.19)$$

where $d_0 = E_0/(2^{1/2}\sigma_0)$, while when insurance is taken out, then the probability, P_1, of making a profit is

$$P_1 = \pi^{-1/2} \int_{-d_1}^{\infty} \exp(-u^2)\, du \qquad (12.20)$$

where $d_1 = E_1/(2^{1/2}\sigma_1)$.

If one is to pay insurance then, at the very least, one would like the probability of a profit to increase, i.e. one requires at a minimum that $E_1/\sigma_1 > E_0/\sigma_0$ –which incorporates inequality (12.16) on the amount of insurance to pay versus the coverage obtained.

However, a more stringent condition is usually in force. A corporation will often insist that a project be undertaken only if a given probability, P, exists of a project being profit-making. Thus, if some predetermined value for P is given, the question then reduces to finding the values of fractional coverage, f, and insurance premium, J, so that equation (12.20) will yield the required value P.

A remarkably good practical approximation for equation (12.20) in $0.1<P<0.9$ is that

$$P = 0.5 + d_1/\pi^{1/2} \qquad (12.21a)$$

or, conversely, if P is given, then one has

$$d_1 = \pi^{1/2}\,(P-0.5) \qquad (12.21b)$$

The form (12.21b) is useful for a prescribed corporate mandate on P, because then one requires

$$E_1 = (2\pi)^{1/2}\sigma_1(P-0.5) \qquad (12.22)$$

The approximation (12.22) provides a constraint on the insurance premium, J, and coverage fraction, f, of the catastrophic loss costs and, provided P>50%, automatically requires $E_1>0$ -a “standard” constraint often invoked for decision-making investments. Because both E_1 and σ_1 involve the coverage fraction, f, but because only E_1 involves the insurance premium, J, it is easiest to re-arrange equation (12.22) to provide the relation

$$J = p_s G - C(1-p_{fc}) - p_{fc}(1-f)C_1 - (2\pi)^{1/2}\sigma_1(P-0.5) \qquad (12.23)$$

Thus, one can insert the fractional coverage required and the corporate-mandated probability of a profit into equation (12.23), and so figure the appropriate premium to pay, subject to the constraint $0<J<p_{fc}\, f\, C_1$, of course.

4.1.2 Numerical Illustration

Consider a typical set of parameters for an exploration project. Let the success probability p_s =30%, with normal costs C about a quarter of expected gains (i.e. G = 4C). The catastrophic probability p_{fc} is usually estimated to be very small, typically 1% or less. We carry through the calculations with $p_{fc} = 10^{-2}$ for definiteness. Estimating catastrophic costs C_1 is another matter entirely. We perform the calculations with C_1 = 100G = 400C so that there would be no chance of a profit in the expected value E_0 without insurance because then

$$E_0 = -3.49C<0 \qquad (12.24a)$$

and one also has

$$E_1 = -3.49C + 4fC - J \qquad (12.24b)$$

For the same numerical values one has the variance σ_1^2 given through

$$(\sigma_1^2/C^2) = 3.36 + 9.6x(0.9975-f) + 1584x(0.9975-f)^2 \qquad (12.25)$$

The connection between insurance premium, J, the mandated corporate success probability P, and the fractional coverage f is, from equation (12.23) given by about

$$(J/C) = -2.79 + 4f - (2\pi)^{1/2}(P-0.5)x[3.36 + 9.6x(0.9975-f) + 1584x(0.9975-f)^2]^{1/2} \qquad (12.26a)$$

provided

$$0 < J/C < 4f \quad (12.26b)$$

and inequality (12.26b) is automatically satisfied for a corporate-mandated profit probability greater than 50%. At 100% coverage (f =1) of the catastrophic losses one has an insurance premium of

$$J/C = 1.21 - 4.61\ (P-0.5) \quad (12.27)$$

Thus at 50% probability of profit one should pay no more than 121% of normal costs, C, to cover the likelihood of catastrophic costs (i.e. no more than 0.33% of catastrophic cost estimates).

4.2 Insurance Coverage After Oil is Found

4.2.1 General Considerations

If oil in commercial quantities were to be found then the insurance coverage a corporation would like to take is predicated on the probability of an oil spillage as shown in figure 12.5. The point here is that one must fold this potential cost of insuring against spillage into the assessment of worth of the project ahead of drilling. In this situation two main factors are operative: First is the fact that until (or unless) one finds commercial amounts of oil one does not take out insurance; second is the fact that the corporation must make a decision to undertake the exploration project ahead of knowing whether oil is present or not. So the corporation must assess potential gains, G, and potential clean-up costs, C_1, if an oil spill were to occur, as well as reduction in potential gains due to the insurance premiums, J, the corporation would pay if oil were to be found. And these assessments must all be done at the stage of deciding whether to undertake the exploration project or not. Thus the decision-tree diagram excluding insurance coverage, but including the potential for an oil spill is given in figure 12.5a; while including insurance cover to a fraction, f, of potential oil spill costs, C_1, at an insurance premium, J, is shown in figure 12.5b.

The expected value of an opportunity in the absence of insurance cover is

$$E_0 = p_s[G-C(1-p_1)]-C \quad (12.28)$$

while the variance, σ_0^2, around this mean value is just

$$\sigma_0^2 = p_s\{(1-p_s)[G-C(1-p_1)]^2 + p_1(1-p_1)C_1^2\} \quad (12.29)$$

Note that if $p_1 = 1$ (no catastrophic oil spill), then σ_0^2 reduces correctly to $p_s(1-p_s)G^2$- the usual variance expression.

If an insurance premium, J, is paid with a corresponding catastrophic insurance coverage of fC_1 paid by the insurance company if the catastrophe occurs, then the corresponding mean expected value is E_1 with

$$E_1 = p_s[G-J-(1-f)(1-p_1)C_1]-C \tag{12.30}$$

The variance around this mean value is

$$\sigma_1^2 = p_s\{(1-p_s)[G-J-(1-f)(1-p_1)C_1]^2 + p_1(1-p_1)(1-f)^2C_1^2\} \tag{12.31}$$

The logic pattern is then the same as for the previous example. One calculates the probability, P_0, of a profitable operation in the absence of insurance coverage from

$$P_0 = \pi^{-1/2} \int_{-b_0}^{\infty} \exp(-u^2)\, du \tag{12.32}$$

where $b_0 = E_0/(2^{1/2}\sigma_0)$. One also calculates the probability, P_1, of a profitable venture in the presence of insurance cover in the amount fC_1 at a premium J from

$$P_1 = \pi^{-1/2} \int_{-b_1}^{\infty} \exp(-u^2)\, du \tag{12.33}$$

where $b_1 = E_1/(2^{1/2}\sigma_1)$. For a fixed corporate probability on profit chances, it follows from equation (12.33) that b_1 must then be a fixed dimensionless number, n, say where

$$E_1 = 2^{1/2} n\sigma_1 \tag{12.34}$$

If a probability P_1 of 50% or greater of making some profit is required by the corporation then one requires $E_1>0$, which can be rewritten as the requirement

$$G>C/p_s + J + (1-f)(1-p_1)C_1 \tag{12.35}$$

Once inequality (12.35) is satisfied then the probability P_1 can be approximated in $0.1<P_1<0.9$ by $P_1 = 0.5 + n/\pi^{1/2}$, or alternatively one can write $n = \pi^{1/2}(P_1-0.5)$. For a given value of n (or a given value of P_1) one has

$$E_1^2 = (2n^2/(1+2n^2))(E_1^2+\sigma_1^2) \quad (12.36)$$

Using the fact that E_1 depends on the fraction, f, of insurance coverage obtained and also on the premium, J, paid one can rewrite equation (12.36) in the form

$$J=G-(1-f)(1-p_1)C_1+C[p_s(k-p_s)]^{-1}\{p_sk-(kp_s)^{1/2}[1-(k-p_s)p_1(1-p_1) \times(1-f)^2C_1^2/C^2]^{1/2}\} \quad (12.37)$$

where $k = (2n^2/(1+2n^2))<1$. In this way one relates the fractional coverage, f, and the premium paid, J, to the value n, which is approximately linear in the desired probability of profit P. A premium less than that given by equation (12.36) for the same coverage and other parameters fixed is clearly a bargain. One should not exceed the premium value given by equation (12.36) for a fixed fractional coverage under the conditions stated.

4.2.2 Numerical Illustration

Consider that a success probability of $p_s = 1/3$ obtains, with potential gains, G, of ten times the normal costs, C, i.e. G = 10C.Then in the absence of any potential for catastrophic loss ($p_1 = 0$) the expected value of the opportunity is

$$E_0 = p_sG-C = 7/3C > 0 \quad (12.38a)$$

and the variance is

$$\sigma_0^2 = (2/9)x10^2C^2 \quad (12.38b)$$

so that $E_0/(2^{1/2}\sigma_0) = 0.36$, with the associated probability of making a profit at P = 0.7.

Including a 1% chance (p_1= 0.99) of a major oil spill after production commences, with an associated clean-up cost of C_1 = 1000C, then yields an expected project value of

$$E_1 = p_s[G-C_1(1-p_1)] - C = -C<0 \quad (12.39)$$

and the corresponding variance on E_1 is $\sigma_1^2 = 3{,}301C^2$, so that $E_1/(2^{1/2}\sigma_1) = -0.0123$. In this case the probability of making a profit is reduced to $P = 0.48$.

If insurance is taken out in the amount $J = jC$, and if the insurance cover is 80% of the catastrophic losses should they occur (i.e. $f = 0.8$), then the expected value, E_2, of the project is now given by

$$E_2/C = (5-j)/3 \tag{12.40}$$

indicating that the insurance premium should be less than 5C if the expected value is to remain positive. The variance on E_2 is given by

$$\sigma_2^2 = 2(j-6)^2/9 + 201 \tag{12.41}$$

In this case one has the ratio

$$E_2/(2^{1/2}\sigma_2) = (5-j)/[18x(2(j-6)^2/9 + 201)]^{1/2} \tag{12.42}$$

The largest value of this ratio (and so the greatest probability of making a profit) is when $j = 0$ of course, i.e. if one can get 80% catastrophic insurance coverage for nothing! As the premium, j, of the insurance cover increases, the probability of making a profit systematically decreases until, at $j = 5$, there is only a 50% chance of a profit with 80% catastrophic coverage. Thus, one should not pay an insurance premium of more than 5 times the normal operating costs for 80% catastrophic coverage if one wishes to maintain a greater than 50% chance of making a profit. This result translates into 0.5% of the total catastrophic costs (or 0.625% of the 80% covered costs) as an adequate insurance premium for the corporate constraint of greater than 50% chance of making a profit.

5 Insuring Against the Chance of Potential Regulatory Changes

This particular sort of catastrophic risk has little to do with physical damage done to the environment by an untoward event, but has more to do with assessing the catastrophic damage that could be done financially to a corporation should a regulatory agency change the rules of operation after a corporation has not only committed to develop some area but is, in fact, heavily engaged based on rules in place at the time the development was undertaken.

For example, a waste disposal has been planned and partially constructed following the rules of TASI (1993, Technical Rules for Anthropogenic Waste). These rules incorporate a high technical standard for all constructive parts of the disposal site concerning the sealing of the underground, the monitoring of the disposal as well as deformation of the ground and emission of polluting agents and other objectives. In summary, high investment costs must be paid for preparation of a long-term project (i.e. at least some tens of years) of waste disposal in a controlled and monitored environment. Imagine then a change of political and legislative direction concerning waste treatment: Emergency burial disposal of waste is still permitted but now the legislative authorities will support financially burning of waste. The basic support for the disposal project vanishes. However, the disposal area has been prepared to about half the originally planned size, the mass of waste that is available for disposal will be reduced to about one third or less, the duration of the project will increase by at least a factor two. Insurance costs must be figured out for such a potential eventuality, else the corporation involved could face immediate bankruptcy.

If a regulatory agency changes the rules in a more stringent direction after such an involvement is underway, then the corporation must obey those new rules and so absorb costs associated with the new rules. Such extra costs can be exorbitant and so ruin the corporation. However, what is not clear at the time the decision to develop is undertaken by the corporation, is whether the regulatory agency will change the rules. In most situations one is aware of what the new rules will be, if enacted, but one is unclear whether they will be enacted. Also, while it is often the case that legislative decisions are slow, and so one has time to set optional plans in place, it does happen that legislatures move with blinding speed once in a while; it is certainly best to then have already taken out some form of financial insurance coverage so that a corporation is not left with massive, unanticipated costs. Thus, a fairly good idea is available of what such new rules would cost the corporation but one has only a probability, p_c, they will be enacted.

What the corporation would like to do is to arrange financial support for itself in the eventuality the rules do come into effect. If no change occurs then all the corporation loses is the premium paid, J, but if the new rules do come into effect then the insurance company is required to pick up a fraction, f, of the extra costs, C_1, so that the loss to the corporation is then limited to $J + (1-f)C_1$. Thus given the uncertainty, as measured by the probability, p_c, that the new regulations will come into effect, the expected corporate loss, L, is then

$$L = J(1-p_c)+[J+(1-f)C_1]p_c = J+(1-f)p_cC_1 \quad (12.43)$$

This expected loss has to be compared with the worth of the project to the corporation before the new rules may come into effect but after the corporation pays out for insurance in the amount J. Let this initial worth be V-J, so that V measures the intrinsic project worth before one pays the insurance premium J.

The argument then proceeds in identical fashion to the example given directly above. One puts together a decision-tree as shown in figure 12.6, representing the worth, $V_1 = V-J$, under no change in rules and $V_2 = V-J-(1-f)C_1$ under a change in rules respectively.

The average value E is then

$$E = (1-p_c)V_1 + p_cV_2 \quad (12.44)$$

with a variance

$$\sigma^2 = (1-p_c)V_1^2 + p_cV_2^2 - E^2 \quad (12.45)$$

The probability, P, that one will make a profit, irrespective of whether the rules are maintained or changed, is then given by

$$P = \pi^{-1/2} \int_{-b}^{\infty} \exp(-u^2)\, du \quad (12.46)$$

where $b = E/(2^{1/2}\sigma)$. Again then one asks for a given corporate mandated probability of making a profit. The basic question is: What are the requirements on the premium, J, that one should pay in relation to the fractional coverage, f, that one wishes to have to compensate for the loss L? This problem is mathematically precisely that given above and there is no need to repeat the argument. In short, insurance against enactment of more restrictive regulatory conditions after development has commenced can be addressed with the same basic tools as already set out in the previous case histories.

6 Discussion and Conclusion

Two major concerns have permeated this chapter. First was to provide some quantitative procedures for assessing the worth of being involved in a project given that there are associated risks of a catastrophe occurring. The examples from the tunnel construction /operation industry and also from the hydrocarbon exploration industry were used to show how such risks of catastrophe can be assessed for both natural and anthropogenic catastrophes using a common framework. In addition, the costs of insurance and the cover such insurance provides were also addressed in a quantitative manner. These examples from very different industries show how the generic procedures can be invoked for almost any such measure of worth of a project from any industry, which was the main point of using such dif ferent examples from very different arenas.

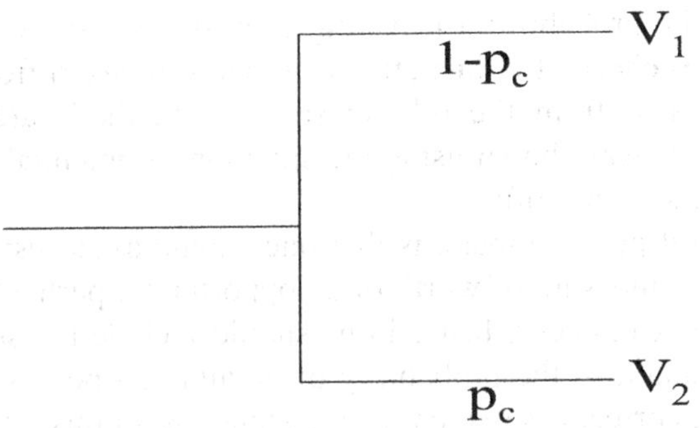

Fig. 12.6. Decision-tree diagram showing the effect of insurance designed to ameliorate the financial risks associated with a potential unilateral change in environmental stringency regulations

Second, clean-up costs and costs associated with enactment of more stringent regulatory changes can be astronomical. Using the methods described in this chapter one can assess the probability that such events can lead to financial ruin. Usually the probabilities of such catastrophic events are small, but because of the high potential losses, inclusion of such events in economic risk models can swamp completely, in an abusive manner, the characteristics of the opportunity that would prevail in the absence of the catastrophic component. This overwhelming nature of catastrophic costs is particularly to be noted if attention is focused only on the expected value as the sole economic arbiter.

The fundamental point being made here is that one can incorporate potential regulatory changes into the worth of a project and can take out insurance against the possibility one could face financial ruin in the event such changes are enacted. This particular aspect is of major importance when one is deciding whether to undertake a venture, such as a waste depository construction, in the face of political "weather" when it is just not clear what the legislature will decide, but it is equally clear that one cannot delay making the decision until the legislature has reached a decision.

Inclusion of catastrophic costs should also allow for the uncertainty on the expected value (measured here by the variance). When such is done then one can properly allow for the risk associated with the catastrophic costs and also make a rational determination of the amount of insurance one should acquire and the amount one should maximally pay for such insurance. The point of the examples given here was to show how one can assess such factors in a systematic way and still honor corporate requirements on the probability of making a profit. While the two major groups of examples chosen to illustrate the points were taken from the tunnel industry and also from the oil industry, the methods are of much greater generality as was shown using the situation of potential regulatory changes as fiscal risks on profit.

The general point to make is that one should avoid using just the expected value as a measure of worth of an opportunity, particularly when catastrophic events can occur. Instead one should include the standard error (or variance) and also the probability of obtaining a positive return in order to provide a correct assessment of the situation. In this way not only can low risk, but high cost, potential clean-up scenarios be addressed but also the amount of insurance premium one should pay to offset such calamitous charges should they eventuate. The illustrations were tailored to sharply focus on the way such events can be handled in an objective, quantitative manner, and so to show how to obtain financial safeguards for corporate profitability when catastrophic events need to be incorporated.

Chapter 13

Integrated Scientific and Economic Uncertainties in Environmental Hazard Assessments: Social and Political Consequences

Summary

This chapter examines the consequences of scientific and economic uncertainties on assessments of remediation of environmental hazards. Several examples from the hydrogeological arena and the oil catastrophe arena provide illustrations of how both the scientific and economic uncertainties are intertwined in contributing to such assessments.

In addition, the political and social consequences of undertaking particular actions, or of not undertaking them, are examined so that one can incorporate the political and legislative decision-making process for any such remediation in a more quantitative manner than is usually done. The determination of loss from clean up costs, and of gains because of future use of a previously environmentally damaged region, are both considered, also from quantitative viewpoints with illustrations. In this way one can obtain an integrated picture of the total effect of environmental hazards and their potential resolutions. One can also determine quantitatively which of the many uncertain scientific, economic and political options need further determination prior to committing to such environmental clean up. The general procedures spelled out can be applied to almost any environmental problem and, as such, provide a consistent framework for a total integrated evaluation.

1 Introduction

Environmental hazards arising from the disposition of chemical, biological, radioactive and toxic wastes are, perhaps, the quintessential problems of this century and the foreseeable future. Humanity produces copious quantities of waste of various sorts. Some of the waste is biological, some is in the form of both household and industrial material, some consists of chemical compounds used in various processes, some is in the form of gaseous emissions that pollute the atmosphere and the earth, some in the form of radioactive waste as by-products, or end-products, of nuclear-driven power plants, submarines, and decommissioned weapons, and some in the form of raw product such as oil and its associated waste.

In a broad sense, the main problems for both waste and for raw product can be classified in about five main categories. The first category arises because material must be transported from a temporary storage or processing facility to a permanent site. This transport requires that decisions be made on the transport vehicles and routes to be used as well as on the material of the transporting containers. For instance, in the case of transport of radioactive waste, the very containers that are used for the transport may eventually become radioactive, and thus the containers themselves may have to be transported and buried along with the primary radioactive waste. In the case of chemical waste transport, the absorption of toxic material to container linings means that such containers may pose problems either by corrosion of the container or by becoming toxic - again requiring that the containers be disposed of properly.

The very act of shipping and transporting hazardous material is fraught with difficulties, due to the desire to avoid population centers where possible, the need to transport the product to a receiving site with the least delay and possibility of danger, and due to the fact that the method of transport can itself be subject to vagaries of the weather, human error, and unforeseen problems - such as bridge collapse, train derailment, or ship breakup, to name but a few.

Assuming one has transported the material successfully to the storage site, be it a waste depository or a storage site for holding product until needed for further use, the second critical issue arises from the location and construction of the site itself. The storage site must be sufficiently well constructed that deposition of the waste material will not compromise either the integrity of the site nor so rapidly fill the site that another must be sought. The site integrity is, arguably, one of the most critical ingredients because external influences, such as water flow, rock fractures, earthquakes, erosion, magmatic activity and the like, should not be sufficiently strong that site integrity is compromised (Fischer 1986). Indeed, geologic, hydrologic, and geochemical conditions at a site should preferably act as natural barriers to potential leakage, supplementing the role of engineering measures (Freeze 1997). The site must be sufficiently far away from population centers that potential leaks will not affect human health, and at locations so that future activities (mining, urban expansion, etc.) will not endanger site integrity (U.S. EPA 1991). In addition, in the case of required long-term storage, well beyond the lifetime of any known political system of government, it becomes of paramount importance that archival records be maintained to advise future generations of the hazardous nature of material contained in a storage site (Keeney 1980).

A third category of problem arises with the requirement of monitoring the hazardous material during both the transport and later storage stages. One must, at the least, have a satisfactory number of monitoring devices of sufficient resolution to detect rapidly any problems that do occur and have contingency plans for action based on the monitoring results. The issues here would seem to be to estimate how many monitoring devices are needed, the length of time over which monitoring should occur (with regular maintenance and replacement being incorporated in the monitoring program), the accuracy and resolution of results reported by such devices, and how measurements performed at different scales can be utilized (Christakos and Hristopulos 1998).

The most critical problems are associated with spillage and leakage, which are treated separately here as two distinct categories. Spillage is usually referred to when, during the acts of transport or after storage, there is a major loss of material from the transporting system or from the storage containers, which has an immediate and significant effect on the surrounding environment. Such types of spillage problems are, perhaps, exemplified by the breakup of a ship carrying crude oil with the immediate loss of the majority of the cargo, resulting in severe, immediate environmental damage, or, perhaps, by the Chernobyl disaster where huge amounts of radioactive material were released to both the atmosphere and into the surrounding ground, thereby causing major contamination problems that required immediate resolution. On the other hand, leakage usually refers to slow loss of material that, while not so hazardous on the short-term, would become so unless action is taken to control and modify the transport or storage system and stop the leakage. Thus, slow leakage from a corroding container filled with toxic chemical waste presents minimal, immediate danger but, unless controlled, will eventually pose a significant long-term hazard (Asante-Duan 1996).

Scientific analyses of environmental hazards are hindered by significant uncertainties. Characterization of present-day conditions at environmental sites is almost always limited because of financial constraints and the complex interaction of physical, chemical, and biological processes that control the transport of contaminants is usually not well understood. This situation is accentuated by the requirement that a repository site needs to provide efficient isolation of the waste from the environment and human population over long periods of time and, correspondingly, of the liability claims that an environmental company may have to face long after the completion of a project. Future geologic events (earthquakes, floods, climate change, etc.), all of which can influence the ability to control and isolate the waste over time, introduce further uncertainties. Moreover, the existing technological solutions are limited in their efficiency, expensive to implement, and can generate by-products that are difficult to control. The situation increases in complexity with litigation issues, where a significant expense can develop, so that funds that could otherwise have been allocated for development and application of new technologies, are used up in attempts to resolve legal issues.

But there is another side to the problem of waste disposal. Eventually a corporation (or corporations) will to have transport, bury and monitor the waste, usually under a contract from a regulatory agency. The contract will usually contain performance criteria to be met but will also contain a price to be paid to the contractor. The question of interest to the corporation is whether the contract is profitable and under what conditions. There is clearly some sort of financial limit depending on the prior estimates of probabilities of transport (with or without spillage), burial (with or without leakage), and on-going monitoring costs. Additionally, regulatory agencies, concerned citizen groups, and political staff attached to a lawmaker, may all wish to become involved in the various components affecting the decision-making at various stages. From the corporate perspective the problem is to ascertain what associated costs make it worthwhile to accept the contract. For an environmental corporation's perspective scientific uncertainties and limitations of the technological solutions are only a subset of the total uncertainties it faces. Changing political, financial, and regulatory conditions constitute other unpredictable components in a project's performance and financial return.

The purpose of this chapter is to explore how scientific and economic considerations, both of which are hindered by significant uncertainties, can be combined in an integrated risk assessment. The political and social issues of decision-making and decision influencing are then discussed within the framework of knowledge of the scientific and economic costs and their uncertainties.

2 Scientific Uncertainties

From the scientific point of view there are several problems that need to be addressed in relation to the transport of hazardous material, the processing of such waste, or the storage of the material in a way that minimizes the potential for spillage or leakage. First, one needs to assess the chemical, biological or radioactive contamination that could occur should waste be released, and to provide remediation solutions to cover such contingencies (Van der Heijde et al. 1986; Nyer 1992; Grasso 1993; Flach and Harris

1997). Second, one needs to plan the transport and storage in such a way that the short or long-term (or both if possible) potential for release is minimized. Thus, in the case of the potential storage site at Yucca Mountain, Nevada for spent radioactive material, one needs to perform major site surveys to check out the current rock fracture pattern, the likelihood of fault and earthquake disturbances that would compromise the integrity of the storage system, the potential for ground water contamination, the geochemical conditions that can accelerate or impede such contamination, and a host of ancillary scientific issues that then provide the best available scientific assessment of the capability of the repository for long term secure storage. This sort of problem has been the subject of major scientific investigation over the years (Hunter and Mann 1992; U.S. D.O.E. 1992; Sandia 1994). In other cases one is, perhaps, not quite so fortunate to have a large body of scientific investigation to hand. The consequence is that, for both short-term and long-term problems, one is more likely to have overlooked or misestimated a potential hazard, with the consequent increase in uncertainty on the capability of the storage system to truly maintain a secure condition (Keeney 1980).

Despite scientific analyses, there are always situations that are difficult to guard against, control, or remediate. Perhaps the most obvious of such problems is the transport of oil by tankers. Here there is little that can be done to safeguard against major storms close to shore that can (and do) easily shipwreck tankers, with the consequent spillage of oil because of disruption of ship integrity. Or again, it would seem difficult to guard against human error as represented by the Exxon Valdez disaster, where a fully laden tanker ripped apart on a submarine reef due to human negligence, spilling its oil contents to the ocean and shore around Valdez, Alaska (Alaska Fish and Game 1989). In other instances one has to deal with old abandoned hazardous landfills and other sites, where leakage has emanated from corroding drums and containers that have been improperly labeled or incompletely burned and, hence, there is very little knowledge of the quantity and nature of the chemicals involved (Magnuson 1980; Malle 1996). The situation is exacerbated by the poor performance of certain commonly used waste site remediation approaches (Canter and Knox 1986). For example, a National Research Council committee review of the performance of pump-and-treat systems at 77 contaminated sites reported that groundwater cleanup goals had been achieved at only 8 sites (MacDonald and Kavanaugh 1995). The U.S. Environmental Protection Agency (EPA) reported that only 14 of 263 Superfund source control projects, for which systematic site remediation solutions were applied, reached completion (Powell 1994).

So, from both the storage and transport points of view, there are not only hazard assessments that need to be made and contingency plans to be developed in response to estimated hazards, but also there is the need for an assessment of the resolution and accuracy of the estimates, and of catastrophic clean-up problems, should such eventualities occur. The point here is that the scientific measurements (from which hazard estimates are made) are limited in number, unevenly distributed in space and time, and have intrinsic uncertainties of their own (Dooge 1997; Christakos and Hristopulos 1998). In addition, the scientific models are themselves uncertain, or require parameters that are poorly known, or do not quite address all the problems that could occur. For instance, groundwater flow and transport models cannot fully account for the spatio-temporal variabilities and uncertainties of the parameters and variables of the problems and, in addition, suffer from the inherent simplification of the complex interaction of physical, chemical, and biological processes that affect the transport of contaminants. The validation of such predictive models in terms of slowly evolving subsurface plumes is currently elusive (Dagan and Neuman 1997). Thus, the ability to provide accurate assessments is limited by data, models, and outside considerations, over which no, or only limited, control exists (Corwin et al. 1999).

But the largest factor, which is often overlooked in scientific assessments, is the financial analysis attendant on any decision made or to be made. Thus, costs of transport, storage, monitoring, leakage remediation, spillage clean up, insurance, catastrophic liability, and the value of bidding on a contract for such transport and storage, need to be intertwined with the scientific assessment. For example, if the cost of undertaking to outfit a tanker with double or triple hulls to offset the chance of an oil spill were to be such that this cost could never be recovered, then there would be little point in taking such action; alternative methods of transport would be searched for that would allow a profit to be made.

Only a government has essentially unlimited financial resources, so any time a contractor is involved in either a transport or storage opportunity, there is always the question from the corporate side of whether the project could prove profitable, or whether the project is so risky that, even with maximum insurance and available technology, the likelihood of spillage and/or leakage is so large as to increase the probability that the company will lose huge amounts of capital. Under such conditions, it would be a brave, or foolhardy, company that would accept such risk without a limited liability clause written into the contract.

The purpose of this part of the chapter is to provide some quantitative methods and associated examples to address the above problems. Illustrations are furnished to show how one can account for the uncertainty in both the scientific and economic estimates in order to reach a rational decision of the worth of a project. In this way, the methods guide the scientific and economic components of a project to increase either the amount of scientific information - and also determine what type of scientific information to increase – or, perhaps, accept the available scientific information but to improve resolution - and to then determine which components of the scientific information need improvement. Also, one must provide a clear set of criteria to determine what is the rationale for the required improvements in narrowing scientific and economic uncertainties while, at the same time, addressing the worth of collecting more information. For example, if such information would do little to help resolve the economic picture or clarify questions of scientific uncertainty, which may be limited primarily by model assumptions rather than by the quality or quantity of the available data, then an intense site characterization may not be the appropriate action - an example of such a situation will be given later.

3 Economic Uncertainties

The evaluation and management of risks in business projects are of increasing importance to environmental firms because of the magnitude of liability costs and of cost overruns. Environmental projects in particular are burdened by the consequences of extreme events for which very little historical data exists and which can significantly alter the outcome of a project. This lack of information severely limits the capabilities to assess the probabilities and consequences of distinct alternatives, a situation that is further accentuated by the complex interaction of physical, chemical, geographical, socio-economic and other factors that enter into decision analyses of environmental projects.

Environmental risks are inherent in most business operations and can have substantial ramifications on balance sheets and, consequently, on shareholder equity if they are not properly addressed (Voorhees and Woellner 1997). Most corporations involved in activities that can affect the environment face the threat of legal and financial liabilities that may result from poor environmental management practices, improper waste disposal methods, changes in environmental regulations, and inadequate risk management (coupled with the potential loss of reputation). Thus, for example, in 1997, the U.S. Environmental Protection Agency (EPA) levied fines of about $169 million against corporations for violations of environmental laws and, furthermore, referred 278 criminal and 426 civil cases (involving corporate compliance assurance programs) to the U.S. Department of Justice (Telego 1998). During the same period, corporate liability in the

United States had reached $250 billion (Merkl and Robinson 1997), with accrued liability for environmental risks related to real estate property estimated at $2 trillion. This amount corresponds to around 16-20% of the total value of all property in the United States (Freeman and Kunreuther 1997).

A significant factor that contributes to the above problems is poor understanding of the complex interactions of the scientific controls that influence the outcome of environmental solutions. Consequently it is difficult to quantify scientifically short and long term liabilities, as well as the potential for limited and catastrophic losses. Thus, liability risks can span the whole range of environmental activities: from Phase I and II property assessments, to site investigations and sampling, to feasibility and remediation studies, to permit authorization, and to facility compliance audits and industrial hygiene surveys (Dixon 1996). A survey of 33 environmental engineering firms found that 21 percent of Phase I reports submitted in 1996 either had misestimated uncertainties related to technical aspects of a project or had neglected to document adverse environmental conditions from past use of a property. Accordingly, a realistic estimation of potential liabilities was impossible (Dunn 1997). Internationally, the broad legal implications and economic ramifications of risks from operations in developing countries, and the potential of liability transfer within the legal framework of the United States, are being increasingly recognized by project finance professionals (Drewnowski 1996; de Souza Porto and de Freitas 1996).

In addition to environmental liabilities, cost overruns present another form of risk to clients and firms. Thus, in 1997, the average cost for private sector, environmental remediation projects in the United States was 25-50% over the initial budget (Al-Bahar and Crandell 1990; Diekmann and Featherman 1998). Correspondingly, the cost overruns related to major military, energy, and information-technology projects in the United States, the United Kingdom, and a number of developing countries were between 40-500% (Morris and Hough 1987). Cost overruns arise, primarily, from: (i) external risks, due to modifications in the scope of a project, and changes in the legal, economic and technologic environments; (ii) technical complexity of a project, due to size, duration or technical difficulty; (iii) inadequate project management manifested in the control of internal resources, poor labor relations, and low productivity; and (iv) unrealistic cost estimates because of improper quantification of the uncertainties involved (Yeo 1990; Minato and Ashley 1998). Moorehouse and Millet (1994) found that poor assessment of uncertainties, and their corresponding consequences, together with inadequate staff training, comprised the two primary causes of financial failure in environmental projects. Jeljeli and Russell (1995) found that underestimation of a project budget, improper insurance coverage, and lack of technical expertise constituted the major factors in liability risks.

Historically, the primary risk management technique for business managers has been to transfer financial losses to insurance schemes. Consider as an example the case of environmental waste disposal problems; some form of insurance may be required in at least two cases. First is the possibility of a catastrophic event occurring during transport and/or after burial of the waste; for instance a truck overturning and spilling waste, thereby involving massive and expensive clean-up costs. A corporation would surely like to take out insurance against this possibility. Second is the possibility that the regulatory agencies may consider a unilateral change in the environmental stringency conditions for transport and/or burial of the material after the project is underway. In this case the corporation could be involved in further costs, thereby lowering potential profitability of a pre-existing contract. The corporation would surely like to option against the possibility of such an event occurring prior to the chance that the contract terms could be changed. These two forms of insurance are not equivalent. In the catastrophic loss event situation one would like to pay an insurance premium to cover the unknown catastrophic costs should they occur. In the regulatory stringency conditions situation, one usually knows ahead of time precisely how such more stringent conditions, if enacted, would influence the corporate profitability and one would like to have a contingency option operating that would be activated if and only if the regulatory agency does indeed enact the new more stringent regulations.

The practice to use insurance as sometimes the sole risk management technique has often promoted complacency in the development of preventive procedures to reduce losses. Furthermore, liability insurance may be unavailable or may contain provisions that exclude coverage for potential release of hazardous wastes (Ness 1992). Even when commercial insurance is purchased, the risk is not completely eliminated because of high deductibles, relatively low individual and aggregate coverage limits, and short time limits on claims-made policies (Frano 1991; Paek 1996).

An innovative approach at self-insurance is the risk reserve fund (Architects and Engineers Insurance Company) of 66 major engineering, architectural, and environmental firms, which aims to cover professional liabilities. The majority of environmental claims against this fund have been related to permits, personal injuries, fee disputes, reporting, and asbestos identification (Janney et al. 1996). Despite such creative approaches to risk limitation, the environmental industry continues to be characterized by: (i) high scientific uncertainty and complexity of problems (for example, the movement of a contaminant plume in the subsurface depends on the geologic, hydrologic, geochemical, and biological characteristics of a site, for which very little information is usually available); (ii) hazardous work conditions and the potential for major health impact to operators and the general population; and (iii) strict regulations that do not usually allow for indemnification clauses and that do not rely on standards of liability based on fault and negligence. Thus, unless the cost of liability is addressed by appropriate management techniques, such as risk retention, the danger exists that the environmental field will be occupied by unqualified contractors willing to take high risks (Paek 1996).

Risk retention is based on the argument that potential liability costs of environmental projects can be treated as legitimate business revenues and may be priced into the projects. A corporation involved in multiple environmental projects will then attribute a weighted, risk-based, premium to each project according to its likelihood of producing a liability. The funds gathered in this manner can be managed as a risk reserve fund with the purpose of meeting potential future claims (Frano 1991). The costs of administering the fund may also be priced into the projects (Tietenberg 1998). Such an approach encourages experienced contractors to invest in environmental projects and leads to procedures for rational allocation of risks (according to the characteristics of the different project types) within the resources of a corporation, without resorting to truncation of the effects of claims or penalties through bankruptcy (Larson 1996; Beard 1990). Recently, Paleologos and Fletcher (1999) developed expressions that allow this fund to be used for a number of liability situations, including cases of distributing the cost to projects of similar monetary value awarded in the same fiscal year or in variable time periods, as well as of proposals of different magnitudes initiated in distinctly different years.

The section that follows illustrates how economic uncertainties can be addressed by considering the simplest situation of a contract being offered under fixed regulations, with a single contract offered per project, with no catastrophic loss events, without inclusion of either optional insurance or of a corporate risk tolerance, and without leakage or spillage occurring.

4 Simple Economic Considerations

Consider that a contract is offered at a fixed price, G, for the transport and long-term storage of hazardous waste, including requirements on storage monitoring for a period of t years after emplacement of the waste material. The contract includes, additionally, storage site preparation. A corporation must then determine whether the fixed-price contract can return a profit before it bids on the contract.

The corporation assesses the cost of transportation on a unit of waste transported. This unit may be a volume or weight measure, or a fixed number of Bequerels in the case of radioactive waste, or some other measure depending on the material being shipped. Let the transport cost per unit be C_T, so that in transporting n units, the total transport cost is n C_T. In the disposal component of the project, there are the fixed costs of site preparation, F_S, and the disposal costs, D_C, per unit transported, for a total disposal cost of $D_T \equiv F_S + n\,D_C$. In addition, the contract requires monitoring after storage. This monitoring requirement includes two components: the fixed cost of all of the monitoring devices, M_1, and the on-going costs of monitoring, maintenance, and replacement of failed devices. Let these latter costs be reckoned on a per year basis at m, so that after t years the total monitor cost is $M \equiv M_1 + m\,t$.

A simple calculation of raw potential profit, P, to the corporation after t years is then

$$P_{raw} = G - n\,C_T - D_T - M, \tag{13.1}$$

where all amounts are calculated in fixed year dollars. There is also a discount factor that needs to be included because of corporate salaries, taxes, and corporate overheads for buildings, etc. Take the salaries to be S per year and the total corporate overhead per year to be H. Then, after t years, the raw potential profit is reduced by (S+H) t, yielding a potential profit of

$$P = G\text{-}nC_T - (F_S + nD_C) - (M_1 + mt) - (S+H)\,t. \tag{13.2}$$

If the profit, P, is positive then taxes must be paid. For simplicity, let the tax rate be a fixed fraction, u, of the profit P. Also assume that the n units of waste are transported and buried at a fixed rate, r, per year, so that one can write n = r t. Then the net profit, NPV, to the corporation is given by

$$NPV = (1\text{-}u)\,[G - (F_S + M_1) - t\,[r(C_T + D_C) + m + S+H]], \tag{13.3}$$

provided that NPV > 0.

Inspection of equation (13.3) shows that two factors control NPV and whether it is positive or not. First, note that the total fixed costs of preparing and monitoring the site, F_S+M_1, must be less than the contract price, G, else there is no hope of making a profit. Second, note that even when $F_S+M_1 < G$ the on-going per year costs of unit transportation and storage, monitor maintenance, and salaries plus overhead, will eventually drive the NPV negative after a time, t_*, given by

$$t_* = [G - (F_S + M_1)] / [r(C_T + D_c) + m + S+H]. \quad (13.4)$$

Thus, there is a critical project lifetime, t_*, that returns NPV=0 beyond which the project will result in a net loss. Clearly, in order for the project lifetime to extend longer than the monitoring time, t_M, which is mandated in the contract, two factors must be considered: G must exceed the fixed costs $F_S + M_1$ and, when such is the case, the costs of transport and storage per year plus the cost of yearly maintenance monitoring plus the cost of salaries and corporate overhead must be kept sufficiently small so that $t_* > t_M$. If such is not the case then no profit can accrue and the corporation loses money on the contract.

5 Inclusion of Economic Uncertainties in Project Considerations

Now the problem with this simplistic estimate, notwithstanding the fact that one has ignored all more involved factors, is that the fixed costs and yearly costs are not known ahead of their occurrence. Thus, for instance, until the project is completed one does not know the yearly maintenance cost of monitoring. Or again, until after the disposal site has been prepared, one does not know (as opposed to estimate) the fixed costs of site preparation. For each component entering equations (13.3) and (13.4), the same uncertainty on costs prevails. Generally, one has some relatively good idea of the range of potential costs but the exact value of these costs is not known with precision (Yeo 1990).

Accordingly, there is an associated uncertainty on both the NPV and on the critical project lifetime, t_*. In the case where the NPV is strongly positive and t_* is much greater than the mandated monitoring time, t_M, it is likely that NPV and the difference $t_* - t_M$ will be positive even when the ranges of uncertainty on the parameters are included. However, for situations where NPV and $t_* - t_M$ are only slightly positive, inclusion of the uncertainty in the parameters can end up providing cases where there are negative values for NPV and $t_* - t_M$. Thus the question devolves into one of estimating the probability of obtaining a positive worth, NPV, and a positive $t_* - t_M$ so that the corporation can assess the risk of making a profit on the environmental project.

To illustrate the influence of uncertainties on the NPV and the value of t_*, suppose then that the fixed component, F_S, of the site preparation costs is unknown but is estimated to range from $F_{S,min}$ to $F_{S,max}$ with a most likely value of F_L. For the purposes of the illustration take it that all other parameters are statistically sharp with no uncertainty, i.e., all other costs are precisely known in advance. The actual statistical distribution of F_S is, in general, unknown, so that one has to make some assessment of the statistical uncertainty of F_S based on the three values $F_{S,min}$, $F_{S,max}$ and F_L.

Denote the estimate of the mean value of F_S by $<F>$, and of the variance of F_S, by $<\delta F^2>$. Note that $<\delta F^2> = <F^2> - <F>^2$ where angular brackets are reserved for statistical averages. Inspection of equation (13.3) shows that NPV is linear in F_S, so that one can immediately write the expected value of NPV as

$$<NPV> = (1-u)\,[G-M_1-t[r(C_T+D_c) + m + S+H]] - (1-u)<F>, \qquad (13.5)$$

and the variance in the expected NPV is

$$<\delta NPV^2> = (1-u)^2 <\delta F^2>. \qquad (13.6)$$

Based solely on the mean, $<NPV>$, and variance, $<\delta NPV^2>$, one can write an equivalent Gaussian cumulative probability of obtaining an NPV in excess of a pre-set value V as

$$P(NPV>V)=(2\pi<\delta NPV^2>)^{-1/2}\int_V^{\infty}\exp[-(x-<NPV>)^2/(2<\delta NPV^2>)]dx \qquad (13.7)$$

If one asks for the cumulative probability, P (0), of making a profit greater than zero, the substitution $x = <NPV> + (2<\delta NPV^2>)^{1/2}\,y$ allows equation (13.7) to be written in the simpler form

$$P(0) = \pi^{-1/2} \int_{-b}^{\infty} \exp(-y^2)\,dy, \tag{13.8}$$

where $b =< NPV > / (2 < \delta NPV^2 >)^{1/2} \equiv 1/(\sqrt{2}\,\nu)$.

Thus the probability of making some profit is controlled not only by the expected NPV but also by the standard error of the expected value. A large positive value of b in equation (13.8) means that the ratio of <NPV> to the standard error, $< \delta NPV^2 >^{1/2}$, is large and positive, so that there is a high probability of making some profit, while a small value of b ($b << 1$) implies that the uncertainty on the expected NPV is large, and that the probability, P (0), is close to 0.5, i.e., there is only about a 50% chance of profitability. The risk of the project is then measured by the volatility, ν. If ν is large (positive or negative) then there is considerable uncertainty on P (0), while a small value of ν implies a relatively sharply defined value of P (0), i.e., one with less risk.

Most corporations set a minimum internal risk factor in the form of a minimum acceptable chance, MAC, to make a profit. If the value of P (0) for a project falls below a predetermined value of MAC then the corporation will usually not invest in a project without some compelling reason. Thus P (0) is an extremely valuable measure of worth of a project to a corporation, as is the volatility, ν, of each project.

The particular illustration here has considered only variations in the range of F_S in order to illuminate the general procedure succinctly. In a more pragmatic situation one should, and does, include all variations of all parameters. The same sense of measure can be applied to the lifetime, t_*, to zero profit versus the mandated monitoring time t_M. Here, considering F_S again to be the only uncertain parameter yields the expression

$$<t_*> = [G-M_1 - <F>] / [r\,(C_T + D_c) + m + S + H], \tag{13.9}$$

for the average value of t_*, while the variance $< \delta t_*^2 >$ around the mean, $<t_*>$, is given by

$$< \delta t_*^2 > =< \delta F^2 > / [r(C_T + D_c) + m + S + H]^2. \tag{13.10}$$

Thus, the cumulative probability that the project will have a profitable lifetime in excess of t_M is

$$P(t > t_M) = (2\pi < \delta t_*^2 >)^{-1/2} \int_{t_M}^{\infty} \exp[-(x - < t_* >)^2 / (2 < \delta t_*^2 >)] dx \tag{13.11}$$

Writing $x = < t_* > + [2 < \delta t_*^2 >]^{1/2}\, y$ enables equation (13.11) to be written in the form

$$P(t > t_M) = \pi^{-1/2} \int_{-B}^{\infty} \exp(-y^2)\, dy\,, \tag{13.12}$$

where

$$B = [< t_* > - t_M] / [2(\delta t_*)^2]^{1/2}\,. \tag{13.13}$$

Thus, if $<t_*>$ is greater than t_M then the cumulative probability that the profitable lifetime will exceed the mandated monitoring time, t_M, is greater than 50% (and may be much greater if B>>1). Equally, from a corporate perspective, one could set a minimum chance P_{min} that the lifetime should exceed t_M. In that case one writes

$$P_{min} = \pi^{-1/2} \int_{-B}^{\infty} \exp(-y^2)\, dy\,, \tag{13.14}$$

and then uses the pre-set value of P_{min} to determine B from equation (13.14). In this way one comes up with an inequality relationship between $<t_*>$ and $< \delta t_*^2 >$ such that one can indeed satisfy the minimum chance, P_{min}. Hence one can set limits on the allowed range of fluctuations in F_S so that the corporate objectives can be achieved.

The point about these simple illustrations is to show how the basic tools can be developed and used to assess the worth to a corporation of becoming involved in a project.

Of particular concern are two major components: first, to provide some assessment of the project and the NPV subject to the constraints imposed (in the simple illustration the constraint was that monitoring be carried out for a specific, mandated time after storage); second, to illustrate how to assess the probability of making at least some profit as well as to determine the probability of honoring the mandated constraint when there is uncertainty in parameters entering the cost assessment. In order to show these aspects as simply as possible, all other considerations involved in environmental projects were ignored. In addition, only one parameter (the fixed cost of storage site preparation) was allowed to be uncertain. In general, however, one cannot ignore problems of potential spillage, multiple projects, limited liability insurance, corporate risk aversion, fixed budgets, etc.; nor can one ignore the problems of multi-parameter uncertainty, which may be correlated or independent, in influencing the corporate decision on project involvement.

6 Parameter Uncertainty Effects

The evaluation of expected value, E, variance, σ^2, and cumulative probability P(V), has so far been predicated on statistically sharp values for the parameters in the decision-tree of Figure 13.1, with the uncertainty arising from the probabilities along the different possible paths.

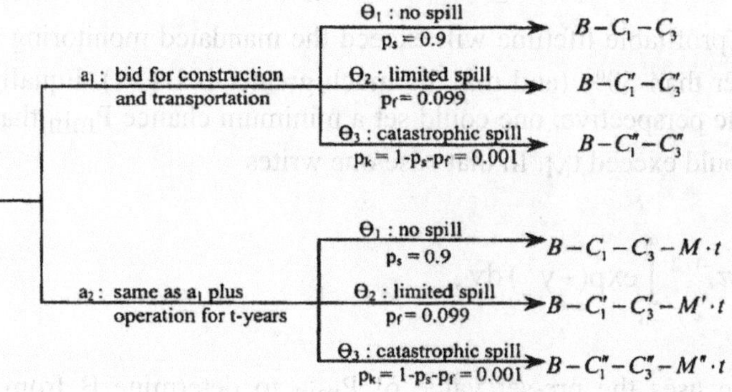

Fig. 13.1. Sketch of the various decision tree possibilities for the monitoring problem given in text

But there is a second form of uncertainty that arises due to the fact that one does not have statistical sharpness on each and every parameter, but rather an estimated range of values within which a parameter can lie.

For instance, one might calculate a success probability, p_S, of transporting wastes without leakage at, say, $p_S = 0.9$, but different methods of estimation and different underlying assumptions can produce a range of estimates of p_S, say $0.8 \leq p_S \leq 0.99$. Similar arguments are relevant for all parameters entering the calculations. Thus, project cost estimates are routinely assigned different error ranges at different stages over the life cycle of a project to account for various contingencies (Yeo 1990).

Two factors are apparent: (i) there is no particularly valid, objective justification for preferring any one value of p_S out of its range over any other value; (ii) one needs to evaluate which ranges of which particular parameters are causing the largest uncertainty in estimates of the expected value E, its variance σ^2, and the cumulative probability P(V). Then one has available the relative importance of each parameter in contributing to system uncertainty, and so one can then determine where to place effort to narrow the ranges of uncertainty of the most important parameters - and with a clear determination of the definition of importance as it relates to the specific system of study.

6.1 Means and Variances

Select at random from the range of each variable parameter. Then, for the ith choice of such a set of parameter values, one can calculate an expected value E_i, and a variance σ_i^2, together with a cumulative probability $P_i(V)$. Clearly, by performing a series of Monte Carlo computations with N total selections, one produces both a mean expected value

$$< E > = N^{-1}\sum_{i=1}^{N} E_i \tag{13.15}$$

and a direct uncertainty $< \rho^2 >$ in $< E >$ due solely to the statistical variations in each parameter given by

$$< \rho^2 > = N^{-1}\sum_{i=1}^{N}(E_i - < E >)^2 . \tag{13.16}$$

In addition, one also has a mean value for the variance, $< \sigma^2 >$, from

$$< \sigma^2 > = N^{-1}\sum_{i=1}^{N}\sigma_i^2 \tag{13.17}$$

and an uncertainty, $< \delta^2 >$, in the mean variance calculated from

$$<\delta^2> = N^{-1}\sum_{i=1}^{N}(\sigma_i^2 - <\sigma^2>)^2. \tag{13.18}$$

Because the direct uncertainty, $<\rho^2>$, in $<E>$ is due to fluctuations in parameter values, while the variance $<\sigma^2>$ is due to the different probabilistic paths, the total uncertainty on $<E>$ is then measured by the variance σ_E^2 where

$$\sigma_E^2 = <\rho^2> + <\sigma^2>. \tag{13.19}$$

Equally, for the cumulative probability P(V) there are several measures of worth. Thus, one can compute a mean cumulative probability, <P(V)>, from

$$<P(V)> = N^{-1}\sum_{i=1}^{N}P_i(V) \tag{13.20}$$

and an associated uncertainty, $<\delta P(V)^2>$ from

$$<\delta P(V)^2> = N^{-1}\sum_{i=1}^{N}(P_i(V) - <P(V)>)^2. \tag{13.21}$$

One can also compute an approximate mean value, $\overline{P}$, using $<E>$ and σ_E in equation (13.8). If the difference

$$|<P(V)> - \overline{P}(V)| \le <\delta(V)^2>^{1/2} \tag{13.22}$$

then it is an accurate enough approximation to use $\overline{P}(V)$, otherwise one must use the ensemble value <P(V)>.

6.2 Relative Importance

In addition to having the ability to compute ensemble averages and their fluctuations, one also requires knowledge of which uncertainty ranges of which parameters are contributing most to the total system uncertainty. This relative importance problem can be addressed simply as follows.

Each parameter in the decision-tree diagram of Figure 13.1 has a mean value within its range of variation. Let a vector **q** denote all of the component parameters, with the jth component, q_j, having a mean value $< q_j >$, with $q_{jmin} \le q_j \le q_{jmax}$ where q_{jmin} and q_{jmax} are, respectively, the minimum and maximum values of q_j.

Suppose that one were to use only the vector of mean values, <**q**>, to calculate the quantities in the previous subsection. Then this choice is precisely the same as choosing specific values of the parameters, so $< \rho^2 > = 0$ and $< \sigma^2 >$ is the same as σ^2 (<**q**>). Now consider the influence of uncertainty in each variable in contributing to variations in σ_E^2, because there is an intrinsic value for σ_E^2 even when the parameters are statistically sharp. The changes in σ_E^2, relative to the value σ^2 (<**q**>) due to variations in each parameter around its mean value, are then determined as follows.

Let all except one, say the jth, of the components of the parameter vector **q** be held at their mean values. This j[th] component of **q** is then varied, randomly, around its mean value. The result is obviously a value for $< \rho^2 >$ and σ_E^2 which is dependent on $< q_j >$, and also on q_{jmin} and q_{jmax}. Denote these values as $< \rho(j)^2 >$ and $\sigma_E(j)^2$. Repeating the process for all components of **q** one can then calculate the relative fractional uncertainty contribution (RC_j) to $< \rho^2 >$ from each q_j as

$$(RC_j) = < \rho(j)^2 > / \sum_{k=1}^{N} < \rho(k)^2 > . \tag{13.23}$$

Equally, one can compute the relative importance, RI_j, to σ_E^2 from

$$(RC_j) = < \sigma(j)^2 > / \sum_{k=1}^{N} < \sigma_E(k)^2 > \tag{13.24}$$

Similarly, for the ensemble mean value, $< E >$ and the cumulative probability, P(V), one can compute relative importance values from, respectively,

$$RI_j(E) = \langle E(j) \rangle / \sum_{k=1}^{N} \langle E(k) \rangle \tag{13.25}$$

and

$$RI_j(P) = P_j(V) / \sum_{k=1}^{N} P_k(V) \tag{13.26}$$

which provide measures of the contributions.

7 Relative Contribution of Uncertain Parameters in Subsurface Hydrology

The heterogeneity of soils and geologic formations is one of the most challenging issues in trying to predict the movement of water and contaminants in the soil. Because of the complexity and heterogeneity of the subsurface environment, the definition of all the properties that describe the dynamics of subsurface physical processes is impossible practically. Additionally, the number of the available field data is always limited due to high cost and technical limitations and, hence, a field cannot be characterized completely. In this section it is shown that application of the concepts presented above is not restricted to economic evaluation of the uncertainties of a project, but can be applied equally well to quantification of the scientific uncertainties of a project. The particular illustration of unsaturated flow in soil formations provides a clear example of how one undertakes such endeavors.

7.1 Introduction

Unsaturated flow in soil formations is a complex process that is not easily understood, even in relatively homogeneous systems. The unsaturated zone is defined as the geologic environment that lies between land surface and the local water table. The upper part commonly consists of the root zone and weathered soil horizons. Soils and bedrock within this zone are usually unsaturated, i.e., their pores are partially filled with water (Hillel 1980; Stephens 1996). One major factor that contributes to this complexity is the spatial variability of soil properties. Field observations have shown that the hydrologic properties of soils vary over several orders of magnitude even in the same formation (Woodbury 1991; Sandia 1994). Thus, the effect of the spatial variability on predictions of water flow and the transport of contaminants is a major focus of scientific investigations. Studies of unsaturated flow within porous media exhibiting random heterogeneities have employed either analytic approximations through linearizations and perturbation methods, or Monte Carlo simulations that have focused on one parameter at a time (Gardner 1958; Mantoglou 1987a, b, c; Yeh 1985a, b, c).

In this section the relative contribution is quantified of the uncertainty and spatial variability of multiple parameters that enter into unsaturated flow problems. First, an investigation is given of several types of probability distributions that may describe the data, and the degree is calculated to which the uncertainty in the distribution type of a physical parameter is contributing to the total system uncertainty. The assumption of the type of distribution is critical because only a small amount of field data is usually available and, even in relatively homogeneous aquifers, these data can be fitted by more than one type of probability distribution (Yeh 1989). The significance for site characterization efforts lies in that, if the exact functional form of the probability distribution is required, then a very dense measurement network needs to be implemented in order to determine unequivocally the statistics of the flow field. Also presented here is the framework that accounts for the uncertainty and the relative contribution and relative importance of various uncertain and/or spatially variable parameters. This aspect has important implications in site characterization and modeling efforts as it allows the focus of time and resources on those factors of a problem that contribute the most and dominate the total system uncertainty.

7.2 Physical Problem

The methodology developed above is applied to the physical problem of one-dimensional infiltration in unsaturated porous media, under constant infiltration rate. For this problem field data are utilized that were collected for the characterization of the flow field at the U.S. Department of Energy site for potential disposition of radioactive wastes, located at Yucca Mountain, Nevada. Figure 13.2 shows the six hydrogeologic units that are considered in the study. These units, from the ground surface to the water table, are: the Tiva Canyon welded (TCw), the Paintbrush nonwelded (PTn), the Topopah Spring welded (TSw), the Topopah Spring vitrophyre (TSv), the Calico Hills nonwelded-vitric (CHnv), and the Calico Hills nonwelded-zeolitic (CHnz).These units exhibit significant differences in their properties and hydraulic behaviors, as illustrated in Tables 13.1 and 13.2 and discussed in detail below.

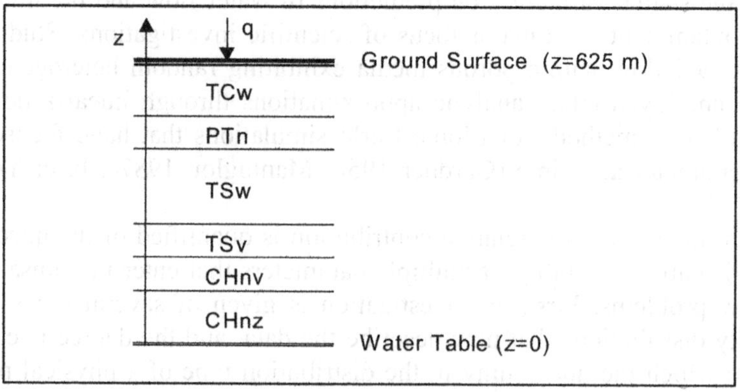

Fig. 13.2. Representation of the various layers for the Yucca Mountain water flow problem

TABLE 13.1. Statistics of the hydrogeologic parameters for different layers (Sandia, 1994)

Rock	Thickness	$\log K_s$		$\log\alpha$		$\log\beta$	
Unit	(m)	E(.)	CV(.)	E(.)	CV(.)	E(.)	CV(.)
TCw	81	-10.90	0.098	-2.094	0.325	0.206	0.259
PTn	39	-7.96	0.202	-1.134	0.636	0.347	0.647
TSw	299	-10.71	0.084	-1.885	0.265	0.233	0.523
TSv	15	-11.00	0.062	-2.624	0.167	0.349	0.539
CHnv	64	-8.99	0.115	-1.644	0.302	0.373	0.614
CHnz	127	-10.79	0.093	-2.270	0.275	0.223	0.540

TABLE 13.2. Minima and maxima of the parameters for different layers (Sandia, 1994)

Rock	K_s	(m/s)	α	(1/m)	β	
Unit	Min	Max	Min	Max	Min	Max
TCw	7.00E-13	4.83E-09	0.0003	0.1338	1.349	2.805
PTn	2.86E-12	2.35E-06	0.0104	1.6990	1.187	11.800
TSw	3.05E-13	5.23E-09	0.0021	0.4244	1.155	5.363
TSv	1.52E-12	6.95E-11	0.0002	0.0077	1.377	4.473
CHnv	5.13E-12	2.92E-07	0.0054	0.3752	1.249	9.888
CHnz	2.37E-14	3.14E-09	0.0004	0.2355	1.184	5.914

The study is concerned with one-dimensional infiltration through the 625m thick unsaturated soil formations depicted in Figure 13.2 (Avanidou 2000). If one assumes, as is commonly done (Sandia 1994), that the infiltration rate is constant at the site, and that steady-state conditions have been attained under gravity-capillary equilibrium, then the hydrologic state between the ground surface and the water table (pressure head and saturation against depth) can be described by the flow equation (Hughson and Yeh 1998)

$$-K(\psi)\left(\frac{d\psi}{dz}+1\right)=q. \tag{13.27}$$

Here q denotes the constant steady-state infiltration rate (m/s), ψ is the pressure head (m), which is positive when the soil is fully saturated and is negative when the soil is partially saturated, and z is the vertical coordinate which is positive in the upward direction. To describe the saturation-pressure head relation, Mualem's model is used:

$$S=\left(1+|\alpha\psi|^{\beta}\right)^{-m} \tag{13.28}$$

where S is the saturation of soil, and α, β, and m are empirical parameters with $m=1-1/\beta$. The unsaturated hydraulic conductivity $K(\psi)$ is then given by (van Genuchten 1980)

$$K(\psi)=K_s\, S^{1/2}[1-(1-S^{1/m})^m]^2, \tag{13.29}$$

where K_s is the saturated hydraulic conductivity (m/s). Thus, the parameter of the problem (unsaturated hydraulic conductivity) depends on the unknown pressure ψ. Through integration of equation (13.27) one obtains the expression

$$\int_0^{\psi} \frac{K(\psi)d\psi}{(K(\psi)+q)} = -(z - z_0), \tag{13.30}$$

where z_0 is the elevation of the water table (pressure equals zero) and z corresponds to a pressure ψ.

TABLE 13.3. Linear Loss function for Monitoring System Project

Number of Contamination Events in T-Years	A(0): Do not Build Monitoring System	A(1): Build Monitoring System
0	0	$C \times T$
1	1×B	$C \times T$
2	2×B	$C \times T$
:	:	:
T	T×B	$C \times T$

Table 13.3 shows the statistical properties of the saturated hydraulic conductivity K_s, and the van Genuchten α and β parameters as tabulated in the Sandia report (1994). The notation E( ) denotes expected value and CV( ) the coefficient of variation. The minimum and maximum values of these parameters are given in Table 13.3 for each hydrogeologic unit (Sandia 1994), illustrating the high degree of variability of the parameters within each layer and among the different layers.

7.3 Theoretical Framework

Consider the parameters K_s, α and β as random variables that obey the same probability distribution. The distributions in the analysis are: Gaussian, lognormal, exponential, uniform, and triangular with the mean, variance and range of each parameter defined in Table 13.2. The first three distributions were chosen because they have been shown to fit data at several sites (Woodbury and Sudicky 1991; Sandia 1994), the uniform distribution because it describes the (common) situation where one has knowledge only of the range within which a parameter lies for a geologic formation, and the triangular because it represents the case where, in addition to the minimum and maximum values, one has information about the most commonly occurring value. By selecting a triplet of values from a specific distribution i for the parameter set (K_S, α, β), and by then performing a series of Monte Carlo computations (Deutsch and Journel 1992) with N total selections, one can create N profiles of ψ with depth, z. At each discretization point of the grid one can then average the N equiprobable values of ψ and obtain the mean pressure head and the variance σ^2 of ψ that apply to this point for a specific distribution i:

$$< \psi >_i = \frac{1}{N}\sum_{j=1}^{N} \psi_j \;; \sigma_i^2 = \frac{1}{N-1}\sum_{j=1}^{N} (\psi_j - < \psi >_i)^2 . \tag{13.31}$$

Now one can calculate, at each point, the global mean $<\psi>_G$, the arithmetic mean of the expected values from the five distributions, as well as the global variance $< \sigma^2 >_G$, the arithmetic mean of the variances obtained from each distribution:

$$< \psi >_G = \frac{1}{5}\sum_{i=1}^{5} < \psi >_i \quad ; < \sigma^2 >_G = \frac{1}{5}\sum_{i=1}^{5} \sigma_i^2 . \tag{13.32}$$

By calculating the quantity $< \rho^2 >_T$,

$$< \rho^2 >_T = \frac{1}{5}\sum_{i=1}^{5} \left(< \psi >_i - < \psi >_G \right)^2 , \tag{13.33}$$

which is just the divergence of the means of the distributions from the global mean, one can obtain the total uncertainty at each point:

$$\sigma_T^2 = < \rho^2 >_T + < \sigma^2 >_G . \tag{13.34}$$

Here $< \rho^2 >_T$ is a measure of the uncertainty in the mean ψ-behavior because of the uncertainty in the type of distribution, and $< \sigma^2 >_G$ is the average fluctuation around the mean ψ-behavior irrespective of distribution.

Now one can examine, for every point, the relative contribution of each distribution i toward the global mean through the expression:

$$RC_p(i) = \frac{\left(<\psi>_i - <\psi>_G\right)^2}{\sum_{m=1}^{5}\left(<\psi>_m - <\psi>_G\right)^2} \tag{13.35}$$

and also calculate the relative importance of each distribution i toward the average variance:

$$RC_{\sigma^2}(i) = \frac{\sigma_i^{\,2}}{\sum_{m=1}^{5}\sigma_m^{\,2}}. \tag{13.36}$$

Finally by calculating the ratios

$$\frac{<\rho^2>_T}{\sigma_T^2} \quad \text{and} \quad \frac{<\sigma^2>_G}{\sigma_T^2} \tag{13.37}$$

one can evaluate to what degree the total uncertainty is dominated by the lack of knowledge in the type of distribution or by the fluctuations around the mean values. A large value of the first ratio indicates that the choice of the probability distribution model is critical in total uncertainty and, hence, more data need to be collected for a clear determination of the shape of the distribution. In contrast, a large value of the second ratio indicates that the fluctuations around the mean ψ-behavior are dominating the total system uncertainty and, hence, the parameters need to be defined more sharply.

Now hold two of the three parameters K_s, α, and β at their mean values and vary the third according to a distribution i with mean, variance, range for that parameter obtained from Table 13.2. By performing Monte Carlo simulations one obtains $<\psi>_{i,k}$, the mean pressure head, and $\sigma_{i,k}^2$, the variance of the pressure head, due to fluctuations in the kth random parameter according to an ith distribution. By repeating the procedure for all parameters one evaluates the relative importance towards the mean pressure of each parameter k for every distribution i:

$$RI_{i,k}^{<\psi>} = \frac{<\psi>_{i,k}}{\sum_{m=1}^{3}<\psi>_{i,m}} \tag{13.38}$$

as well as the relative importance toward the variance of each parameter k and distribution i:

$$RI_{i,k}^{\sigma^2} = \frac{\sigma_{i,k}^2}{\sum_{m=1}^{3} \sigma_{i,m}^2} \tag{13.39}$$

Clearly, this process can be repeated for all distributions (i=1,..., 5) and then, for each parameter k, one can calculate the relative importance toward the mean:

$$RIC_k^{<\psi>} = \frac{\sum_{i=1}^{5} <\psi>_{i,k}}{\sum_{i=1}^{5}\sum_{m=1}^{3} <\psi>_{i,m}}, \tag{13.40}$$

and the variance:

$$RI_{i,k}^{\sigma^2} = \frac{\sum_{i=1}^{5} \sigma_{i,k}^2}{\sum_{i=1}^{5}\sum_{m=1}^{3} \sigma_{i,m}^2}, \tag{13.41}$$

irrespective of distribution. Thus, the above analysis can provide a ranking of the importance of each parameter in the evaluation of the mean and variance of the pressure for a specific distribution and also irrespective of the choice of distribution.

7.4 Results and Discussion

The flow domain was discretized into elements of length of 0.5 m, for a total of 1250 nodes. A constant infiltration rate of 0.1 mm/yr was assumed. The upper boundary was considered as a prescribed flux boundary, whereas the lower boundary was treated as a stationary water table. Because the relationship between the three variables K_s, α and β can not be specified from the limited field data available, it was assumed that the variables were random processes, perfectly correlated, with their statistics given by Table 13.2. The non-linear equation was solved using an iterative scheme, while the integration was performed through an adaptive Newton-Cotes 9-point rule (Press et al. 1987). For each node, and for the parameter set (K_s, α, β), 500 Monte Carlo simulations were performed. Thus, for each choice of distribution 500 equiprobable ψ-profiles were created along the 1250 nodes. This process was repeated for all five types of distributions.

The quantities were then calculated but, in order to simplify the depiction of the results for each distribution, the detailed mean point profiles were averaged over the six formations. This situation represents the case where an evaluation is needed that ranks the different parameters with

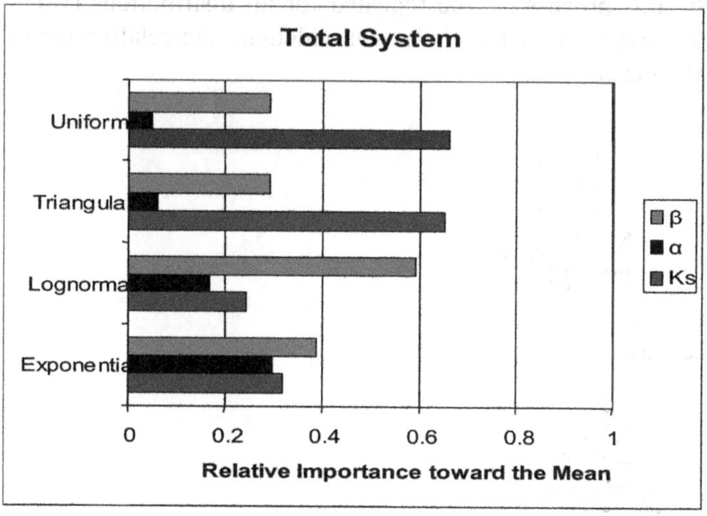

Fig. 13.3. Relative Importance in contributing to the mean value for the total flow system for different distributions of the uncertain parameters given in the Tables

respect to the total system uncertainty. One can also use the results of the simulations to compare detailed point profiles, or depth-average over each layer only, and rank the parameters at each point of the flow field or for each individual layer.

Figure 13.3 plots the contribution of each type of distribution toward the global mean while Figure 13.4 plots the relative importance of each distribution model toward the average variance. The two figures indicate that a Gaussian model assumption for the parameters will produce a mean ψ-profile, as well as fluctuations around this profile, that are significantly larger than those for any other statistical model. Hence, these results demonstrate that the assumption of a particular distribution is critical in the prediction of the mean hydraulic behavior and the fluctuations about this mean in an unsaturated zone field. This conclusion is supported by the values of the ratios where the first term, which measures the importance of parameter assumptions to total uncertainty, is 75.3%, whereas the second term, which measures the contribution of fluctuations, is 24.7%.

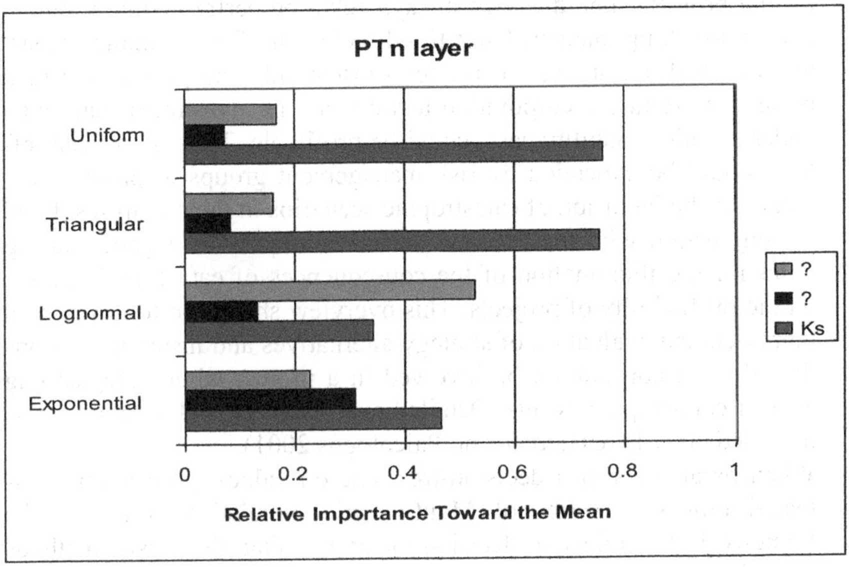

Fig. 13.4. Relative Importance in contributing to the mean value for the total flow system for different distributions of the uncertain parameters given in the Tables

7.5 Conclusions

The following conclusions can be drawn from this section:

1. The criterion of the maximum expected monetary value (MEMV) is widely used in decision-making analyses of environmental and gas and oil projects to select a preferred action as the one returning the maximum expected return. Sole use of this criterion may lead to erroneous decisions in the presence of uncertainty.
2. The MEMV fails to differentiate the economic consequences of limited and catastrophic failures in a project. The use of additional statistical measures, such as standard error, volatility and cumulative probability, provides insight into the selection process, illuminating the implications of each case. Thus, equations that provide expressions for the volatility and the probability of profit, respectively, should be used in conjunction with the MEMV criterion to analyze a project.

3. Consideration of a low-probability, high-cost (catastrophic) event in a decision analysis can have the effects of: (i) reversing the expected return from a positive to a negative value for a range of contract awards of a project; (ii) significantly increasing the standard error and volatility, and (iii) substantially reducing the probability of success.
4. Consideration of catastrophic scenarios in risk analyses of environmental projects should almost always never be performed in a routine decision-making manner. Unjustified inclusion of catastrophic scenarios into such an analysis can alter substantially the perspective of a project and guide a corporation away from an investment that, even under a limited liability case, could be profitable. Thus, significant effort should be expended by risk management groups to properly account for the inclusion of catastrophic scenarios in their analyses. Used in conjunction with the MEMV, such efforts provide a good starting point for the illumination of the consequences of catastrophic events on the profitability of projects. This overview should be followed by a more detailed evaluation of strategy alternatives and insurance options that allow a corporation be involved in a project while safeguarding against catastrophic events. Detailed exposition of such alternatives is provided elsewhere (Lerche and Paleologos 2001).
5. When parameters of a decision-tree take on values lying within estimated ranges, a series of Monte Carlo simulations provides the framework for relevant decision-making, using the basic methods given here as a framework.
6. The use of the concept of the relative uncertainty can guide selection of those parameters that dominantly influence the total system uncertainty, thus allowing one to concentrate resources on efforts to minimize the range in such dominant parameters. Such efforts may be used to illuminate the critical components of a project.

8 Bayesian Decision Criteria

8.1 Optimal Expected Value Bayes Decision

The objective of this section is to provide an exposition of some more advanced concepts of the Bayesian decision theory and to illustrate the use of this theory in risk analyses of environmental projects (see figure 13.5). This objective is accomplished with the presentation of a simple example, which has the advantage of providing close-form solutions that can be easily reproduced by the reader. The procedure in the section follows closely the development by Duckstein et al. (1978) for the case of drainage or not of a plot of land. Duckstein et al. (1978) have provided a simple, self-contained guide that can be used to analyze hydrologic and economic risk, the worth of additional information, and the sensitivity of the decision-making in scientific and economic assumptions. The procedure is applied here in the case of monitoring at a waste disposal site.

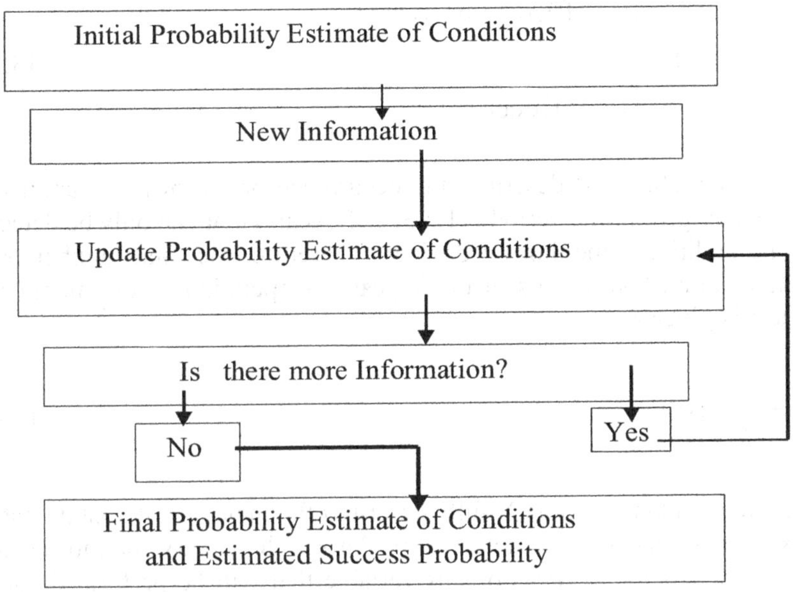

Fig. 13.5. Flow diagram for estimating probability based on Bayesian updating of a priori estimates using additional information

Consider again the problem of toxic (or radioactive) waste disposal. In addition to the monitoring network that is mandated by federal or other requirements to operate at a repository site a decision-maker may be considering whether or not to install additional monitoring devices at an annual cost C. Such a situation may arise when components of the repository, that under normal conditions would not have been monitored, are governed by enough geologic uncertainty and present a potential for contamination that consideration is given to extra monitoring of some regions. Cost C represents the total annual cost of the additional monitoring devices that includes the cost of installation and maintenance over T years. Here uncertainties related to the amortization factor are not considered. The time horizon T corresponds to the lifetime of the additional monitoring network and the decisions that can be made are A(0) corresponding to the action "no additional monitoring system," and A(1) that corresponds to the action "build additional monitoring system." Designate by E an extreme contamination leak event that would not have been detected without the presence of the additional monitoring devices. The remediation cost of a contamination leak is projected to be B. Define the binary random variable by

$$X(i) = \begin{cases} 0 & \text{if E does not occur} \\ 1 & \text{if E occurs.} \end{cases} \tag{13.42}$$

Variable X(i) describes the occurrence or not of a contamination leak in any given time period i=1,2,3,...,T (years) that can only be detected by the additional monitoring devices. Therefore, the total number N of contamination leak events in the T-years of operation of the monitoring system is given by

$$N = \sum_{i=1}^{T} X(i). \tag{13.43}$$

Thus, the random variable N fully describes the incidents of contamination leaks in T-years that are detected by the supplementary monitoring network and it is the analysis of this variable that will be performed subsequently.

Assuming that only one contamination leak per year gets detected by the additional monitoring network the probability of an event E is given by

$$P(X=1) = p, \tag{13.44}$$

and the probability that no leak occurs by

$$P(X=0)=1-p. \tag{13.45}$$

Here, for reasons of simplicity it is assumed that the proposed additional system would result in early detection of all contamination leaks that get undetected by the regular monitoring system. Because the variable X(i) is a Bernoulli variable, N in equation (13.43) is binomial (Benjamin and Cornell 1970) and the probability that j contamination leak events will occur in T-years of operation of the monitoring network is given by

$$P(N=j)=\binom{T}{j}p^{j}(1-p)^{T-j}. \tag{13.46}$$

Here j takes the values j=0,1,2,...,T, and the combination of j out of T depicted at the right hand side of equation (13.46) equals T! / (j! (T-j)!) with $j!=1\times 2\times ...\times j$ and 0!=1.

The financial losses incurred as a result of the different actions or decisions made are encoded in a function that is designated as the loss or objective function. For this study if action A(1) to build the monitoring system is followed the financial loss over T-years is given by the value of C×T. The values of the loss function can be tabulated in a table of discrete losses such as Table 13.4.

TABLE 13.4 Loss table for two states of nature

State of Nature	A(0): Do not Build Monitoring System	A(1): Build Monitoring System
S1: $BN < CT$	BN	CT
S2: $BN > CT$	BN	CT

These losses clearly depend both on the action followed (i.e., on the action space A) and the number of contamination events that may occur (i.e., on variable N).

Alternatively, the loss function of this problem can be written in a compact form as

$$L(A,N)=\begin{cases} BN & \text{if } A=A(0) \\ CT & \text{if } A=A(1). \end{cases} \tag{13.47}$$

The loss function in equation (13.47) or Table 13.4 has been assumed to vary linearly with N. Indeed, such dependency will depend on the particular problem at hand. Later a quadratic form of the loss function is investigated and comparisons are drawn between the two loss function models.

The expected value of the loss function with respect to the variable N is termed in the Bayesian decision theory as the goal or risk function (Raiffa and Schlaifer 1961; Duckstein et al. 1978; Berger 1985). For action A(0) the goal function is given by

$$G(A(0), p) = E^N[L(A(0), N)] = E^N[B\,N] = B\,T\,p, \tag{13.48a}$$

where in the last equality the mean value of a Binomial variable N, E(N) = T p, was utilized. For action A(1) the goal function is given by

$$G(A(1), p) = E^N[L(A(1), N)] = E^N[C\,T] = C\,T. \tag{13.48b}$$

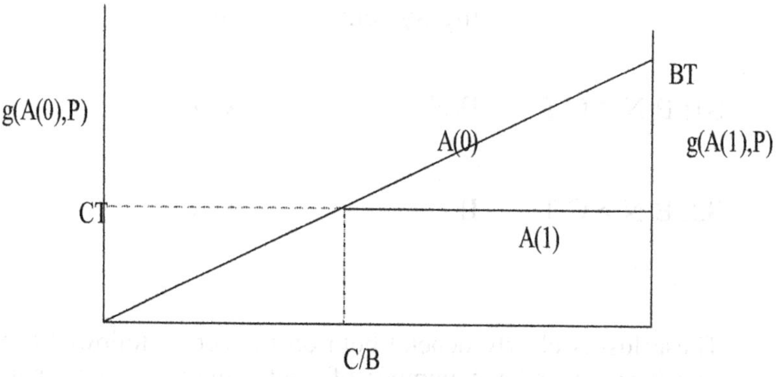

Fig. 13.6. Optimum decision for monitoring problem

Equations (13.48a) and (13.48b) indicate what one risks to lose on average if actions A(0) and A(1) are taken, respectively. In a standard benefit-risk analysis one would proceed then to evaluate which of the two decisions minimizes the goal function or even consider other measures of uncertainty . This procedure can be easily performed in this case with the use of figure 13.6.

Here, the goal functions that correspond to actions A(0) and A(1) are plotted on the left and right y-axes, respectively, whereas the probability p is plotted on the x-axis. Figure 13.6 indicates (heavy solid line segments) the regions where the optimal action minimizes the goal function. To find the probability value where a change in a decision is made it suffices to set the two goal functions equal to each other:

$$G(A(0), p) = G(A(1), p) = B\,T\,p = C\,T$$
$$\Rightarrow p = C / B. \quad (13.48c)$$

Thus, the ratio C / B provides the break point where a change from the decision A(0): "no additional monitoring system," to A(1): "build additional monitoring system" is recommended.

The particular application of a Bayesian analysis in this problem is the recognition that the probabilities p, and (1-p) are themselves uncertain, and hence they need to be considered as random variables. We proceed to discuss the estimation of a subjective prior probability density function (pdf) of the variable p (Duckstein et al. 1978).

It is possible that based on past experience from similar sites the operator of the repository facility may be able to provide a rough estimate on how often he expects a contamination leak to occur. Assume that his answer is that he expects one serious leak every four years. Based on this information one can construct the prior pdf of p through the use of the conjugate prior distribution which for the case of the binomial distribution studied here is the Beta distribution (Benjamin and Cornell 1970)

$$f_p(p) = Be(t,r,p) = \frac{(t-1)!}{(r-1)!(t-r-1)!} p^{r-1}(1-p)^{t-r-1}, \quad (13.49)$$

with mean and variance given by

$$m = r/t, \quad \text{and} \quad \sigma^2 = r(t-r)/t^2(t+1). \quad (13.50)$$

Discussion of the use of conjugate functions for the development of prior pdf is beyond the scope of this tome and can be found in several standard textbooks dealing with the Bayesian decision theory (Raiffa and Schlaifer 1961; Berger 1985). Utilizing the numbers r=1, t=4 provided by the operator of the site allows one to evaluate the prior distribution of the probability p as

$$f_p(p) = \frac{3!}{0!2!} p^0(1-p)^2 = 3(1-p)^2. \quad (13.51)$$

The Bayes risk R is defined (Duckstein et al. 1978; Berger 1985) as the expected value of the goal function in equations (13.49a) and (13.49b) with respect to the unknown parameter p. The Bayes risk is then given by

$$R(A) = \begin{cases} R(A(0)) = BT\int_0^1 3p(1-p)^2\,dp = BT/4 \\ R(A(1)) = CT. \end{cases} \quad (13.52)$$

R(A(0)) in equation (13.52) can be evaluated by integration or simply by noticing that $R(A(0))=E^{f(p)}[BT\ p] = BT\ E^{f(p)}[p] = BT / 4$ by virtue of m=1/4. The optimal Bayes decision is obtained by the action that minimizes the Bayes risk function. For example, if B=40 and C=8 then A(0) returns the value of R(A(0))=10 T, whereas A(1) returns the value of R(A(1))=8 T. Hence for these values A(1) should be selected and additional monitoring devices should be installed. Because the risk function involves a linear and a constant part, respectively, construction of a graph similar to figure 13.6 can be easily made and then the optimal regions can be defined according to:

$$\text{If } B\,T < 4\,C\,T \text{ then } A^* = A(0); \text{if } B\,T > 4\,C\,T \text{ then } A^* = A(1). \quad (13.53)$$

Here, A* designates the optimal action for each particular case.

8.2 Expected opportunity loss.

Table 13.4 or equation (13.47) depicting the loss function of the problem can be presented as follows: The first column of Table 13.4 indicates the two possible states of nature (i.e., the two possible situations that may occur in terms of costs) and the two other columns tabulate the losses incurred under these two possible scenarios. It is straightforward from this table to construct the regret table as follows:

TABLE 13.5. Regret for monitoring problem

State of Nature	A(0): Do not Build Monitoring Network	A(1): Build Monitoring Network
S1: $B\,N < C\,T$	0	$C\,T - B\,N$
S2: $B\,N > C\,T$	$B\,N - C\,T$	0

Table 13.5 tabulates the extra loss that would have been incurred if a particular action were followed given that one had perfect knowledge of the true state of nature. Thus if it were known that the remediation costs were less than the cost of the additional monitoring system (B N < CT), no opportunity would have been lost by following action A(0) (i.e., no regret), but on the other hand if A(1) was followed an extra amount (C T – B N), not required by the sistuation, would have been used. Similar arguments apply for the case where the cost B N > C T.

One should notice here that N entering Table 13.5 is a random variable and hence a more appropriate measure of the regret is through the expected value of the loss functions, defined as the goal functions. Utilizing these two equations one can easily construct the opportunity loss (OL) table as follows:

The opportunity loss provides a measure of the average loss incurred by following a certain action because of imperfect information on the true state of nature. Alternatively, it provides a numerical value of the worth of obtaining more information on the probability of a contamination leak event p.

The values of Table 13.6 can be presented in closed form according to:

$$OL(A(0),p) = \begin{cases} 0 & \text{if } 0 \le p \le C/B \\ (Bp - C)T & \text{if } C/B \le p \le 1 \end{cases} \tag{13.54a}$$

$$OL(A(1),p) = \begin{cases} (C - Bp)T & \text{if } 0 \le p \le C/B \\ 0 & \text{if } C/B \le p \le 1. \end{cases} \tag{13.54b}$$

TABLE 13.6. Opportunity loss for monitoring problem

State of Nature	A(0): Do not Build Monitoring Network	A(1): Build Monitoring Network
S1: B N < C T	0	C T − B N
S2: B N > C T	B N − C T	0

The expected opportunity loss (EOL) can now be evaluated by calculating the expected values of the opportunity losses given in equations (13.54a) and (13.54b) with respect to the prior distribution of the probability of a leak.

$$\mathrm{EOL}(A(0)) = 3\int_{C/B}^{1}(Bp-C)T(1-p)^2\,dp = 3T\left(\frac{B}{12}-\frac{C}{3}+\frac{C^2}{2B}-\frac{C^3}{3B^2}+\frac{C^4}{12B^3}\right), \tag{13.55a}$$

$$\mathrm{EOL}(A(1)) = 3\int_{0}^{C/B}(C-Bp)T(1-p)^2\,dp = 3T\left(\frac{C^2}{2B}-\frac{C^3}{3B^2}+\frac{C^4}{12B^3}\right). \tag{13.55b}$$

The values of B=40 and C=8 result in

$$\mathrm{EOL}(A(0)) = 4.096\ T, \text{ and } \mathrm{EOL}(A(1)) = 2.096\ T. \tag{13.56}$$

Therefore, for this problem and values of the parameters B, and C the worth of obtaining more information about the probability of leak p is 2.096T or in other terms this is the amount that on average one would be willing to pay to define the contamination potential. Comparison of the optimum EOL(A(1)) = 2.096T with the Bayes risk function R(A(1)) = 8T indicates that the expected opportunity loss represents 26.2% of the optimum Bayes risk. For the particular values of the parameters B, and C chosen here the cost of uncertainty appears to be a significant component of the total project costs.

8.3 Influence of the length of the observation record on decision-making

Assume now that the operator's answer is that he would expect two contamination leak events in eight years, or alternatively, in four additional years of observation an extra contamination event was observed. Although the ratio of contamination events per number of years remains the same as that utilized previously, the numbers r=2, t=8 provided by the operator of the site have the effect of modifying the prior distribution of the probability p. Thus

$$f_p(p) = \frac{7!}{5!1!} p^1 (1-p)^5 = 42\, p\,(1-p)^5 . \qquad (13.57)$$

Using equation (13.57) and the values of B=40 and C=8 yields

$$\mathrm{EOL(A(0))} = 1680\mathrm{T} \int_{C/B=0.2}^{1} (p-0.2)\, p\,(1-p)^5\, dp \approx 3.3\mathrm{T}, \qquad (13.58a)$$

$$\mathrm{EOL(A(1))} = 1680\mathrm{T} \int_{0}^{0.2} (0.2 - p)\, p\,(1-p)^5\, dp \approx 1.3\mathrm{T}. \qquad (13.58b)$$

Equations (13.58a) and (13.58b) indicate that extending the length of the observation record results in a reduction of the expected opportunity loss, for action A(0) from an EOL(A(0)) of about 4.1T to a value of 3.3T, and for action A(1) from an EOL(A(1)) of about 2.1T to a value of approximately 1.3T. In other words, the largest the observation record the more one could estimate the true state of nature and hence the lesser would the regret be for each economic action.

8.4 Influence of the form of the loss function on decision-making

Consider now a loss function of the form

$$L_2(A,N) = \begin{cases} BN^2 & \text{if } A = A(0) \\ CT & \text{if } A = A(1). \end{cases} \tag{13.59}$$

In equation (13.59) the cost for adding the supplementary monitoring system has remained the same (CT), but for the decision A(0), not to augment the monitoring network, the consequences have become severe. Thus, the losses in equation (13.59) increase rapidly the more contamination events get undetected by the regular monitoring system, effectively penalizing the decision not to install the additional devices. In practice, such a loss function can be used to simulate the situation where an increasing number of contamination events (and associated reme\diation costs) points to inadequacy of a regular monitoring program (because of poor selection of the mandated devices' locations, underestimation of the uncertainty characterizing the site etc.).

The goal function for action A(0) is given by

$$G_2(A(0), p) = E^N[L_2(A(0), N)] = E^N[B\,N^2] = B\,Tp\,(Tp+1\text{-}p). \tag{13.60a}$$

Here we have used that $Var(N)=E[\,N^2]\text{-}(E[N])^2$, and that the mean and variance of the Bernoulli distribution are given by $E[N]=Tp$ and $Var(N)=Tp(1\text{-}p)$. For action A(1) the goal function is given again by

$$G_2(A(1), p) = E^N[L_2(A(1), N)] = E^N[C\,T] = C\,T. \tag{13.60b}$$

The probability p* where a change in action is recommended is obtained by equating the goal functions in equations (13.60a) and (13,60b) yielding the quadratic equation

$$p^{*2}\,(T-1) + p^* - \frac{C}{B} = 0\,, \tag{13.61}$$

which for the values of B=40 and C=8 and a time period of T=10 years returns the value of p*=0.1.

The Bayes risk for action A(0) is given now by

$$\begin{aligned} R_2(A(0)) &= E^{f_p(p)}[G(A(0)),p] = E^{f_p(p)}[BTp\,(Tp+1-p)] = \\ &= BT^2\,E^{f_p(p)}[p^2] + BT\,E^{f_p(p)}[p] - BT\,E^{f_p(p)}[p^2], \end{aligned} \tag{13.62}$$

where the notation $E^{fp(p)}[\ ]$ stands for the expected value with respect to the prior distribution. For this distribution the mean and variance are given as m=1/4 and $\sigma^2 = 0.0375$ (for r=1 and t=4). Substituting the relation $E^{f_p(p)}[p^2] = \sigma^2 + m^2 = 0.1$ into equation (13.34) yields

$$R_2(A(0)) = 0.1\ BT\ (T+1.5). \tag{13.63a}$$

The Bayes risk for action A(1) is given by

$$R_2(A(1)) = E^{f_p(p)}[G(A(1)),p] = E^{f_p(p)}[CT] = CT. \tag{13.63b}$$

For the values of B=40 and C=8, under a quadratic loss function, action A(0) returns a value of $R_2(A(0))$=4T(T+1.5) and a $R_2(A(1))$=8T. The time when the expected losses of both actions are equal to each other can be found by equating equation (13.63a) to (13.63b):

$$0.1BT(T+1.5) = CT \Rightarrow 4T+6 = 8 \Rightarrow T = 0.5. \tag{13.64}$$

Hence for the case where B=40 and C=8 the optimal action is defined according to

$$\text{If } T < 0.5 \text{ then } A^*=A(0);\ \text{if } T > 0.5 \text{ then } A^*=A(1). \tag{13.65}$$

In general the optimal action is determined through the following relations:

$$\text{If } 0.1\ B\ (T + 1.5) < C \text{ then } A^*=A(0); \text{if } 0.1\ B\ (T + 1.5) > C \text{ then } A^*=A(1). \tag{13.66}$$

One should notice that the previous optimal action was determined by relating the costs B and C only, whereas, under the quadratic loss modelinvolves, additionally, the time horizon T of the supplementary monitoring system. Thus, under a linear loss model, action A(1), to install a supplemantary system, is to be followed only when the ratio of the costs of remediation to those of the monitoring system exceeds the value of four. In contast, under a quadratic loss model, action A(1) is recommended if B/C > 10 / (T+1.5). Table 13.7 indicates the value of the ratio B/C for the two loss models for different time periods. For the linear loss model B/C remains fixed over the years, whereas for the quadratic loss model the criterion of how high the contamination costs must be in relation to the cost of the monitoring system in order to install additional devices is relaxed with the length of operation.

TABLE 13.7. Installation of monitoring network: Criteria for different loss models

Time period T (years)	Linear Loss Model: Value of B/C favoring	Quadratic Loss Model: Value of B/C favoring
1	4	4
2	4	2.9
3	4	2.2
4	4	1.8
5	4	1.5
:	:	:
10	4	0.9

The loss function for the case of a quadratic loss function is depicted in Table 13.8, while the Regret table of the quadratic loss case is Table 13.9 that tabulates the loss that is incurred by each action because information that would have revealed the true state of nature is not available. Thus, if it were known for certain that monitoring would be more expensive than cleaning up a contamination leak then action A(1) expends ($C\,T - B\,N^2$) more than what is needed to address the situation. On the other hand, if contamination costs turn out to be trully higher than monitoring costs then action A(1) is utilizing resources efficiently (no regret).

TABLE 13.8. Loss table for quadratic model

State of Nature	A(0): Do not Build Monitoring System	A(1): Build Monitoring System
S1: $B N^2 < C T$	$B N^2$	$C T$
S2: $B N^2 > C T$	$B N^2$	$C T$

TABLE 13.9. Regret for quadratic model

State of Nature	A(0): Do not Build Monitoring Network	A(1): Build Monitoring Network
S1: $B N^2 < C T$	0	$C T - B N^2$
S2: $B N^2 > C T$	$B N^2 - C T$	0

The opportunity loss table (OL) for this case is obtained by taking the expected value of all quantities entering table 13.9. Solving the inequalities in the first column of table 13.10, yields a probability p* where a change in action is recommended. Using the probability p* allows presentation of the relations in table 13.9 according to:

$$OL_2(A(0),p) = \begin{cases} 0 & \text{if } 0 \le p \le p^* \\ [Bp(Tp + 1 - p) - C]T & \text{if } p^* \le p \le 1 \end{cases} \qquad (13.67a)$$

$$OL_2(A(1),p) = \begin{cases} [C - Bp(Tp+1-p)]T & \text{if } 0 \le p \le p^* \\ 0 & \text{if } p^* \le p \le 1. \end{cases} \quad (13.67b)$$

TABLE 13.10. Opportunity loss for quadratic model

State of Nature	A(0): Do not build Monitoring Network	A(1): Build Monitoring Network
S1: Bp (Tp+1-p) < C	0	[C – Bp (Tp+1-p)] T
S2: Bp (Tp+1-p) > C	[Bp (Tp+1-p) – C] T	0

The expected opportunity loss (EOL) can now be evaluated for the quadratic loss model by calculating the expected values of the opportunity losses given in equations (13.67a) and (13.67b) with respect to the prior distribution of the probability of a contamination leak in equation (13.51).

$$EOL_2(A(0)) = 3\int_{p^*}^{1} [Bp(Tp+1-p) - C)]T(1-p)^2\,dp, \quad (13.68a)$$

$$EOL_2(A(1)) = 3\int_{0}^{p^*} [C - Bp(Tp+1-p)]T(1-p)^2\,dp. \quad (13.68b)$$

The values of B=40, C=8, and T=10 return the value p*=0.1 which, in turn, yields

$$EOL_2(A(0)) = 30 \int_{p^*=0.1}^{1} (360p^2 + 40p - 8)(1-p)^2 \, dp \approx 393, \qquad (13.69a)$$

$$EOL_2(A(1)) = 30 \int_{0}^{p^*=0.1} (8 - 40p - 360p^2)(1-p)^2 \, dp \approx 13.37. \qquad (13.69b)$$

For the linear model and a time period of ten years the expected opportunity losses were found to be EOL(A(0)) ≈ 40, and EOL(A(1)) ≈ 20.96, respectively.

Therefore, for this problem and values of the parameters B, and C the worth of obtaining more information about the probability of leak p is 2.096T ; this is the amount that on average one would be willing to pay to define the contamination potential. Comparison of the optimum EOL(A(1)) = 2.096T with the Bayes risk function R(A(1)) = 8T indicates that the expected opportunity loss represents 26.2% of the optimum Bayes risk. For the particular values of the parameters B and C chosen here the cost of uncertainty appears to be a significant component of the total project costs.

9 Conclusion

The purpose of this section of the chapter has been to illustrate, with particularly simple examples, how one can go about using the occurrence of particular events to update the initial probability estimates. This Bayesian aspect of environmental risk assessment is particularly powerful in attempting to improve knowledge of the likely causes of the environmental hazard. In such a way one can continually up-date contingency plans for remedial action based on a most likely cause. In addition, one can assess the need to put such plans into action based on the actual events that have occurred and their use to improve the estimates of further disastrous events occurring.

While the example here has been tailored to illustrate simply the Bayes' procedure, it is clear that more than two scenarios can be handled equally well. The dominant theme is to have some a priori way of estimating the probability of a given scenario being correct and then, within the framework of each scenario, to provide an a priori estimate of a particular type of event occurring or not occurring. Once these two aspects of the environmental risk problem have been addressed adequately, it is then a relatively simple matter to update the likelihood of further events occurring and also to update the probability that a given scenario is valid using the Bayesian method.

Illustration of this particular goal, and how to achieve it practically, was the objective of this section. A more extensive discussion on the Bayesian theory can be found in several textbooks and articles, this section's aim was to motivate decision-making groups on the use of this powerful method.

10.Political Decisions

Despite the advice that is achievable scientifically on the ways to proceed with environmental remediation based on the methods above, the political powers who control approval of what to do often make decisions based on very different criteria, some of which may even be logical. First there is uncertainty in the scientific assessment so that any political decision can make use of this uncertainty to decide in favor of an alternative that provides the minimum acceptable chance of a successful remediation, rather than a complete remediation. Second, there is the economic cost of any sort of remediation and the associated uncertainties on such costs. A political decision will, in general, opt for the lowest cost solution that may not be the most satisfactory choice because a minimum cost decision may end up not performing a remediation that is unassailable in its success. Third, a political decision is often determined by laws being enacted to ensure compliance with the avowed wishes of the legislature, whether the majority of the population wishes it or not. Indeed, in connection with the American Congress, USA Today in its editorial of 16 October 2003, has written:

"When Congress tackles national problems, its tendency is to take the course of least resistance or greatest pressure. From its refusal to touch Social Security to its eagerness to create new corporate tax breaks, Congress has shown repeatedly that it listens to those groups that speak the loudest. Senior citizens are plenty vocal, both through their muscular lobbies, such as the AARP, and their voting clout. This helps explain lawmakers' preoccupation with crafting an expensive Medicare drug benefit to ease the health care burdens faced by a slice of seniors. But it doesn't excuse it. The singular focus on expanding the benefits for the 38.5 million individuals who already are covered by one of the nation's most affordable and dependable health plans is dangerous in two respects. It diverts political capital and precious taxpayer resources from more pressing national health-coverage concerns.

In 2002, 43.6 million people lacked health insurance for the entire year, an increase of more than 2 million from 2001, according to a Census Bureau study released last month. The reason for the jump is clearly documented. Faced with double-digit increases in health insurance premiums for a third year in a row, private companies with coverage are dropping it, and others are refusing to offer it. The increase in the number of uninsured would have been even greater had government programs such as the Medicaid health plan for the poor not picked up more than 1 million people.

The same study shows a stark contrast between the growing ranks of people without insurance and the guaranteed benefits for people in Medicare. Those 65 and older are covered for just $57.80 per month. While the plan doesn't cover prescription drugs, its price is considerably less than individuals pay for comparable private plans if they are fortunate enough to have one.

Yet Congress is concentrating its energies on further enhancing benefits for seniors. The Senate and House each have passed Medicare drug plans and are trying to work out a compromise measure by the year's end. In spite of pronouncements of concern about the rise in health care costs and number of uninsured, lawmakers have failed to put broader solutions on the calendar.

In fact, the push for a $400 billion Medicare drug plan during the next decade reduces the chances for more far-reaching legislation. Among the reasons:

Lost time. Debate on the drug bill has prevented Congress from completing work on other pressing health issues. Last spring, it set aside $50 billion to address the problems of the uninsured. Yet the money remains unspent because lawmakers have not agreed on what to do with it, if anything. Also stalled is President Bush 's proposal to revamp Medicaid coverage for the poor. He wants to cap spending on the rapidly growing program in the future, but for now give cash-strapped states more federal aid and discretion in allocating the money. Congressional leaders see little chance of taking up those or other politically contentious issues on health coverage now that a presidential election is approaching.

Lost money. The $400 billion reserved for the Medicare drug bill takes away funding that could go toward broader health coverage. In fact, lawmakers are under pressure to cut spending elsewhere because the drug benefit's steep price would increase a yawning budget deficit already projected to approach an unprecedented $500 billion.

Supporters of the drug bill say it is necessary to halt price increases that are overwhelming for seniors. Yes, drug prices are shooting up. Prescriptions for popular medicines such as Zocor, a high-cholesterol treatment, or Norvasc, for hypertension, cost more than $100 per month, nearly twice Medicare's premium. And the outdated Medicare program is in need of revamping to match reforms common in private plans, from drug coverage to cost efficiencies. That not only leaves seniors on limited incomes liable for huge drug bills, but also ignores the potential savings that could come with an overhaul of the program.

Even so, the drug bill would address only part of that problem. And it wouldn't even do that very well. Some seniors who exchange the new Medicare plan for the drug coverage they now get through former employers or separate health policies would see their out-of-pocket expenses rise, the Senate Budget Committee reports. And as many as 37% of seniors who now get a drug benefit from former employers could lose it, according to the Congressional Budget Office. The reason: Companies could use the new plan to justify dropping their benefit for retirees. The $400 billion drug benefit also relies on accounting gimmicks that underestimate the huge cost taxpayers likely would bear.

Addressing the issue of the uninsured will be expensive and complicated. Just extending coverage to children and young adults up to age 25 would cost $53.4 billion a year, according to an Emory University study.

Ignoring the problem won't make it go away, which is why a national conversation about it is long overdue. That conversation can't be heard, however, when it's drowned out by all of the noise generated by a flawed senior drug bill."

The point about this example is to show that a legislature responds to pressure from affluent groups (or vocal-enough groups) that may have little to do with either the majority of the population influenced by a particular problem or that may be interested in self. Thus the general political solution to a problem is colored and tempered by this involvement-and environmental problems are subject to the same sort of legislative influence.

To illustrate how one can estimate the impact on an environmental problem of legislative partisan actions, consider the situation where one is concerned that something could go wrong with an environmental clean-up. For instance, if a project is saddled with the small chance of a catastrophic blow-out while drilling through a gas-bearing dump, and the associated high clean-up costs, then it is all too easy to take what could have been a very worthwhile environmental project and downgrade it to less than worthwhile. Equally, even if a blowout scenario is not included while drilling, a similar catastrophic loss condition is often envisioned by the legislature: environmental clean-up starts and, once into production, there is the chance of a contaminant spill with, again, enormous environmental clean-up costs. Accordingly, politicians often use the catastrophic loss excuse to get out of (or never get into) environmental clean-up projects. Thus political bias can lead to a re-evaluation that takes, on average, a very good project and makes it seem very poor.

But the point about all of these political biases is that if only the expected value (average value) is used to quantify the risk, then one is completely overlooking the fact that there is also an uncertainty on the expected value, which ameliorates the "catastrophic loss" effect. The purpose of Chapter 12 was precisely to show how this amelioration operates in practice. Two case histories were outlined: "catastrophic failure" during (a) the environmental assessment; and (b) the development assessment.

The purpose of the case histories given in Chapter 12, and their numerical illustrations, was to show how such allowances can be made by specific example.

It is strongly recommended that politicians should avoid using just the expected value as an absolute measure of opportunity worth. Instead one should, more correctly, include the standard error, and also the probability of obtaining a positive return (for the corporation contracted by the government to undertake the environmental clean-up project) from an environmental project. In this way, low risk, but high potential clean-up cost scenarios are correctly included. And the cases here were tailored to maximize the illustration of the way such factors are to be handled, so that political decisions to proceed with an environmental project can be made without pejorative phraseology that adds little to a rational assessment of costs and worth.

11 Social Bias and Social Consequences

The difficulties in arranging that an environmental clean-up and subsequent use of the cleaned area conform to the wishes, and demands, of society are painfully clear. Often, even with the best will in the world, a clean up proposed and undertaken can lead to a strong emotive reaction from fractions of society. Such reactions can be in terms of where one should deposit radioactive waste from nuclear power plants, where and how one should decommission old oil platforms or old oil tankers, or even whether one should clean up contaminated land at all or just leave it fallow. The diverse nature of humanity guarantees that there will always be some group or groups of people who disagree, sometimes violently, with any proposed scheme. Yet the same such groups are often loath to propose any other alternatives, presumably because they do not have any that are any better than those already proposed. It is all too easy to criticize without the concomitant will to be pro-active in suggesting alternatives.

There is also the scare factor that can be so trivially enhanced by the press and also by the criticizing groups themselves. It is so easy to promulgate the ultimate disaster scenario without being aware, or at least not actively making others aware, that such scenarios are less likely to happen than being struck by a meteor and lightning at the same time. The point here is that the bias introduced into the beliefs of society concerning the dangers of environmental clean-up is not balanced by an appreciation of the potential benefits, a demagoguery type of behavior.

At the same time, such watchdog groups do provide a balance of a different sort. If handled correctly, they make it clear to the greater public that there are negative (and positive) consequences for any action undertaken, including no action at all. This broad level education and information service can be used to select environmental solutions that make the most sense to the greatest number of people involved, as well as making aware to the authorities who propose clean-up schemes just what is tolerable to society and what is not.

Presumably, one will always have such conflict of interest arguments. Yet, in the end, some sort of remedial action needs to be taken with environmental problems if we are not wallow in our generated filth, with long-term detriment to humanity as a whole. This sort of ultimate perspective of accumulated environmental problems is almost self-defining in terms of seeing that some sorts of solutions are needed to vast arrays of environmental problems. The alternative is an ever-increasing pollution of the one and only planet we call home, with nowhere else to go, leading to death and sickness on massive scales.

The scientific and economic quantitative procedures painted above, while not providing a universal panacea to the total of all environmental problems, have as an advantage that they can at least be addressed in terms of quantifiable actions, rather than through emotive passionate arguments that do not include the facts of the problems.

12 Conclusions

There are no easy solutions to the environmental problems that humanity has visited upon itself over the decades, nor are there any easy solutions to the environmental problems that humanity continues to generate today. But there are remedial actions that can be taken provided humanity is prepared to absorb the costs of such undertakings. There are also lessons to be learnt from past environmental problems so that, hopefully, in the future less environmentally damaging effects will be produced. It may be that this hope is forlorn unless we learn to control the sheer mass of humankind and their education as an environmentally friendly people. Until such a time arrives, the best we can likely hope for is to try to integrate the scientific, economic, political, and social aspects of environmental clean-up scenarios in a more rational approach than heretofore seems to have been the case. And that is, perhaps, the quintessential message that this chapter, and the whole volume too, have been trying to promote through case histories and logical procedures outlining how one can go about attempting to resolve environmental problems.

References

Alaska Fish and Game (1989) Special Oil Spill Issue, 21, July-August 1989

Al-Bahar JF, Crandall KC (1990) Systematic risk management approach for construction projects. *Construction Engineering and Management, ASCE*, 116: 533-546

Alpers CN, Blowes DW (1994) Environmental geochemistry of sulphide oxidation. ACS Symposium Washington DC Series 550

American Factfinder (2002) U.S. Census Bureau Online Data, http://factfinder.census.gov

Asante-Duan DK (1996) *Management of Contaminated Site Problems*, Lewis, Boca Raton, FL

Aslanbog I, Höster R, Meyer FH (1979) Umweltschäden an Straßen bäumen in Hannover.- Das Gartenamt, 28: 364 – 376

Avanidou T (2000), Infiltration in stratified soils: Relative importance of-parameters and model variations. M.S. Thesis, University of South Carolina

Beard RT, (1990) Bankruptcy and Care choice, RAND Journal of Economics 21: 23-28

Bellmann HJ; Pilot J, Rösler HJ (1977) Untersuchungen zur Petrographie und Genese von Karbonatkonkretionen im braunkohlenführenden Oligozän der Leipziger Bucht.- Z.f.angew. Geol., 23:334 – 341

Bellmann HJ, Starke R (1990), Ausbildung und Lithogenese des marinen Oligozaens der Leipziger Bucht. Z. angew. Geol. 25124-128

Bierns de Haan S (1991) A review of the rate of pyrite oxidation in aqueous systems at low temperatures, Earth-Science Reviews, 31: 1 – 10

Benjamin JR, Cornell CA (1970) *Probability, Statistics and Decision for Civil Engineers*, McGraw-Hill, New York

Berger JO (1985) *Statistical Decision Theory and Bayesian Analysis*, Springer Verlag, New York

Blatt H (1997) Our Geologic Environment. Prentice Hall Publishing, p 144-147

Canter LW, Knox RC (1986) *Ground water pollution control,* Lewis, Chelsea MI

Carstensen A, Pohl W (2000) Long-term stability of overburden dump slopes in recultivated lignite mines of Central Germany – the prediction of maximal pore water pressure. In: Bromhead E, Dixon N, Ibsen ML (eds.), Landslides in Research, Theory and Practice, Proceeding of the 8th Symposium on Landslides at Cardiff, Vol. 1, Telford, pp 221-226

Caruccio FT, Hossner LR, Geidel G (1988) Pyritic materials: acid drainage, soil acidity, and liming. In: Hossner LR (ed.), Reclamation of surface-mined lands, CRC Press, Boca Raton, pp. 159-189

Cerling TE, Solomon DK, Quade J, Bowman JR (1991) On the isotopic composition of carbon in soil carbon dioxide, Geochim. Cosmochim. Acta, 55: 3403 – 3405

Cesnovar R, Pentinghaus H (2000) Flooding of central German lignite mining dumps may cause severe groundwater contamination. In: Rammelmair D, Mederer J, Oberthuer T, Heimann RB, Pentinghaus H (eds.), Applied Mineralogy, pp 495-498

Christakos G, Hristopoulos DT (1998) *Spatiotemporal environmental health modeling: A tractatus stochasticus,* Kluwer Academic Publishers, Boston

Christoph G (1995) Modellrechnungen zur Entwicklung des Hufeisensees in der Naehe der subhydrischen Deponie Halle - Kanena (Hufeisensee).- UFZ-Bericht 4/1995, 81-93, Bad Lauchstaedt, Leipzig [ISSN 0948-9452]

Cole KW (1988) *Ground Engineering Building over Abandoned Shallow Mines: A Strategy for the Engineering Decisions on Treatment* (reference as given in Scott and Statham, 1998)

Corwin DL, Loague K, Ellsworth TR (eds.), (1999) Assessment of Nonpoint Source Pollution in the Vados Zone, American Geophysical Union, Geophysical Monograph 108

Dagan G, Neuman SP (eds.) (1997) *Subsurface Flow and Transport: A Stochastic Approach*, Cambridge University Press, Cambridge

Deines P (1980) The isotopic composition of reduced organic carbon.- In: Handbook of Environmental Isotope Geochemistry, 1. The Terrestrial Environment. (eds. P. Fritz and J.Ch. Fontes) p329 - 406. Elsevier, Amsterdam

De Souza Porto MF, de Freitas CM (1996) Major chemical accidents in industrializing countries: The socio-political amplification of risk, Risk Analysis 16: 19-29

Deutsch CV, Journel AG (1992) GSLIB: Geostatistical Software Library and User's Guide, Oxford University Press, Oxford

Dickmann JE, Featherman DW (1998) Assessing Cost Uncertainty: lessons from environmental restoration projects, Journal of Construction Engineering and Management 124: 445-451

Dixon SA (1996) Lessons in Professional Liability: DPIC's Loss Prevention Handbookfor Environmental Professionals, DPIC Co., Monterey, CA

Dohrmann H (2000) Untersuchungen zu hydraulischen und hydrochemischen Prozessen im Initialstadium bindiger Mischbodenkippen des Mitteldeutschen Braunkohlenreviers. Beispiel: Tagebaukippe Zwenkau / Cospuden.- Dissertation, Univ. Leipzig; UFZ-Bericht 34 / 2000, 136 p Leipzig

Dooge JC (1997) Scale problems in hydrology, In Reflections on Hydrology, Scienceand Practice, N. Buras (ed.), American Geophysical Union, Washington, DC, p 85-143

Duckstein L, Kisiel CC (1971) Efficiency of hydrological data collection systems: Roles of Type I and II errors, Water Resources Bull. 7: 592-604

Duckstein L, Krysztofowicz R, Davis D (1978) To build or not to build: A Bayesian analysis, Journal of Hydrological Sciences 5: 55-68

Dunn JH (1997) Environmental liability: New insurance policies can help protect investors against unseen risk, *Property Management*, 62: 54-60

Drewnowksi S (1996) Evaluating and managing environmental risk, Energy Economist 182: 10-14

Evangelou VP (1998) Pyrite chemistry: The key for abatement of acid mine drainage. In: Geller W, Klapper H, Salomons W (eds.), Acidic mining lakes, Springer, Berlin, Heidelberg, New York, pp 197-222

Feller W (1957) *An Introduction to Probability Theory and Its Applications*. Vols. 1 and 2, John Wiley and Sons, New York

Fernald EA, Purdum ED, (1998) Water Resources Atlas of Florida, Institute of Science and Public Affairs, p 57-62

Fischer JN (1986) Hydrological Factors in thee Selection of Shallow Land Burial for the Disposal of Low-Level Radioactive Waste, U. S. Geological Survey Circular 973, Washington, DC

Flach GP, Harris MK (1997) *Integrated hydrogeological model of the General Separations area (U), Vol. 2: Groundwater flow model,* Westinghouse Savannah River Co., Savannah River Site, WSRC-TR-96-0399, Rev. 0

Florida Department of Environmental Protection Website, Sinkhole Data (www.dep.state.fl.us/geology/forms/sinkholereport/sinkreportform .html)

Frano AJ (1991) Pricing hazardous-waste revisited, Journal of Management in Engineering 7: 428-440

Freeman PK, Kunreuther H (1997) *Managing Environmental Risk through Insurance*, Kluwer Academic Press, Boston

Freeze RA (1997) Groundwater contamination: Technical analysis and social decision making, *Reflections on Hydrology, Science and Practice*, N. Buras (Ed.), American Geophysical Union, Washington, DC, p 148-180

Gamboa S (2003) Flawed data used in nerve gas report, The State Newspaper, South Carolina, 31 May

Geller W, Klapper H, Salomons W (eds.), (1998) Acidic mining lakes, Springer, Berlin, Heidelberg, New York

Gardner WR (1958) Some steady-state solutions of the unsaturated moisture flow equation with application to evaporation from a water table, Soil Sciences 85: 228-232

Geologischer Dienst Nordrhein-Westfalen (1988) Abgedeckte Geologische Karte M 1/ 25 000, aus Blatt Bochum

Glaeser HR (1995) Geophysikalische Untersuchungen zum Nachweis von Fliesswegen - Fallstudie Hufeisensee.- UFZ-Bericht 4/1995, 94-102, Bad Lauchstaedt, Leipzig [ISSN 0948-9452]

Glaesser W (1994) Umweltgeologische Untersuchungen *in* Mitteldeutschland als Beitrag aktuogeologischer Prozesse.- Altenbg. nat.-wiss. Forsch., 7, 236-249; DEUQUA-Meeting 1994, Leipzig, Altenburg

Glaesser W (1995a) Der Einfluss des Braunkohlebergbaus auf Grund- und Oberflächenwasser Hydrogeologische Untersuchungen in Mitteldeutschland. Geowissenschaften, Bd. 13, H. 8/9, Ernst & Söhne, 291-296, Berlin

Glaesser W (1995b) Das gestörte Grundwasserregime im Südraum von Leipzig. Beträge zu Lehre und Forschung, Hochschule für Technik, Wirtschaft und Kultur (HTWK), Sonderheft 1: 6-8

Glaesser W, Lazik D, Witzke, T (1997) Versuche zur Simulation der Sekundärmineralbildung auf Kippenmaterial eines Braunkohlentagebaus im nördlichen Weißelsterbecken. Ber. d. Dt. Min. Ges., Bh. 1; Eur. J. Min., H. 9: 121

Glaesser W, Nitsche P, Lerche I (2004) Carbon Dioxide Development in Aerobic Parts of Lignite Mining Dumps: The Influence of Rising Groundwater in the Cospuden/Zwenkau Dump I. Observations and Inferences. J. Environmental Geosciences (in press)

Gleason JD, Kyser TK (1984) Stable isotope composition of gases and vegetation near naturally burning coal, Nature, 307: No. 5947, 254 – 247

Grasso D (1993) *Hazardous Site Remediation*, Lewis, Boca Raton, FL

Haas H; Fisher DW; Thorstenson DC; Weeks EP (1983) $^{13}CO_2$ and $^{14}CO_2$ measurements on soil atmosphere in the sub-surface unsaturated zone in the western Great Plains of the US, Radiocarbon, 25: No.2, 301 – 314

Hillel DW, (1980) *Fundamentals of Soil Physics*, Academic Press, Orlando

Hiller A, Moller, Trettin R (1988) Zum Einfluß kohleführender Schichten auf Kohlenstoffhaushalt und Isotopenzusammensetzung des Grundwassers unter besonderer Beruecksichtlgung natuerlicher Oxidationsprozesse der Braunkohle.- Studie; 2fl - Bericht; 76 - 124; Leipzig

Hughson DL,. Yeh T.-C (1998) A geostatistically based inverse model for three dimensional variably saturated flow, Stoch. Hydro. and Hydraul. 12: 285-298

Hunter RL, Mann CJ (1992) *Techniques for Determining Probabilities of Geological Events and Processes*, Oxford University Press, Oxford

Iijima M, Ito K, Horizoe H, Noguchi Y, Tazaki Y, Shindo Y, Koide H (1993) Characteristics of Carbon Dioxide Solubility in Water, Kagaku Kogaku Ronbunshu, 19: 914-918

Janney JR, Vince CR, Madsen JD (1996) Claims analysis from risk-retention professional liability group, Journal of Performance of Constructed Facilities 10: 115-122

Jeljeli MN, Russell JS (1995) Coping with uncertainty in environmental construction:Decision analysis approach, *Construction Engineering and Management, ASCE*, 121: 370-380

Johnson CA, Thornton I (1987) Hydrogeological and chemical factors controlling the concentrations of Fe, Cu, Zn, and As in river systems contaminated by acid mine drainage. Water Research 21: 359-365

Keeney RL, (1980) *Siting energy facilities*. Academic Press, New York

Kirmer A, Mahn EG (1996) Verschiedene Methoden zur Initiierung von naturnaher Vegetationsentwicklung auf unterschiedlichen Böschungsstandorten in einem Braunkohletagebau – erste Ergebnisse, Verh. Ges. Oekologie 26: 377-385

Kölling M (1990) Modellierung geochemischer Prozesse in Sickerwasser und Grundwasser.- Beispiel: Die Pyritverwitterung und das Problem saurer Grubenwässer, Ber. FB Geowissenschaften Univ. Bremen 8: 135 p Bremen

Landesoberbergamt Nordrhein-Westfalen (2000) Tagesbrüche Oberflächennaher Bergbau, Blatt 2590 5698 Muttental, M 1/5 000, Dortmund

Langhoff K (1971) Bekampfung von Baumschäden durch Erdgas in Den Haag, Amsterdem und Lübeck.- Garten und Landschaft 81: 20 -21

Larson BA (1996) Environmental policy based on strict liability: Implications of uncertainty and bankruptcy, Land Economics 72: 33-42

Lempp C, Lerche I, Paleologos E (2003) Environmental Concerns: Catastrophic Events and Insurance, Energy Explor. Exploit. 20, (in press)

Lerche I, Bagirov EB (1999) Impact of Natural Hazards in Oil and Gas Extraction:The South Caspian Basin , Plenum Press, New York 352 p

Lerche I, Foth H (2003) Quantitative Risks of Death and Sickness from Toxic Contamination, Environmental Geosciences (in press)

Lerche I, MacKay JA, (1999) Economic Risk in Hydrocarbon Exploration Academic Press, 403 p

Lerche I, Paleologos E (2001) Environmental Risk Analysis , McGraw-Hill Book Co., New York, 437 p

MacDonald JA, Kavanaugh MC (1995) Superfund: The cleanup standard debate, *Policy and Planning*, 55-61

MacKay JA (2003) The More Zones, The Merrier, AAPG Explorer June Issue, 24-25

Maiss M, Walz V, Zimmermann M, Ilmberger J, Kinzelbach W, Glaesser W (1998) Experimentelle Tracerstudien und Modellierungen von Austauschprozessen in einem meromiktischen Restsee (Hufeisensee).- UFZ-Bericht 1/1998, 232 p Heidelbereg, Bad Lauchstaedt, Leipzig (ISSN0948-9452)

Malle KG (1996) Cleaning up the River Rhine, Scientific American 274: 70-75

Magnuson E (1980) The poisoning of America, Time 116: 58-69.

Mantoglou A, Gelhar LW (1987a) Stochastic modeling of large-scale transient unsaturated flow systems, Water Resources Research 23: 37-46

Mantoglou A, Gelhar LW (1987b) Capillary Tension head variance, mean soil moisture content, and effective specific soil moisture capacity of transient unsaturated flow in stratified soils, Water Resources Research 23: 47-56

Mantoglou A, Gelhar LW (1987c) Effective hydraulic conductivities of transient unsaturated flow in stratified soils, Water Resources Research 23: 57-68

Mathess G (1990) Lehrbuch der Hydrogeologie. Bd. 2, 2. Aufl., Gebr. Borntraeger Berlin

Maund JG, Eddleston M (eds.), (1998) Geohazards in Engineering Geology, Geological Society, London, Engineering Geology Special Publications, 15

McKnight TL (1999) Physical Geography: A Landscape Appreciation. Prentice Hall 5th ed., p 533-536Merkel B, Eichinger L, Uoluft P (1992) Same Aspects of Modelling the Carbon System in the Unsaturated Zone. In: Progress in Hydrogeochemistry (eds. Matthes G, Frimmel FH, Hirsch P, Schulz HD, Usdowski E) p 428 -438, Springer-Verlag, Berlin

Merkl A, Robinson H (1997) Environmental risk management: Take it back from lawyers and engineers, *McKinsey Quarterly*, 14: 150

Milliman JD (1974) *Marine Carbonates Part 1*, Springer-Verlag, New York

Mills AL (1985) Acid mine waste drainage: microbial impact on the recovery of soil and water ecosystems. In: Tate RL, Klein DA (eds.), Soil reclamation processes, Dekker, pp 35-80, New York

Minato T, Ashley DB (1998) Data-driven analysis of “cooperative risk” using historical cost-control data, Journal of Construction Engineering and Management 124: 42-47

Moorhouse DC, Millet RA (1994) Identifying causes of failure in providing geotechnical and environmental consulting services, *Management in Engineering, ASCE*, 10: 56-64

Morris PWG, Hough GH (1987) *The anatomy of major projects*, Wiley, New York

Moses CO, Nordstrom DK, Herman JS, Mills AL (1987) Aqueous pyrite oxidation by dissolved oxygen and ferric iron, Geochim. Cosmochim. Acta 51: 1561-1571

Ness A (1992) A contracting for environmental remediation, Construction Business Review 2: 20-75

New Rash of Sinkholes Tells Same Old Story St. Petersburg Times, Pasco Section, June 23, 2002, p 1-3

Nordstrom DK (1982) Aqueous pyrite oxidation and the consequent forming of secondary iron minerals. In: Hossner LR (ed.), Acid sulfate weathering, Soil Science Society of America, Chapter 3, pp 37-56

Normann JM, Kucharik CJ, Gower ST, Baldocchi DO, Crill PM, Savage K, Striel RG (1997) A comparison of six methods for measuring soil surface carbon dioxide fluxes, J. Geophys. Res. - Atmosphere, 102: 28771 – 28777

Nyer EK, (1992) *Practical Techniques for Groundwater and Soil Remediation*, Lewis, Boca Raton, FL

Oliver P (2003) Big polluter gets off lightly, The State Newspaper, South Carolina, 31 May

Odum HT (1998) Environment and Society in Florida. Lewis Publishers, Boca Raton, p 119-123

Paek JH (1996) Pricing the risk of liability associated with environmental clean-up projects, Trans. Am. Assoc. Cost Engineers CC4.1-CC4.5

Paleologos EK, Fletcher C (1999) Assessing risk retention strategies for environmental project management, Environmental Geosciences 6: 130-138

Pasco County Office of Emergency Management Website, Sinkhole Checklist, (www.pascocountyfl.net/oem/sinkhole.asp)

Pflug W (1998) Braunkohlentagebau und Rekultivierung, 1068 p Springer, Berlin, Heidelberg, New York

Pinellas County Economic Development Website, www.pced.org/live_work_play/

Powell DM (1994) Selecting innovative cleanup technologies, EPA resources Chemical Engineering Progress, 33-41

Powell L (2002) Digital Topographic Data and the Detection of Incipient Sinkhole Formation in Central Florida p 1-4

Prein A (1993) Sauerstoffzufuhr als limitierender Faktor für die Pyritverwitterung in Abraumippen, Mitt. Inst. Wasserwirtsch., Hydrologie und landwirtsch. Wasserbau Univ. Hannover 79, 126 p

Prenzel J (1985) Verlauf und Ursachen der Bodenversauerung, Z. dt. geol. Ges., 136: 293-302

Press WH, Flannery BP, Teukolsky SA, Vetterling WT (1987) *Numerical Recipes: The Art of Scientific Programming*, Cambridge University Press, Cambridge

Raiffa H (1968) Decision Analysis: introductory lectures on choices under uncertainty Random House, New York.

Raiffa H, Schlaifer R (1961) *Applied Statistical Decision Theory*, Harvard University Press, Cambridge, MA

Sandia Report (1994) *Total system performance assessment for Yucca Mountain-SNL second iteration*, Sandia National Laboratory Technical Report, SAND93-2675

Schreck M, Nishigaki M, Gläser HR, Christoph G, Grosswig S, Hurtig E, Kasch M, Kohn K (1998) Determination of heat production zones at open-cast mine dump, ASCE Journal of Environmental Engineering, 124: 7, 546 – 651

Scott MJ, Statham I (1998) Development advice maps: mining subsidence, In: Maund JG, Eddleston M (eds.) Geohazards in Engineering Geology, Geological Society, London, Engineering Geology Special Publications, 15, 391-400

Showstack R (2003) Reliability of current U. S. modeling of atmospheric plumes questioned, EOS 8: Number 24, p226, 17 June 2003

Sinkholes, www.fiu.edu/~whitmand/Research_Projects/C_Fl_Sinkhole.html

Sinkhole Filled; S.R. 52 Reopens, Tampa Tribune, North Tampa Section, February 24, 2003, p 3

Sinkhole Surprises Homeowners. Tampa Tribune, Pasco Section, March 27, 2003, p 1-2

Sinkhole Swallowing Up Business. St. Petersburg Times, Pasco Section, October 15, 2000, p 1-2

Sinkhole Swallows 20 Feet of Property on Peak Street. Tampa Tribune, Pasco Section, March 28, 2003, p 4

Singer PC, Stumm W (1970) Acid mine drainage. The rate determination Step, Science 167: 1121 – 1123

South West Florida Water Management District, (www.swfwmd.state.fl.us/emer/sinkhole/sinkpage.html)

Speck RC, Bruhn RW (1995) Non-uniform mine subsidence ground movement and resulting surface-structure damage, Environmental and Engineering Geoscience, 1: 61-74

Stadtspiegel Bochum Nr. 61, 29th July 2000

Stephens DB (1996) *Vadose Zone Hydrology*, Lewis, Boca Raton, FL

Strauch G, (compiler) (1996) Untersuchung an einer anaeroben Deponie zu Stofftransport, Stoffumsetzungen und Wechselbezeichnungen zwischen Deponie und Grund-bzw Oberflaechenwaessern, Umweltforschungszentrum Leipzig-Halle GmbH, Report, 269 p

Stumm W; Morgan JJ (1981) Aquatic chemistry. An introduction emphasizing chemical equilibria in natural waters, 780 p John Wiley & Sons, New York

Tampa Chamber of Commerce Website: Housing and Communities, www.tampachamber.com/gipr-tahc_housing.html

TASI (Technische Anleitung zur Verwertung, Behandlung und sonstigen Entsorgung von Siedlungsabfaellen) (1993) Bundesanzeiger, Cologne, Germany

Taylor BE, Wheeler MC, Nordstrom DK (1984) Stable isotope geochemistry of acid mine drainage: Experimental oxidation of pyrite, Geochim. Cosmochim. Acta, 48: 2669 – 2678

Telego DJ (1998) A growing role: Environmental risk management, *Risk Management*, 6: 19-21

Thomsen RO, Lerche I (1997) Relative Contributions to Uncertainties in Reserve Estimates, Marine Petrol. Geol. 14: 65-74

Thorstenson DC, Weeks EP, Haas H, Fisher DW (1983) Distribution of gaseous $^{12}CO_2$, $^{13}CO_2$, and $^{14}CO_2$ in the sub-soil unsaturated zone of the western US Great Plains, Radiocarbon, 25: No.2, 315 – 346

Tietenberg TH (1998) *Environmental Economics and Policy*, 2nd Edition, Addison-Wesley, Reading, MA

US DOE (1994) Total System Performance Assessment-1993: An Evaluation of the Potential Yucca Mountain Repository, DOE Report B00000000-01717-2200-00099-Rev.01

US EPA (1991) Is your drinking water safe? EPA Report 570-9-91-0005

USGS Website wwwga.usgs.gov/edu/earthgwsinkholes/html

van der Heijde P, Bachmat Y, Bredehoeft J, Andrews B, Holtz D, Sebastian S (1986) *Groundwater Management: The Use of Numerical Models*, American Geophysical Union, Water Resources Monograph 5

van Genuchten MT (1980) A closed-form equation for predicting the hydraulic conductivity of unsaturated soils, Soil Science Journal 44: 892-898

Voorhees J, Woellner RA (1986) *International Environmental Risk Management*, Lewis, Boca Raton, FL

Waltham AC (1989) Ground Subsidence, Blackie and Son Ltd., London, 202 p

Warren JE (1978) The development decision for frontier areas: The North Sea, Eur. Offshore Pet. Conf, Exhib., London

Warren JE (1978) The Evaluation of Oil and Gas Plays, SPE 17554

Warren JE (1979) Basin evaluation; Society of Petroleum Engineers, Economics and Evaluation Symposium, Dallas, February

Warren JE (1981) The Development Decision: Frontier Areas, SPE 9558

Warren JE (1983a) A Strategic Exploration Model, SPE 11442

Warren JE (1983b) The Development Decision: Value of Information, SPE 11312

Warren JE (1988) The Evaluation of Oil and Gas Plays, SPE 17554

Westdeutsche Allgemeine Zeitung (WAZ) Nr. 147, 28th June 2001-10-16

Whittaker BN, Reddish DJ (198) Subsidence: occurrence, prediction and control, Elsevier, New York, 528p

Wiegand U (2002) Hydro- und geochemische Prozesse in oberflächennahen Kippensedimenten des Tagebaus Zwenkau.- Dissertation, Univ. Leipzig, UFZ-Bericht 6/2002, 111 p., 39 Anlagen, Leipzig

Wiegand U, Lazik D, Glaesser W (2000) Mineralogical development of dump sediments in Central German lignite pits, In: Rammelmair D, Mederer J, Oberthuer T, Heimann RB, Pentinghaus H (eds), Applied Mineralogy, pp 701-704

Wilson R, Couch EAC (2001) Risk-Benefit Analysis (Second Edition), Harvard University Press, Harvard, 370 p

Wisotzky F (1994) Untersuchungen zur Pyritoxidation in Sedimenten des Rheinischen Braunkohlenreviers und deren Auswirkungen auf die Chemie des Grundwassers, Bes. Mitt. Dt. Gewässerkundl. Jb., 58: 153 p, Essen

Wisotzky F (1996) Hydrochemische Reaktion im Sicker- und Grundwasserbereich von Braunkohlentagebaukippen, Grundwasser 2: 129-136

Woodbury AD, Sudicky EA (1991) The geostatistical characteristics of the Borden aquifer, Water Resources Research 27: 533-546

Yeh T.-C, Gelhar LW, Gutjahr AL (1985a) Stochastic analysis of unsaturated flow in hetereogeneous soils I. Statistically isotropic media, Water Resources Research 21: 447-456

Yeh T.-C, Gelhar LW, Gutjahr AL (1985b) Stochastic analysis of unsaturated flow in hetereogeneous soils II. Statistically anisotropic media with variable α, Water Resources Research 21: 457-464

Yeh T.-C, Gelhar LW, Gutjahr AL (1985c) Stochastic analysis of unsaturated flow in hetereogeneous soils III. Observations and applications, Water Resources Research 21: 465-472

Yeh T.-C (1989) One-dimensional steady state infiltration on heterogeneous soils, Water Resources Research 25: 2149-2158

Yeo KT (1990) Risks, classification of estimates, and contingency management, Management in Engineering, ASCE, 6: 458-470

Index